Birkhäuser

Monografie Matematyczne
Instytut Matematyczny Polskiej Akademii Nauk (IMPAN)

Volume 74
(New Series)

Founded in 1932 by
S. Banach, B. Knaster, K. Kuratowski,
S. Mazurkiewicz, W. Sierpinski, H. Steinhaus



Volumes 31–62 of the series
Monografie Matematyczne were published by
PWN – Polish Scientific Publishers, Warsaw

Sergey Kislyakov • Natan Kruglyak

# Extremal Problems in Interpolation Theory, Whitney-Besicovitch Coverings, and Singular Integrals

Birkhäuser

Sergey Kislyakov
Steklov Mathematical Institute
St. Petersburg Branch
St. Petersburg
Russia

Natan Kruglyak
Department of Mathematics
Linköping University
Linköping
Sweden

ISBN 978-3-0348-0468-4 ISBN 978-3-0348-0469-1 (eBook)
DOI 10.1007/978-3-0348-0469-1
Springer Basel Heidelberg New York Dordrecht London

Library of Congress Control Number: 2012950867

Mathematics Subject Classification (2010): 26B35, 41A10, 41A63, 42B20, 42B25, 42B35, 42B70, 42C40, 46E30, 46E35



Printed on acid-free paper

Springer Basel is part of Springer Science+Business Media (www.springer.com)

# Contents

# Preface

The acquaintance of a newcomer with real interpolation often starts (and fairly often also finishes) with the Marcinkiewicz theorem about the Lorentz scale $L^{p,q}$, see, e.g., [SW], [Z], or the introductory material in [BL]. It can be seen from the classical proof that this subtle result is related to careful splitting of a function in two parts. More advanced expositions show (though this may be hidden behind technicalities; see, e.g., [BL] except the introductory material) that, in fact, not too many splittings are required. Specifically, the *entire* content of the Marcinkiewicz theorem stems from the solution of *only one* extremal problem.

**Problem 0.1.** *Suppose* $0 < p < \infty$, $f \in L^p$, *and* $t > 0$. *Find the distance from* $f$ *to the* $t$*-ball of* $L^\infty$ *in the* $L^p$*-metric, i.e., the quantity*

$$\mathrm{dist}_{L^p}(f,\, B_t(L^\infty)) = \inf\{\|f - g\|_{L^p} :\ g \in L^\infty, \|g\|_{L^\infty} \le t\}.$$

Luckily, this problem can be resolved, moreover, a *minimizer*, i.e., a function $g$ at which the infimum is attained, can be exhibited immediately:

$$g = \frac{f}{\max(1, t^{-1}|f|)}. \tag{0.1}$$

The splitting mentioned above is going to be $f = g + (f - g)$.

In general, real interpolation is in intimate relationship with similar extremal problems, but for more or less arbitrary couples of spaces $(X, Y)$ in place of $(L^p, L^\infty)$. However, for general couples the solution may be not as easy. Typically, there is no way to explicitly find a minimizer for the *distance functional* $\mathrm{dist}_X(f,\, B_t(Y))$. In this book we shall show that, in recompense, *explicit* construction of *near-minimizers*, i.e., of functions at which the distance in question is attained roughly (up to a multiplicative constant), is often possible.

Largely, a near-minimizer gives as much information as an exact minimizer, and, if well-chosen, it may behave even better. We shall describe a certain unified algorithm for finding near-minimizers with important additional properties. The method applies to surprisingly many couples $(X, Y)$ of function spaces (and for many of them no explicit near-minimizers were available previously), and is based on some far-reaching refinements of a procedure invented by Calderón and

Zygmund [CZ] for the needs of the theory of singular integral operators. Vaguely speaking, these refinements are related to the case when $Y$ is a space of *smooth* functions, e.g., a Sobolev or Lipschitz class. In this case, we say that a function undergoes a *smooth version of the Calderón–Zygmund decomposition.* As in the classical case, a system of cubes adjusted to one another in some intricate way is required for this procedure.

Usually, families of cubes with special properties arise from *covering theorems.* We discuss classical covering theorems (in particular, those due to Besicovitch and Whitney), and then prove several *joint* refinements of these results. These refinements apply to families of cubes possessing a certain type of smoothness, and yield coverings with the entire list of nice features available in the two classical statements named above. Moreover, a numerical characteristic (the so-called $\alpha$-capacity) of the outcoming covering is controlled in terms of a similar quantity for the initial family of cubes.

This being done, smooth Calderón–Zygmund decompositions and, with them, near-minimizers can be written out almost immediately. It should be mentioned that, even in the scarce situations where an exact minimizer is available, very often it does not fit any longer if we deviate even slightly from the initial setting. This is the case of the function given by (0.1): though simple, this formula is "too rigid". Near-minimizers constructed by our methods are much more flexible: we prove that they remain near-minimizers after the action of a singular integral operator (see the Introduction below for more details).

Our principal aim in this book is to give a complete, unified, and polished treatment to this stuff, covered before only partly by journal publications (see, e.g., [K1], [K0], [KK]). However, in Part 1 we present a substantial background material (whence its title). For instance, classical covering theorems are discussed there, and very basic facts about spaces of smooth functions and singular integrals are exposed. This material is included only to the extent required for more or less independent presentation of the main topic. Some original results pertaining to this main topic appear already in Part I, but, generally, this part is more easy reading than Part II, where the principal line is developed fully.

It shoud be noted, however, that we did not aim at an anywhere near detailed exposition of interpolation theory when discussing the background. When talking about "interpolation", we mostly mean a careful decomposition of a function in two parts, and leave interpolation spaces (and operators acting in them) apart. More advanced results about interpolation are employed from time to time, so that the reader familiar with them will feel more comfort, but, strictly speaking, this knowledge is not critical for the understanding of the main body of the book.

# Introduction

The present book consists of two parts. As has already been mentioned, the first part is preparatory for the more technical second part and may be used as a brief introduction to several important topics in analysis, such as classical covering theorems, spaces of smooth functions, or singular integrals. However, when discussing these subjects, we do not aim at completeness and, in general, do not claim originality (except for scarce points where otherwise is stipulated explicitly). Our purpose was merely to make the exposition of the main line more or less self-contained, for the reader's convenience. We have restricted ourselves to the "auxiliary" facts strictly necessary for that (the epithet "basic" can be used equally well instead of "auxiliary"). Fortunately, the required material has not turned out to occupy much space.

At the same time, we do start this main line itself already in the first part of the book. Among other things, we make some important observations, some of which have not received due emphasis before. The central observation (however simple and obvious it can seem when stated) is the fact that the so-called "good" part $f_t$ in the Calderón–Zygmund decomposition

$$f = f_t + (f - f_t)$$

is a near-minimizer for the $E$-functional (the same as the distance functional) of a function $f \in L^1$ with respect to the couple $(L^1, L^\infty)$, and moreover, we have the equivalence (see (1.48))

$$E(t, f; L^1, L^\infty) = \inf_{\|g\|_{L^\infty} \le t} \|f - g\|_{L^1} \approx \sum_{i \in I} |Q_i| ,$$

where $\{Q_i\}_{i\in I}$ is the family of cubes arising under the construction of the "good" function $f_t$. Specifically, we recall that

$$f_t = \sum_{i \in I} c_i \chi_{Q_i} + f \chi_{(\mathbb{R}^n \setminus \cup Q_i)},$$

where $\{Q_i\}_{i\in I}$ is the family of maximal dyadic cubes satisfying $\frac{1}{|Q_i|}\int_{Q_i} |f| \ge t$, and the constants $c_i$ are chosen from the condition $\int_{Q_i}(f - c_i) = 0$. In this connection, we mention a somewhat related formula for the $K$-functional due to Peetre, see [Pe1].

The next observation (probably, first made in [JRW]) is an intrepretation of the Holmstedt formula (a well-known formula in interpolation theory) in the language of near-minimizers. Together with the above fact, it implies that, after a power-type change of the parameter $t$, the classical Calderón–Zygmund decomposition provides also a near-minimizer for the distance functional with respect to the couple $(L^1, L^p)$ for $p < \infty$.

Also in the first part of the book, we show how covering theorems can be used to construct smooth decompositions of Calderón–Zygmund type. Originally, this possibility was observed by Stein [S1], and then the idea was used by Stein and Ch. Fefferman in [FS1]. We describe explicitly the properties of cubes required for construction of smooth Calderón–Zygmund decomposition and show how such systems of cubes (Whitney–Besicovitch coverings, or WB-coverings for short) arise. Fairly involved WB-coverings will play an important role in the second part of the book. The class of WB-coverings is distinguished here for the first time.

As an application, also in Part I, we prove the (nontrivial) theorem about the stability of smooth near-minimizers for the couple $(L^1, L^\infty)$ under the action of sigular integral operators. Some important elements of the proof are used also in the treatment of subtler stability theorems in Part II.

In connection with the above comments, the question arises as to whether the classical Calderón–Zygmund construction can be refined to do the same job for couples of function spaces other than $(L^1, L^\infty)$ (or $(L^1, L^p)$ with finite $p$). Surely, we are interested in refinements for which, in particular, the "good" function $f_t$ is a near-minimizer for the $E$-functional, and a certain analog of the above formula for the $E$-functional holds true. By the way, such an extension may be useful not only in Fourier analysis on Euclidean spaces (the original Calderón–Zygmund procedure is an important tool in it), but also in inverse problems of image processing, because near-minimizers for various couples of Banach spaces are an indispensable ingredient in that theory, see, e.g., [SGGHL]. (This theme will be touched upon very slightly in Chapter 6.)

As has already been mentioned, the first step of the classical Calderón–Zygmund procedure is the construction, for a given function $f$ and a parameter $t > 0$, of a certain family $\{Q_i\}_{i\in I}$ of cubes; after that $f$ is approximated by the "piecewise polynomial" function $g = \sum\limits_{i\in I} c_i\chi_{Q_i}$. For the moment, the polynomials have zero degree — the $c_i$ are constants of roughly best local approximation to $f$ on $Q_i$, because

$$\int_{Q_i} |f - c_i| \leq 2 \inf_c \int_{Q_i} |f - c| \,.$$

In view of this, it seems natural to try to extend the Calderón–Zygmund construction to couples of function spaces whose norms can be described in terms of local polynomial approximation. There are plenty of such spaces, see, e.g., Triebel's book [T]. Many deep results concerning them were obtained by Brudnyi in the 1970s, see the survey [Br1] (unfortunately, not all of these results were published at that time).[1] In particular, Brudnyi showed (see [Br]) that, when we deal with local approximation in $L^p$, a description in question is available for variation spaces and Besov spaces and, moreover, even for the Sobolev spaces $\dot{W}_q^k$ if $q > 1$ and the Sobolev embedding $\dot{W}_q^k \subset L^p$ occurs, which is equivalent to the inequality

$$\frac{k}{n} + \frac{1}{p} - \frac{1}{q} \geq 0,$$

where $n$ is the dimension (of $\mathbb{R}^n$). So, it is natural to look for Calderón–Zygmund type decompositions for couples of the form $\left(L^p, \dot{W}_q^k\right)$ with the above restriction on the parameters.

The second part of the book is devoted to this and other related problems. The most difficult matter is the construction of an appropriate family $\{Q_i\}_{i\in I}$ of cubes. The reason is that, in the case of the couple $\left(L^p, \dot{W}_q^k\right)$, the "good" function $f_t \in \dot{W}_q^k$ must be smooth to a certain extent, and so, instead of a piecewise polynomial approximant, we "glue" an approximant from polynomials with the help of a smooth partition of unity. As has already been said, the existence of such a partition of unity requires special properties of the family of cubes. Moreover, even finer properties will be needed to ensure due norm estimates. An appropriate family of cubes is produced with the help of new covering theorems (so-called controlled covering theorems), which provide subcoverings with a blend of properties that occur in the classical covering results by Whitney and Besicovitch.

It should be noted that merely taking Whitney cubes (as Stein did, and as it was done in the paper [BSh] in a setting related to Stein's) would not work here, if for no other reason than the fact that we must deal sometimes with coverings of the entire $\mathbb{R}^n$, which cannot be the case for Whitney coverings.

Now we discuss some applications of the techniques in question. It is convenient to start with two basic definitions. As was already mentioned in the Preface (which was written primarily for nonexperts), we perceive real interpolation as a theory of splitting a function into two parts ("big and small", "good and bad", etc.), which are going to belong to different spaces. The "quality" of such a splitting is measured in terms of various functionals. The distance functional (alias, $E$-functional) mentioned in the Preface and at the beginning of this Introduction is among them, but often the use of it is not quite convenient. More practical is to employ the $K$-functional defined below. Surely, it is related to the distance functional. This statement can be given a precise form; see formula (5.21) and Proposition 5.10, which show (by the way) that the relationship is not so transparent as it might be supposed.

[1] See also the quite recent monograph [BB].

For a couple of Banach spaces, the $K$-functional is introduced by the formula

$$K(t,x;X_0,X_1) = \inf_{x_1\in X_1} (\|x-x_1\|_{X_0} + t\,\|x_1\|_{X_1}).$$

It is in terms of this functional that the interpolation spaces of the real method are usually defined, see (5.8). Along with the $K$-functional, it is natural to consider a slightly more general (and sometimes more convenient) $L_{p,q}$-functional, defined for $0 < p,q < \infty$ by the formula

$$L_{p,q}(t,x;X_0,X_1) = \inf_{x_1\in X_1} (\|x-x_1\|_{X_0}^p + t\,\|x_1\|_{X_1}^q),\quad t>0.$$

The following problem has long been known in interpolation theory (see, e.g., p. 484 of the book [BK]): to find the $K$-functional for the couple $\left(\dot W_{p_0}^{k_0}, \dot W_{p_1}^{k_1}\right)$. In essence, the problem can be attributed to classical questions, but the answer had existed only in two particular cases, namely, for $p_0 = p_1$ (Peetre, see [Pe0]) and for $k_0 = k_1$ (DeVore–Scherer, see [DeVSch]; see also [CM] for a different proof).

Now it should be explained what precisely we mean by the symbol $\dot W_q^k$ (here $q \ge 1$). This is the standard notation for the *homogeneous* Sobolev space defined as the closure of the set of infinitely differentiable functions on $\mathbb{R}^n$ in the seminorm

$$\|g\|_{\dot W_q^k} = \sup_{k_1+\dots+k_n=k} \left\| \frac{\partial^k}{\partial^{k_1}x_1\partial^{k_2}x_2\dots\partial^{k_n}x_n} g(x)\right\|_{L^q}.$$

In the present book, not only do we give (among other things) a fairly simple formula for the $L_{p,q}$-functional for the couple $(L^p,\,\dot W_q^k)$, but also construct near-minimizers, i.e., functions at which the infimum in the definition of the $L_{p,q}$-functional is roughly attained.

In connection with the comments above, a general remark is in order. Surely, the knowledge of a near-minimizer always leads to the calculation of a $K$- or $L$-functional (up to equivalence). Generally speaking, this calculation *itself* is a difficult extremal problem. Despite the fact that the study of real interpolation has a rather long history (first publications appeared in the late 50s and early 60s of the 20th century), this problem has been resolved only for a relatively small number of couples. There are subtle methods of calculating such functionals (up to equivalence, of course) based, e.g., on Littlewood–Paley decompositions (see [BL] or, for example, [FJ] to represent more recent time) or wavelet expansions (see [BDKPW]). However, these methods provide *explicit* near-minimizers only in exceptional cases (at least, the authors do not know how to adapt them for this purpose in a systematic way). But this can be achieved by refinements of the Calderón–Zygmund decomposition, as we shall see in the present book.

We shall work under the restriction

$$\frac{k}{n} + \frac{1}{p} - \frac{1}{q} > 0 \tag{0.2}$$

(unlike the situation in Brudnyi's result, we require the strict inequality). We start the description of the main results with merely giving an explicit formula for the $L$-functional in question; it is convenient to slightly postpone explicit expressions for near-minimizers.

Throughout, by "a cube" we mean a closed cube with faces parallel to coordinate hyperplanes. Suppose that $p \neq q$. Given $f \in L^p = L^p(\mathbb{R}^n)$ and a number $t > 0$, we denote by $\Omega_{f,t}$ the set of cubes $Q$ satisfying

$$\frac{1}{|Q|^\alpha} \int_Q |f(x) - f_Q(x)|^p \, dx \geq t^p, \tag{0.3}$$

where

$$\alpha = 1 + \frac{k}{n(\frac{1}{p} - \frac{1}{q})} \tag{0.4}$$

and $f_Q(x)$ stands for a polynomial of degree strictly smaller than $k$ that provides the best approximation of $f$ on $Q$ in the $L^p$-metric:

$$\int_Q |f(x) - f_Q(x)|^p \, dx = \inf_{\deg P < k} \int_Q |f(x) - P|^p \, dx.$$

From (0.3) it follows that, whenever $\pi$ is a subfamily of $\Omega_{f,t}$ that consists of mutually disjoint cubes, we have

$$t^p \sum_{Q \in \pi} |Q|^\alpha \leq \|f\|_{L_p}^p < \infty.$$

Thus, the *$\alpha$-capacity* of the family $\Omega_{f,t}$ is finite. The $\alpha$-capacity is defined by the formula

$$|\Omega_{f,t}|_\alpha = \sup_\pi \Big(\sum_{Q \in \pi} |Q|^\alpha\Big),$$

where the supremum is taken over all subfamilies $\pi$ of mutually disjoint cubes in $\Omega_{f,t}$. The notion should not be confused with the $\alpha$-capacity of a *set* in potential theory.

**Theorem 0.2.** *Under the conditions $p \neq q$ and* (0.2), *we have the following equivalence* (*with constants independent of $f \in L^p$ and $t > 0$*):

$$L_{p,q}(t^{p-q}, f; L^p, \dot{W}_q^k) = \inf_{g \in \dot{W}_q^k} \left(\|f - g\|_{L^p}^p + t^{p-q} \|g\|_{\dot{W}_q^k}^q\right) \approx t^p \, |\Omega_{f,t}|_\alpha \,. \tag{0.5}$$

Recall that we have already mentioned a similar (though much less involved) formula in the case of the couple $(L^1, L^\infty)$. Specifically, the distance functional can be expressed in terms of the volumes of cubes arising in the *classical* Calderón–Zygmund decomposition; see (1.48).

Now, the proof of Theorem 0.2 provides also a near-minimizer for the functional (0.5); the construction will be outlined later in this Introduction, as has already been promised.

If we put formally $p = q$, we encounter degeneration: the parameter $\alpha$ becomes infinite. However, in this case the following recipe applies.

**Theorem 0.3.** *Split $\mathbb{R}^n$ into congruent cubes $Q_i$ with side $t$ and put $K_i = \frac{3}{2}Q_i$. Then*

$$K(t^k, f; L^p, \dot{W}_p^k) = \inf_{g \in \dot{W}_q^k} (\|f - g\|_{L^p} + t^k \|g\|_{\dot{W}_q^k}) \approx \left( \sum_{i \in I} \int_{K_i} |f - f_{K_i}|^p \right)^{\frac{1}{p}}.$$

We signalize immediately that these results can be extended to couples in which *both* components are Sobolev spaces. We state a counterpart of Theorem 0.2. For a multiindex $k$, denote by $D^k f$ the corresponding partial derivative of a function $f$.

**Theorem 0.4.** *Suppose $p \neq q$, $m > l$, and*

$$\frac{m - l}{n} + \frac{1}{p} - \frac{1}{q} > 0. \tag{0.6}$$

*Then the quantities $L_{p,q}(t^{p-q}, f; \dot{W}_p^l, \dot{W}_q^m)$ and $\sum_{|k|=m-l} L_{p,q}(t^{p-q}, D^k f; L^p, \dot{W}_q^{m-l})$ are equivalent.*

The second quantity is provided by Theorem 0.2. This "shift in the smoothness scale" is possible by stability theorems. We are going to discuss them in some more detail a bit later.

The above statements can be supplemented with formulas for distance functionals (alias, $E$-functionals)[2] for the couples $\left(L^p, \dot{C}_p^{s,k}\right)$, where $\dot{C}_p^{s,k}$ is the Morrey–Campanato space built on the basis of $L^p$. This is the space determined by the seminorm

$$\|f\|_{\dot{C}_p^{s,k}} = \sup_Q \frac{1}{|Q|^{\frac{s}{n}}} \left( \frac{1}{|Q|} \int_Q |f(x) - f_Q(x)|^p \, dx \right)^{\frac{1}{p}}, \quad s \in \left( -\frac{n}{p}, k \right]. \tag{0.7}$$

As previously, $f_Q(x)$ stands for a polynomial of degree strictly smaller than $k$ that provides the best approximation to $f$ on $Q$ in the $L^p$-metric, and $k = 0, 1 \ldots$. The parameter $s$ (if it is positive) indicates smoothness. For instance, if $0 < s \leq 1$ and $k = 1$, the space $\dot{C}_p^{s,1}$ coincides with the homogeneous Lipschitz class $\mathrm{Lip}_s$ (this is nontrivial, though widely known; see Chapter 1 for the proof). If $s = 0$, we arrive at the very important class BMO of functions with bounded mean oscillation. In this connection, we mention the paper [BSh] (already quoted) by Bennet and

[2]Heuristically, the distance functional in question (see Theorem 0.5) can be viewed as the limiting case ($q = \infty$) of the $L_{p,q}$-functionals treated above. Indeed, if we formally put $q = \infty$ in the expression $\|u - y\|_X^p + t^{p-q}\|y\|_Y^q$ (the infimum of which over $y \in Y$ is equal to $L_{p,q}(u, t^{p-q}; X, Y)$), we see that the second term is infinite or equals $t^p$ if $\|y\|_Y \geq t$, and is zero otherwise. So, for $q = \infty$, the infimum in question becomes $\inf_{\|y\|_Y < t} \|u - y\|_X^p$.

Sharpley, who in fact constructed a near-minimizer for the couple $(L^1, \mathrm{BMO})$ by using local approximations and the Whitney lemma. As in the case of Sobolev spaces, the Whitney lemma is insufficient for nonzero smoothness.

**Theorem 0.5.** *Define the family $\Omega_{f,t}$ of cubes by condition* (0.3) *with*

$$\alpha = 1 + \frac{sp}{n}. \tag{0.8}$$

*Then, with constants independent of $f \in L^p$ and $t > 0$, we have the equivalence*

$$\inf_{\|g\|_{\dot{C}_p^{\alpha,k}} \le t} \|f - g\|_{L^p}^p \approx t^p \left|\Omega_{f,t}\right|_\alpha .$$

A glance at the formulas shows that the parameter $\alpha$ in Theorems 0.3 and 0.5 runs through all nonzero real numbers. It is somewhat curious to analyze its behavior in more detail. The expression (0.8) coincides with (0.4) if we put $q = \infty$ and $s = k$ in the latter. Next, the case of $\alpha < 0$ corresponds to the couples $(L^p, \dot{W}_q^k)$ with $q < p$. Positive values of $\alpha$ correspond to the couples $(L^p, \dot{C}_p^{s,k})$ if $0 < \alpha \le \frac{kp}{n} + 1$ and again to $(L^p, \dot{W}_q^k)$ but with $q > p$ if $\alpha > \frac{kp}{n} + 1$. It should be noted that in Chapter 6 we shall establish covering theorems of two different sorts (opposite or dual to each other in a sense): controlled extension theorems and controlled contraction theorems. The proofs of the above results involve controlled extension for $\alpha > 0$ and controlled contraction for $\alpha < 0$. We feel that this distinction is more than a mere technicality.

There are two exceptional values, $\alpha = \infty$ and $\alpha = 0$, not covered by (0.8) and (0.4). The value $\alpha = \infty$ corresponds to $p = q$, see Theorem 0.3 for the result in this case. It might be supposed that the case of

$$\alpha = 1 + \frac{k}{n(\frac{1}{p} - \frac{1}{q})} = 0$$

corresponds to the couple $(L^p, \dot{W}_q^k)$ for a limit exponent:

$$\frac{k}{n} + \frac{1}{p} - \frac{1}{q} = 0.$$

However, things are more complicated. Unfortunately, some information is available here only in the case of dimension 2. This is discussed in Chapter 11. The construction presented there does stem from the algorithm for the couples $\left(L^p, \dot{W}_q^k\right)$ after a limit passage, but Sobolev spaces disappear "in a natural way" in the limit: the algorithm in question provides a near-minimizer for the couple $(L^p, V)$, where $V$ is a certain two-dimensional analog of the space of functions with bounded $(p/2)$-variation. If $p = 2$, we obtain the couple $\left(L^2, \mathrm{BV}\right)$, which is important in connection with image processing.

Before passage to algorithms that yield near-minimizers, we mention that certain *unbounded* operators may take some specific near-minimizers to near-minimizers. To better explain the nature of the problem, first we consider the case of an operator $T$ *bounded* on the couple $(X_0, X_1)$. If $x_t$ is a near-minimizer for the $L_{p,q}$-functional of $x$ with parameter $t$, then

$$\|Tx - Tx_t\|_{X_0}^p + t\,\|Tx_t\|_{X_1}^q \le c(T)(\|x - x_t\|_{X_0}^p + t\,\|x_t\|_{X_1}^q) \le c(T)L_{p,q}(t, x; X_0, X_1). \tag{0.9}$$

So, for $Tx_t$ to be a near-minimizer of $L_{p,q}(t, Tx; X_0, X_1)$, it suffices that

$$L_{p,q}(t, x; X_0, X_1) \le cL_{p,q}(t, Tx; X_0, X_1). \tag{0.10}$$

Now, we put a question.

**Problem 0.6.** *If we discard the assumption that $T$ is bounded (in practice this means that $T$ is defined on $X_0$ and $X_1$, but may fail to send these spaces into themselves), when is it still possible to deduce from* (0.10) *that $Tx_t$ is a near-minimizer for $L_{p,q}(t, Tx; X_0, X_1)$?*

Whenever the answer is in the positive, we say that the near-minimizer in question is *$T$-stable.*

The main result in this direction is that this question is answered in the positive for the *near-minimizers constructed in this book* if we choose singular integral operators and their relatives for the role of $T$. It should be noted that this is new even for the most well-studied couples of the form $(L^1, L^p)$. (A related result in the paper [Bo], which served for us as a motivation, is somewhat weaker. See Subsection 1.4.2.) We call such statements *stability theorems* for near-minimizers.

We discuss two sample results in order to show the power of stability theorems. In the first of them, we again return for a short while to the distance functional.

Let $F = (F_1, \dots, F_n)$ be a gradient vector field all components of which belong to $L^1$. We want to approximate it by a vector field $G = (G_1, \dots, G_n)$ with components lying in the ball of radius $t > 0$ in the space $\mathrm{Lip}_\alpha$. The question is whether the approximation becomes worse if we restrict ourselves to *gradient* fields $G$.

It turns out that it does not.

**Theorem 0.7.** *Suppose $\nabla f \in (L^1)^n$. Then there exists a function $g$ such that $\|\nabla g\|_{(\mathrm{Lip}_\alpha)^n} \le ct$ and*

$$\|\nabla f - \nabla g\|_{(L^1)^n} \le c \inf_{\|G\|_{(\mathrm{Lip}_\alpha)^n} \le \frac{t}{c}} \|\nabla f - G\|_{(L^1)^n}$$

*with a constant $c \ge 1$ independent of $f$ and $t > 0$.*

The proof of this fact is based on a stability theorem for the couple $(L^1, \mathrm{Lip}_\alpha)$ (here we mean spaces of vector-valued functions). The operator "$T$" from the above discussion is hidden in the statement. We disclose that this is the orthogonal projection $Q$ of the space $(L^2)^n$ onto its closed subspace of gradient vector fields. The theorem remains true if we replace $L^1$ with $L^p$, $1 < p < \infty$, but in that form it is much less involved because $Q$ is bounded on $L^p$ and on $\mathrm{Lip}_\alpha$. (By the way, these boundedness results are not immediate, so that the statement is nontrivial still.) Next, $Q$ can be applied to $n$-tuples of $L^1$-functions, but the image may fall out of $L^1$, i.e., $Q$ is unbounded on $L^1$. As to the relative of (0.10) for the distance functional (which, of course, must be ensured in any case), it is true indeed in the $L^1$-setting, because we must check it for the element $x = \nabla f$, and then $Qx = x$.

The second example is Theorem 0.4 above. If we put $l = 1$ in it for simplicity, that statement is a consequence of another stability theorem with the same projection $Q$. Again, it is the *special construction of near-minimizers discussed in this book* that makes them stable under $Q$.

After so much praise we have paid to our algorithm of finding near-minimizers, the time is ripe to sketch it in some detail. It consists of three steps (see Figure 1). At the first step, after analyzing local approximations to a function $f$, we construct the "initial family" $\{Q_x\}$ of cubes. At the second step, we apply a certain covering theorem to this family, which results in a new family $\{K_i\}_{i\in I}$ of cubes with special properties. At the third step, the family $\{K_i\}_{i\in I}$ is used to construct a partition of unity, and to write out a near-minimizer with its help.

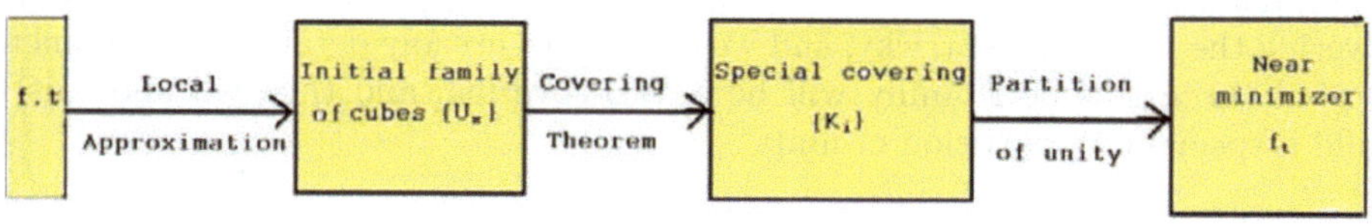


Figure 1: Three steps of the algorithm of constructing a near-minimizer.

Returning to our favorite $L_{p,q}$-functionals and Sobolev spaces, we signalize that the algorithms in question differ somewhat in the cases of $p < q$ and $p > q$. Below we describe the algorithm for the couples of the form $(L^p, \dot{W}_q^k)$ under the restriction

$$\frac{1}{p} < \frac{1}{q} < \frac{k}{n} + \frac{1}{p}, \tag{0.11}$$

which is equivalent to (0.2) for $q < p$.

So, given $f \in L^p$ and a number $t > 0$, we want to construct a function $f_t$ such that

$$\begin{aligned}\|f - f_t\|_{L^p}^p + t^{p-q}\,\|f_t\|_{\dot{W}_q^k}^q &\le c \inf_{u\in \dot{W}_q^k} (\|f - u\|_{L^p}^p + t^{p-q}\,\|u\|_{\dot{W}_q^k}^q) \\ &= cL_{p,q}(t^{p-q}, f; L^p, \dot{W}_q^k)\end{aligned} \tag{0.12}$$

with a constant $c > 0$ independent of $f \in L^p$ and $t > 0$.

In order to construct the initial family $\{Q_x\}$ of cubes (the first step of the algorithm), for every point $x \in \mathbb{R}^n$ we consider the function

$$\varphi_x(r) = \left( \frac{1}{|Q(x,r)|^\lambda} \int_{Q(x,r)} \left| f(x) - f_{Q(x,r)} \right|^p dx \right)^{\frac{1}{p}}, \quad \lambda = 1 + \frac{k}{n(\frac{1}{p} - \frac{1}{q})},$$

where $f_{Q(x,r)}$ is a polynomial of degree strictly smaller than $k$ that realizes the best approximation to $f$ on $Q(x,r)$ in the $L^p$-metric.

Since $\lambda < 0$ under the restrictions (0.11), the function $\varphi_x(r)$ is continuous and monotone nondecreasing, and its values range through all positive numbers. We note that, since no polynomials belong to $L^p(\mathbb{R}^n)$, the function $\varphi_x(r)$ cannot vanish identically.

Therefore, for every $x \in \mathbb{R}^n$ there exists $r = r_x$ (see Figure 2) such that

$$\varphi_x(r_x) = \left( \frac{1}{|Q(x,r)|^\lambda} \int_{Q(x,r)} \left| f(x) - f_{Q(x,r)} \right|^p dx \right)^{\frac{1}{p}} = t.$$

For the role of the initial family $\{Q_x\}$, we take the family $\{Q(x, r_x)\}$.

At the second step, with the help of a *controlled covering* theorem, out of the initial family we construct a new special family $\{K_i\}_{i \in I}$ of cubes. The controlled covering theorem is most tricky, and we do not discuss any details at the moment. We only note this new family will be a WB-covering, and this will allow us to build a special $C^\infty$-partition of unity

$$\sum_i \psi_i = 1$$

on $\mathbb{R}^n$. Among other things, the functions of this partition satisfy $\operatorname{supp} \psi_i \subset K_i$ and

$$\left| D^k \psi_i \right| \le \frac{c}{|K_i|^{\frac{|k|}{n}}}, \quad D^k = \frac{\partial^k}{\partial^{k_1} x_1 \dots \partial^{k_n} x_n},$$

where $c = c(k,n)$, $k = (k_1, \dots, k_n)$, $|k| = \sum_{i=1}^n k_i$.

Finally, a near-minimizer for the $L_{p,q}$-functional (0.12) is defined by the formula

$$f_t = \sum_i f_{K_i} \psi_i,$$

where $f_{K_i}$ stands for a polynomial of degree strictly smaller than $k$ that gives a nearly best approximation to $f$ on $K_i$ in the $L^p$-metric.

It should be noted that the algorithm is much simpler if $p = q$. We describe it. We cover $\mathbb{R}^n$ by a system $Q_i$ of cubes with disjoint interiors and with edge length $t$. Next, we put $K_i = \frac{3}{2} Q_i$. Starting with this family $\{K_i\}$, it is easy to construct a smooth partition $\sum_i \psi_i = 1$ of unity with the properties described above (in this

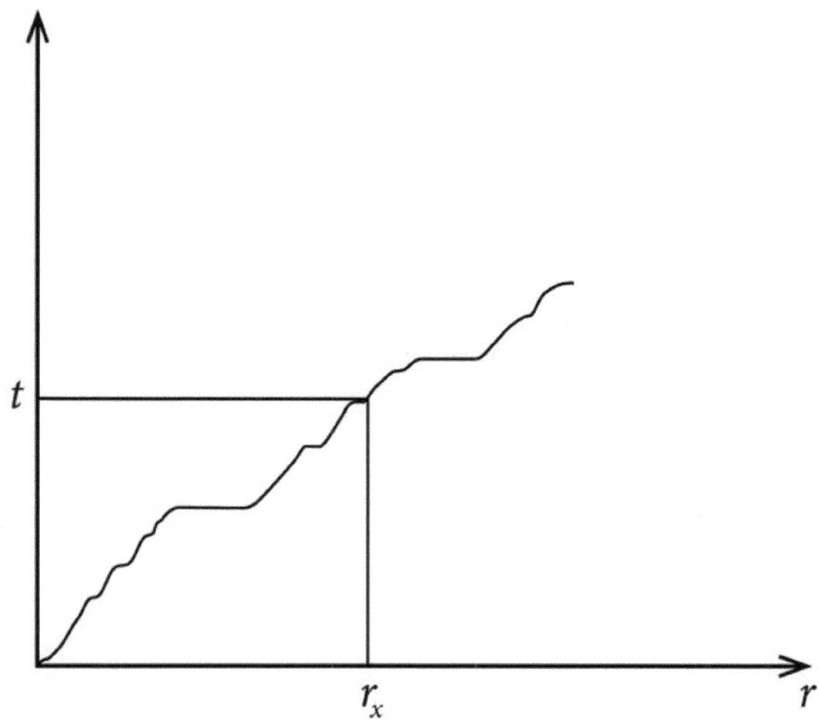


Figure 2: The graph of $\varphi_x$ and the point $r_x$.

case, all functions $\psi_i$ can be obtained from one by shifts). A near-minimizer for the $K$-functional

$$K(t^k, f; L^p, \dot{W}_p^k) = \inf_{u \in \mathrm{BV}} (\|f - u\|_{L^p} + t^k \|u\|_{\dot{W}_p^k})$$

is defined by the formula

$$f_t = \sum_i f_{K_i} \psi_i.$$

To finish the introduction, we list formally and very briefly the topics included. The first part "Background" deals with the classical Calderón–Zygmund decomposition and its smooth version (we present several well-known applications of these constructions, which are quite instructive), classical covering theorems, some information on spaces of smooth functions and on singular integrals. Two initial stability theorems are also included. We also show how the concepts and methods of real interpolation can be used to relate to one another near-minimizers for various functionals of interpolation theory and various couples of spaces. For the most part, detailed proofs are presented. The second part "Advanced theory" is devoted to refinements of the Calderón–Zygmund decomposition for couples different from $(L^1, L^\infty)$ (mostly, a space of smooth function is involved). This leads to constructions of near-minimizers for various functionals and to stability theorems, in the spirit described above in this Introduction. These refinements are obtained with the help of new covering theorems, called "controlled covering theorems". The term reflects the fact that the resulting covering is controlled by a certain family of cubes fixed beforehand. Originally, the most part of these results were obtained by N. Kruglyak in the 1990s; later, stability theorems were proved by the authors jointly.

The results will be presented for function spaces on $\mathbb{R}^n$ but, mostly, the arguments can be adjusted to convex domains in $\mathbb{R}^n$ with minor changes. See [K1], where the exposition was given for a cube.

# Definitions, notation, and some standard facts

Basically (though not entirely), this section plays the part of a guide. We hope it will enable the reader to quickly find the precise form of some definitions, or some important notation and references in the case of need. Some proofs are included for completeness.

## Geometry

$Q$ denotes a closed cube in $\mathbb{R}^n$ with edges parallel to coordinate axes;
$r(Q)$ is the radius of $Q$ (half its edge length);
$Q(x,r)$ is the cube in $\mathbb{R}^n$ centered at $x$ and of radius $r$;
$\lambda Q$ stands for the cube with the same center as $Q$, but of radius

$$r(\lambda Q) = \lambda r(Q);$$

$|\Omega|$ is the volume of a set $\Omega$;
$\|\cdot\|$ is the norm in $\mathbb{R}^n$ defined by the formula

$$\|x\| = \max_{i=1,\dots,n} |x_i|, \quad x = (x_1, \dots, x_n);$$

$D$ is the set of dyadic cubes in $\mathbb{R}^n$. A cube $Q$ is said to be dyadic if it is the translate of a cube of the form $[0, 2^m]^n$, $m \in \mathbb{Z}$, by a vector of the form $2^m k$, where $k \in \mathbb{Z}^n$.

**Definition 0.8.** A collection of pairwise nonintersecting cubes will be called a *packing*.

**Definition 0.9.** A collection $\{Q_i\}_{i\in I}$ of cubes is said to be *almost disjoint* if for every $i, j \in I$, $i \neq j$, we have

$$\min\left(r(Q_i), r(Q_j)\right) < \|x_i - x_j\|,$$

where $x_i, x_j$ are the centers of $Q_i$ and $Q_j$.

Geometrically, this means that the center of the bigger cube lies outside the smaller.

**Definition 0.10.** A family $\{Q_x\}_{x\in\Omega}$ of cubes (where each $Q_x$ is centered at $x$) is said to be Lipschitz with constant $q$ if

$$|r(Q_x) - r(Q_y)| \leq q\,\|x-y\|$$

for every $x, y \in \Omega$.

**Definition 0.11.** A family $\{Q_x\}_{x\in\Omega}$ of cubes is said to be strongly Lipschitz with constant $q$ if the function

$$F(x) = \begin{cases} r(Q_x), & x \in \Omega, \\ 0, & x \in \mathbb{R}^n \setminus \Omega \end{cases}$$

satisfies the Lipschitz condition with constant $q$:

$$|F(x) - F(y)| \leq q\,\|x-y\|\,, \quad x, y \in \mathbb{R}^n.$$

**Definition 0.12.** A family $\{K_i\}_{i\in I}$ of cubes is called a WB-covering of a set $\Omega \subset \mathbb{R}^n$ if it has the following properties:

- the Whitney property:

$$\bigcup_{i\in I}\Big(\frac{1}{1+\delta}K_i\Big) = \bigcup_{i\in I} K_i = \Omega;$$

- finite multiplicity:

$$\sum_{i\in I} \chi_{K_i} \leq M;$$

- strong engagement:

$$\text{if } K_i \cap K_j \neq \varnothing, \text{ then } |K_i \cap K_j| \geq \varepsilon \max(|K_i|\,, |K_j|), \qquad i, j \in I,$$

  where the positive constants $\delta$, $M$, and $\varepsilon$ do not depend on $i, j \in I$.

In many cases a family of cubes with the above properties will emerge in no relationship with any set $\Omega$ prescribed beforehand; then we still use the term "a WB-covering" meaning that the family covers its union.

**Definition 0.13.** The $\alpha$-capacity of a family $\{Q\}$ of cubes is the quantity

$$|\{Q\}|_\alpha = \sup_\pi \sum_{Q\in\pi} |Q|^\alpha < \infty,$$

where the supremum is taken over all subfamilies $\pi$ in $\{Q\}$ formed by pairwise nonintersecting cubes.

## Spaces

Let $E$ be a measurable subset of $\mathbb{R}^n$ of positive measure. As usual, $L^p(E)$ is the space of functions on $E$ determined by the norm

$$\|f\|_{L^p} = \left(\int_E |f(s)|^p \, ds\right)^{\frac{1}{p}},$$

with the usual modification for $p = \infty$;

$L^{1,\infty}(E)$ ("the weak $L^1$") is the space of functions on $E$ determined by the quasinorm

$$\|f\|_{L^{1,\infty}} = \sup_{t>0} t \cdot |\{x \in E : |f(x)| > t\}| \, .$$

We write simply $L^p$, etc. if $E = \mathbb{R}^n$.

If $\Omega$ is an open subset of $\mathbb{R}^n$, the symbol $L^p_{\text{loc}}(\Omega)$ denotes the space of functions that are locally in $L^p$; it consists of all functions $f$ on $\Omega$ such that every $x \in \Omega$ possesses a neighborhood $Q \subset \Omega$ for which the integral $\int_Q |f(s)|^p \, ds$ is finite. If $p = 1$, we talk about *locally integrable* functions on $\Omega$; again, we mean $\mathbb{R}^n$ if $\Omega$ is not indicated.

The (*homogeneous*) Sobolev space $\dot{W}^k_q = \dot{W}^k_q(\mathbb{R}^n)$ is defined to be the set of all distributions $g$ on $\mathbb{R}^n$ whose $k$th derivatives belong to $L^q(\Omega)$. This space is endowed with the natural seminorm

$$\|g\|_{\dot{W}^k_q} = \sup_{k_1+\cdots+k_n=k} \left\| \frac{\partial^k}{\partial^{k_1} x_1 \partial^{k_2} x_2 \dots \partial^{k_n} x_n} g(x) \right\|_{L^q} .$$

BMO is the space of functions on $\mathbb{R}^n$ determined by the seminorm

$$\|f\|_{\text{BMO}} = \sup_Q \frac{1}{|Q|} \int_Q |f(s) - f_Q| \, ds, \;\; f_Q = \frac{1}{|Q|} \int_Q f(s) \, ds. \tag{0.13}$$

$\mathcal{P}_k$ is the space of algebraic polynomials of degree strictly smaller than $k$. We agree that $\mathcal{P}_0 = \{0\}$.

Finally, the Morrey–Campanato spaces $\dot{C}^{s,k}_p$ are defined for $s \in (-n/p, k]$, $k$ being a nonnegative integer. If $p < \infty$, the space $\dot{C}^{s,k}_p$ is determined by the seminorm (cf. (0.7))

$$\|f\|_{\dot{C}^{s,k}_p} = \sup_Q \inf_{g \in \mathcal{P}_k} \frac{1}{|Q|^{\frac{s}{n}}} \left( \frac{1}{|Q|} \int_Q |f(x) - g(x)|^p \, dx \right)^{\frac{1}{p}}, \;\; s \in \left(-\frac{n}{p}, k\right]. \tag{0.14}$$

For $p = \infty$ (this implies that $0 \le s < k$), the seminorm is defined by

$$\|f\|_{\dot{C}^{s,k}_\infty} = \sup_Q \inf_{g \in \mathcal{P}_k} |Q|^{-s/n} \operatorname{ess\,sup}_Q |f - g|. \tag{0.15}$$

## Functionals of real interpolation

Let $X_0$, $X_1$ be two Banach spaces that form a Banach couple (usually, this means that they are embedded continuously in some linear topological space, though this requirement can be relaxed somewhat).

**Definition 0.14.** For $x \in X_0 + X_1$ and $t > 0$, we introduce the $K$-functional

$$K(t, x; X_0, X_1) = \inf_{x_1 \in X_1} (\|x - x_1\|_{X_0} + t\,\|x_1\|_{X_1}).$$

When $x$ is fixed, this will be viewed as a function of $t$ only and called the $K$-functional of the element $x$.

We do not hesitate to use the same definition if, say, $X_1$ is only a seminormed space, such as BMO or the homogeneous Sobolev space $\dot{W}_k^q$.

Though the definition of (near-)minimizers for the $K$-functional is clear from the preceding, we write it out.

**Definition 0.15.** An element $x_t \in X_1$ is called a near-minimizer for the $K$-functional of an element $x \in X_0 + X_1$ at the point $t$ if

$$\|x - x_t\|_{X_0} + t\,\|x_t\|_{X_1} \leq cK(t, x; X_0, X_1),$$

where $c$ is independent of $x$ and $t$. If $c = 1$, then we talk about an exact minimizer.

**Definition 0.16.** For $x \in X_0 + X_1$, the $L_{p,q}$-functional of $x$ is defined by the formula

$$L_{p,q}(t, x; X_0, X_1) = \inf_{x_1 \in X_1} (\|x - x_1\|_{X_0}^p + t\,\|x_1\|_{X_1}^q).$$

The definition of a near-minimizer for this functional is much similar to the preceding case, and we omit it.

**Definition 0.17.** For $x \in X_0 + X_1$, the $E$-functional (or the distance functional) is defined as follows:

$$E(t, x; X_0, X_1) = \operatorname{dist}_{X_0}(x, B_{X_1}(t)) = \inf_{u \in B_{X_1}(t)} \|x - u\|_{X_0}.$$

Here $B_{X_1}(t)$ stands for the ball of radius $t > 0$ and centered at 0 in $X_1$:

$$B_{X_1}(t) = \left\{x \in X_1 : \|x\|_{X_1} \leq t\right\}.$$

As to minimizers, here it is natural to allow even more flexibility than in the preceding two cases.

**Definition 0.18.** Suppose $t > 0$ is such that $\operatorname{dist}_{X_0}(x, B_{X_1}(t)) > 0$. An element $x_t \in X_1$ is called a near-minimizer for the distance functional $\operatorname{dist}_{X_0}(x, B_{X_1}(t))$ (alias $E$-functional) of an element $x \in X_0 + X_1$ if

$$\|x_t\|_{X_1} \leq ct \text{ and } \|x - x_t\|_{X_0} \leq c \operatorname{dist}_{X_0}\left(x, B_{X_1}\left(\frac{t}{c}\right)\right).$$

Here $c$ must not depend on $x$ and $t > 0$.

## Theorems

**Theorem 0.19** (Lebesgue theorem). *Suppose $f$ is a locally integrable function on $\mathbb{R}^n$. Then*

$$f(x) = \lim_{|Q|\to 0,\, x\in Q} \left( \frac{1}{|Q|} \int_Q f(s)\, ds \right)$$

*for almost every $x \in \mathbb{R}^n$.*

Recall that $\mathcal{P}_k$ stands for the space of algebraic polynomials of degree strictly smaller than $k$.

**Theorem 0.20** (Approximation by a Taylor polynomial). *Let $f$ be an infinitely differential function, and $f_T$ the Taylor polynomial for $f$ relative to the center of $Q$ and of order $k-1$. Then*

$$\left(\int_Q |f - f_T|^p\right)^{\frac{1}{p}} \le c\,|Q|^{\frac{1}{p}+\frac{k}{n}} \max_{|m|=k} \|D^m(f)\|_{L_\infty(Q)} \, .$$

*Here $c$ is independent of $Q$ and $f$.*

The proofs of the following two theorems will be given at the end of this section.

**Theorem 0.21** (Markov inequality). *If $Q$ is a cube and $P \in \mathcal{P}_k$, then*

$$\left\| \frac{\partial^l}{\partial^{l_1} x_1 \partial^{l_2} x_2 \dots \partial^{l_n} x_n} P \right\|_{L^\infty(Q)} \le c \frac{1}{|Q|^{\frac{l}{n}}} \|P\|_{L^\infty(Q)} \, ,$$

*where $l = l_1 + \dots + l_n$ and the constant $c > 0$ depends only on the dimension $n$ and the bound $k$ for the polynomial's degree.*

**Theorem 0.22** (On the growth of a polynomial). *Suppose $G$ is a convex domain and $P \in \mathcal{P}_k$. If a set $\Omega \subset G$ satisfies*

$$\frac{|\Omega|}{|G|} \ge \varepsilon > 0,$$

*then*

$$\|P\|_{L^\infty(G)} \le c\,\|P\|_{L^\infty(\Omega)} \, ,$$

*with a constant $c > 0$ depending only on $k$, $\varepsilon$, and the dimension $n$.*

For $f \in L^1_{\mathrm{loc}}(\mathbb{R}^n)$, the Hardy–Littlewood maximal function $Mf$ is defined by

$$(Mf)(x) = \sup_{x\in Q} \frac{1}{|Q|} \int_Q |f|, \; x \in \mathbb{R}^n. \tag{0.16}$$

We remind the reader of well-known estimates for $Mf$: $\|Mf\|_{L^s} \le C_{s,n}\|f\|_{L^s}$, $1 < s \le \infty$, and $\|Mf\|_{L^{1,\infty}} \le C_n\|f\|_{L^1}$, see, e.g., [S].[3] More involved is the following fact, to be employed in Part 2.

[3] These two estimates will be proved in passing in the sequel.

**Theorem 0.23** (Fefferman–Stein maximal theorem; see [FS]). *Let $1 < p, q < \infty$. Then*

$$\left\| \left( \sum_{i=1}^{\infty} (Mf_i)^q \right)^{\frac{1}{q}} \right\|_{L^p} \leq c \left\| \left( \sum_{i=1}^{\infty} |f_i|^q \right)^{\frac{1}{q}} \right\|_{L^p}.$$

## Two improper proofs

We do not mean "incorrect". Any proper proof of Theorems 0.21 and 0.22 should give some idea of the dependence of the constant $c$ on $k$, $\varepsilon$, and $n$. But nowhere in what follows shall we need the knowledge of this dependence. If we do not need any explicit control of $c$, fairly simple arguments are available. For simplicity, in Theorem 0.22 we assume that $G$ is also a cube $Q$ (this is the only case we shall need).

Now, by shifts and dilations the two theorems in question are reduced to the case of the cube $Q_0 = [-1/2, 1/2]^n$ of unit volume. For $Q_0$, in the case of Theorem 0.21 we consider the differential operator $T : f \mapsto \frac{\partial^l}{\partial^{l_1} x_1 \ldots \partial^{l_n} x_n} f$. Then Theorem 0.21 simply claims that, on the *finite-dimensional* space $\mathcal{P}_k$ with the norm of $L^\infty(Q_0)$, this operator is bounded (as a mapping to $L^\infty(Q_0)$ in our situation). This is a commonplace. Theorem 0.22 is more involved, because, even for $Q_0$, we must ensure the dependence of $c$ only of the measure rather than on the position of the subset $\Omega$. This can be done by the following argument due to Yu. Brudnyi.

Suppose for some $\varepsilon > 0$ there is a sequence of subsets $\Omega_m$ of $Q_0$ with measure of at least $\varepsilon$ each, and a sequence of polynomials $p_m$ in $\mathcal{P}_k$ such that $\max_{x \in Q_0} |p_m(x)| = 1$ but $a_m = \sup_{x \in \Omega_m} |p_m(x)| \to 0$. Since the unit sphere of the finite-dimensional space $\mathcal{P}_k$ is compact, we may assume that the sequence $p_m$ tends uniformly to some polynomial $p$ with $\max_{x \in Q_0} p(x) = 1$. Now the set $\{x \in Q_0 : p(x) = 0\}$ has zero Lebesgue measure, so it is contained in a relatively open set $G \subset Q_0$ of measure strictly smaller than $\varepsilon$. But then each $\Omega_m$ intersects the set $F = Q_0 \setminus G$, so $\min_{x \in F} |p_m(x)| \leq a_m$. Since $F$ is closed, this implies that $p$ has a zero in $F$, a contradiction.

However, we shall need more information on the behavior of the constant $c$ in case $\Omega$ is also a cube. Moreover, we shall need the following subtler fact.

**Lemma 0.24.** *For any two cubes $K \subset K_1$, and a polynomial $P$ of degree at most $k$, we have*

$$\|P\|_{L^\infty(K_1)} \leq c \left( \frac{|K_1|}{|K|} \right)^{\frac{k}{n}} \left( \frac{1}{|K|} \int_K |P| \right)$$

*with a constant $c$ depending only on $k$ and the dimension $n$.*

*Proof.* The claim is an easy consequence of Lemma 0.25 below. Consider a cube $Q$ with center $y$ and a polynomial $P$ of degree $l$. Then, clearly, $P$ can be expanded in the powers of the variables $x_i - y_i$: $P(x) = \sum_{|m| \leq l} a_m (x_1 - y_1)^{m_1} \ldots (x_n - y_n)^{m_n}$.

**Lemma 0.25.** *With some constants $b$ and $B$ depending only on $l$ and $n$, we have*

$$b\|P\|_{L^\infty(Q)} \le \sum_{|m|\le l} |a_m||Q|^{\frac{|m|}{n}} \le B\left(\frac{1}{|Q|}\int_Q |P|\right).$$

To verify this, we can again argue "improperly": if $Q = [-1/2, 1/2]^n$, the existence of $b$ and $B$ is merely the equivalence of all norms on a finite-dimensional space. Then the case of a general $Q$ follows by dilations and shifts. □

# Part I

# Background

The nature of this part is reflected by its title. Our aim here is to quickly introduce the notions and concepts required in the sequel, to prove basic statements about them, and to discuss their interrelation in the light of real interpolation.

In Chapter 1, we describe the classical procedure invented by Calderón and Zygmund, which splits a function in two parts with good properties. We discuss some applications of this procedure and relate it to real interpolation; also, we prove our first stability result for near-minimizers (in a rather elementary setting). Chapter 2 is a brief introduction to the theory of singular integral operators (for us, they serve as "raw material" for stability theorems). In Chapter 3 we prove classical covering theorems due to Besicovitch, Wiener, and Whitney, describe a smooth version of the Calderón–Zygmund decomposition, and present our first serious stability statement. In Chapter 4 we discuss some basic facts about homogeneous spaces of smooth functions and prove that, typically, singular integral operators are bounded on them. Chapter 5 is a brief summary of the real interpolation theory in its relationship with near-minimizers.

In two final Chapters 6 and 7, we dwell on two topics related to the theory under study but staying somewhat apart. In Chapter 6 we give a brief account of the use of near-minimizers for $L$-functionals in ill-posed problems, particularly in those arising in image processing. In Chapter 7, we discuss a different method of producing near-minimizers. This method is not related to decompositions of Calderón–Zygmund type. It is applicable only in the framework of certain spaces of analytic functions of one variable, but leads to very powerful results. The reader interested only in nonclassical Calderón–Zygmund procedures may pass directly to Chapter 8 after Chapter 5.

# Chapter 1

# Classical Calderón–Zygmund decomposition and real interpolation

We start with the Riesz rising sun lemma, and then pass to its many-dimensional substitute, i.e., the Calderón–Zygmund lemma. We discuss some applications of the latter, the John–Nirenberg and Campanato inequalities being among them. To relate the matter to real interpolation, we observe that the so-called "good" part of the Calderón–Zygmund decomposition is a near-minimizer for the couple $(L^1, L^\infty)$. Next, in a rather elementary setting, we prove that certain near-minimizers are stable under the action of certain unbounded operators that are nevertheless "bounded far away" from the support of the function to which they are applied. They will be called *long-range regular operators.*

## 1.1 Riesz rising sun lemma and the Calderón–Zygmund procedure

### 1.1.1 Riesz rising sun lemma

In 1932, F. Riesz (see [R]) suggested a simple proof of the Hardy–Littlewood maximal theorem based on the following lemma, the geometric interpretation of which has led to its name.

**Lemma 1.1** (Rising sun lemma). *Given a function $f \in L^1(\mathbb{R})$ and a number $t > 0$, there exist at most countably many finite closed intervals $Q_i = [a_i, b_i]$ $(i \in I)$ with disjoint interiors such that*

$$\frac{1}{b_i - a_i} \int_{a_i}^{b_i} |f(s)|\, ds = t, \qquad i \in I, \tag{1.1}$$

*and*

$$\left\| f\chi_{\mathbb{R}\setminus \bigcup Q_i} \right\|_{L^\infty} \le t. \tag{1.2}$$

*Proof.* Consider the function

$$F(x) = \int_0^x |f(s)|\, ds - tx. \tag{1.3}$$

Since $f \in L^1$, it is clear that $F(x)$ is continuous and satisfies

$$\lim_{x\to-\infty} F(x) = \infty, \qquad \lim_{x\to\infty} F(x) = -\infty. \tag{1.4}$$

Consider the graph of $F(x)$ and perceive it as a mountain chain illuminated by rising sun (the rays arrive from the right and are parallel to the $x$-axis); see Figure 1.1. Let $\Omega$ denote the projection to the $x$-axis of the "shadowed" pieces of the graph.

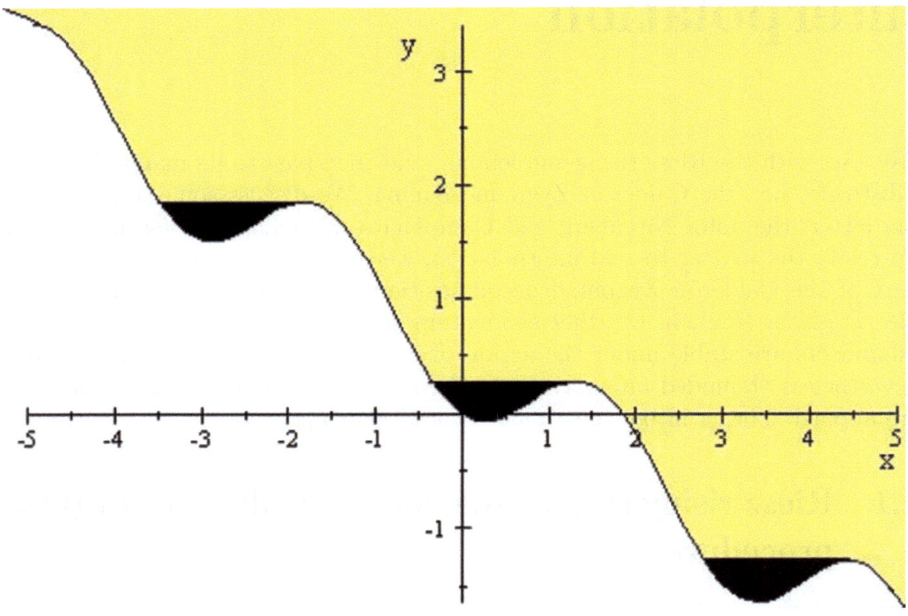


Figure 1.1: The graph of $F(x)$ illuminated by light rays parallel to the $x$-axis and arriving from the right.

Now, $x \in \Omega$ means that there exist $y$ with

$$y > x \text{ and } F(y) > F(x).$$

Since $F$ is continuous, we see that $\Omega$ is open and, consequently, it splits into at most countably many open intervals $(a_i, b_i)$; by (1.4), all of them are finite. At the ends of each interval, $F(x)$ takes equal values, that is, $F(a_i) = F(b_i)$. Thus,

$$\int_0^{a_i} |f(s)|\, ds - ta_i = \int_0^{b_i} |f(s)|\, ds - tb_i,$$

which implies (1.1).

We prove (1.2). If $x$ lies outside $\Omega$, then the point $(x, F(x))$ of the graph is not shadowed, which means that $F(x) \geq F(x+h)$ for every $h > 0$, implying

$$\int_0^{x} |f(s)|\, ds - tx \geq \int_0^{x+h} |f(s)|\, ds - t(x+h).$$

Consequently,

$$\frac{1}{h}\int_x^{x+h} |f(s)|\, ds \leq t$$

for $h > 0$; by the Lebesgue theorem (Theorem 0.19), we have $|f(x)| \leq t$ a.e. outside $\Omega$, which proves (1.2). □

### 1.1.2 Calderón–Zygmund lemma

In 1952, for the purposes of the theory of singular integrals, A. Calderón and A. Zygmund (see [CZ]) proved the following statement, which can be viewed as a multidimensional analog of the rising sun lemma. We refer the reader to "Definitions, notation, and some standard facts" after the Introduction for the notion of dyadic cubes, to be used in the proof.

**Lemma 1.2** (Calderón–Zygmund). *Given a finction $f \in L^1(\mathbb{R}^n)$ and a number $t > 0$, there exists an at most countable family $Q_i$ $(i \in I)$ of dyadic cubes with mutually disjoint interiors and such that*

$$t \leq \frac{1}{|Q_i|}\int_{Q_i} |f(s)|\, ds < 2^n t, \qquad i \in I, \tag{1.5}$$

$$\left\| f\chi_{\mathbb{R}^n \setminus \bigcup Q_i} \right\|_{L^\infty} \leq t. \tag{1.6}$$

Note that (1.6) is a precise analog of (1.2), but instead of equality in (1.1) we only have equivalence in (1.5).

*Proof.* In the set $D$ of dyadic cubes, consider the subset $D_{f,t}$ of all cubes satisfying the left inequality in (1.5):

$$D_{f,t} = \left\{ Q \in D : \frac{1}{|Q|}\int_Q |f(s)|\, ds \geq t \right\}. \tag{1.7}$$

For the role of $\{Q_i\}_{i \in I}$, we can take the family of maximal cubes in $D_{f,t}$ (a cube $Q \in D_{f,t}$ is *maximal* if it is not included in any other cube of the family $D_{f,t}$).

If $Q_i$ is a maximal cube in $D_{f,t}$, take the dyadic cube $\tilde{Q}_i$ containing $Q_i$ and of volume $2^n |Q_i|$. Then $\tilde{Q}_i$ is not in $D_{f,t}$, whence

$$\frac{1}{|Q_i|}\int_{Q_i} |f(s)|\, ds \le \frac{2^n}{\left|\tilde{Q}_i\right|}\int_{\tilde{Q}_i} |f(s)|\, ds < 2^n t.$$

Thus, $Q_i$ satisfies (1.5). To prove (1.6), first we observe that, since $f \in L^1$, for every cube $Q \in D_{f,t}$ we have

$$|Q| \le \frac{1}{t}\int_Q |f(s)|\, ds \le \frac{1}{t}\|f\|_{L^1}. \tag{1.8}$$

This volume restriction implies that an arbitrary cube $Q \in D_{f,t}$ is included in some maximal cube $Q_i \in D_{f,t}$. We see that if $x \in \mathbb{R}^n \setminus \bigcup Q_i$, then any dyadic cube $Q$ containing $x$ is not in $D_{f,t}$, whence

$$\frac{1}{|Q|}\int_Q |f(s)|\, ds < t.$$

Now, the Lebesgue theorem (see Theorem 0.19) shows that $|f(x)| \le t$ for a.e. $x \in \mathbb{R}^n \setminus \bigcup Q_i$, which is (1.6). □

We mention a "more constructive" form of the same procedure in the original exposition by Calderón and Zygmund. They started with splitting $\mathbb{R}^n$ into congruent dyadic cubes of volume greater than $\frac{1}{t}\|f\|_{L^1}$. This ensures the inequality $\frac{1}{|Q|}\int_Q |f(s)|\, ds < t$ for each of them; also, it is easily seen that any member of $D_{f,t}$ is included in one of these cubes (see (1.8)). At the next step, any such cube underwent splitting in $2^n$ congruent dyadic cubes, among which those belonging to $D_{f,t}$ (i.e., satisfying $\frac{1}{|Q|}\int_Q |f(s)|\, ds \ge t$) were selected. The procedure was repeated consecutively for the remaining cubes (no cubes selected at some stage underwent further splitting). Clearly, this leads to the same family $\{Q_i\}_{i\in I}$ of cubes as above. An experienced reader will readily recognize here a stopping time argument from the theory of martingales.

It should be noted that in place of the *standard* dyadic grid $D$ we could work with any grid $D_1$ obtained from $D$ by dilation and shift. A useful application of this is as follows: if $Q$ is an arbitrary cube, we can choose $D_1$ in such a way that it contains $Q$.

### 1.1.3 Calderón–Zygmund decomposition

Amazingly, the simple geometric lemmas discussed above lead to deep results in Fourier analysis. A bridge to many of them is splitting a function $f \in L^1(\mathbb{R}^n)$ in a "good" and a "bad" part.

As before, suppose that we are given a function $f \in L^1(\mathbb{R}^n)$ and a number $t > 0$. Taking the family $\{Q_i\}_{i\in I}$ from the Calderón–Zygmund lemma, we introduce the "good" part of $f$ by the formula

$$f_t = \sum_{i\in I} \left( \frac{1}{|Q_i|} \int_{Q_i} f(s)ds \right) \chi_{Q_i} + f\, \chi_{\mathbb{R}^n \setminus \bigcup Q_i}. \tag{1.9}$$

The "bad" part is then given by

$$f - f_t = \sum_{i\in I} \left( f - \frac{1}{|Q_i|} \int_{Q_i} f(s)ds \right) \chi_{Q_i}. \tag{1.10}$$

The Calderón–Zygmund decomposition of $f$ (at the level $t$) is defined to be

$$f = f_t + (f - f_t)\,. \tag{1.11}$$

With this decomposition, we relate two linear operators $P_{f,t}$ and $I - P_{f,t}$ given by

$$P_{f,t}(h) = \sum_{i\in I} \left( \frac{1}{|Q_i|} \int_{Q_i} h(s)ds \right) \chi_{Q_i} + h\chi_{\mathbb{R}^n \setminus \bigcup Q_i} \tag{1.12}$$

and

$$(I - P_{f,t})(h) = \sum_{i\in I} \left( h - \frac{1}{|Q_i|} \int_{Q_i} h(s)ds \right) \chi_{Q_i}. \tag{1.13}$$

They (and especially their analogs in subtler settings) will be of much use in what follows. Clearly,

$$P_{f,t}(f) = f_t, \; (I - P_{f,t})f = f - f_t. \tag{1.14}$$

The Calderón–Zygmund decomposition possesses a series of simple but important properties. We collect them in the lemma below. In order to state it, we introduce the notation

$$B_{L^\infty}(t) = \{g \in L^\infty : \|g\|_{L^\infty} \le t\} \tag{1.15}$$

(this is the ball of radius $t$ and centered at zero in $L^\infty$) and define the distance functional

$$\operatorname{dist}_{L^1}(f, B_{L^\infty}(t)) = \inf_{g\in B_{L^\infty}(t)} \|f - g\|_{L^1}\,. \tag{1.16}$$

Also, we refer the reader to "Definitions, notation, and some standard facts" after the Introduction for a definition of the space BMO.

**Lemma 1.3.** *The Calderón–Zygmund decomposition has the following properties.*

(CZ1) (A capacity estimate.) *The volumes of the cubes $Q_i$, $i \in I$, obey the inequality*

$$\sum_{i\in I} |Q_i| \le \frac{2}{t} \operatorname{dist}_{L^1}(f, B_{L^\infty}(\tfrac{t}{2})). \tag{1.17}$$

(CZ2) (Splitting the "bad" function into blocks.) *Put*

$$b_i = \left(f - \frac{1}{|Q_i|}\int_{Q_i} f(s)ds\right)\chi_{Q_i} \qquad i \in I. \tag{1.18}$$

*Then*

$$f - f_t = \sum_{i\in I} b_i \tag{1.19}$$

*and all the blocks* $b_i$*,* $i \in I$*, satisfy*

$$\operatorname{supp} b_i \subset Q_i, \qquad \int_{\mathbb{R}^n} b_i(s)ds = 0, \qquad \|b_i\|_{L^1} < 2^{n+1}t\,|Q_i|. \tag{1.20}$$

(CZ3) (Approximation property.) *We have*

$$\|f_t\|_{L^\infty} \le 2^n t; \tag{1.21}$$

$$\|f - f_t\|_{L^1} \le 2^{n+2}\operatorname{dist}_{L^1}(f, B_{L^\infty}(\frac{t}{2})). \tag{1.22}$$

(CZ4) (Operator estimates.) *We have*

$$\|P_{f,t}\|_{L^1\to L^1} \le 1, \qquad \|I - P_{f,t}\|_{\mathrm{BMO}\to L^1} \le \frac{2}{t}\operatorname{dist}_{L^1}(f, B_{L^\infty}(\frac{t}{2})). \tag{1.23}$$

*Proof.* (CZ1) Let $g \in B_{L^\infty}(\frac{t}{2})$. The left inequality in (1.5) shows that

$$\begin{aligned}\|f-g\|_{L^1} &\ge \sum_{i\in I}\int_{Q_i}|f-g| \ge \sum_{i\in I}\int_{Q_i}|f| - \sum_{i\in I}\int_{Q_i}|g| \\ &\ge t\sum_{i\in I}|Q_i| - \frac{t}{2}\sum_{i\in I}|Q_i| = \frac{t}{2}\sum_{i\in I}|Q_i|.\end{aligned}$$

Since $g \in B_{L^\infty}(\frac{t}{2})$ is arbitrary, we obtain (1.17).

(CZ2) This is an immediate consequence of (1.10) and the right inequality in (1.5).

(CZ3) Inequality (1.21) follows from the definition (1.9) and the Calderón–Zygmund lemma. By (1.19) and (1.20), we see that

$$\|f - f_t\|_{L^1} < 2^{n+1}t\sum_{i\in I}|Q_i|.$$

It remains to apply (1.17).

(CZ4) The inequality $\|P_{f,t}(h)\|_{L^1} \le \|h\|_{L^1}$ is a consequence of the definition (1.12). If $h \in \mathrm{BMO}$, then (1.13) implies

$$\begin{aligned}\|(I - P_{f,t})h\|_{L^1} &= \sum_{i\in I}\int_{Q_i}\left|h(y) - \frac{1}{|Q_i|}\int_{Q_i}h(s)\,ds\right|dy \\ &\le \|h\|_{\mathrm{BMO}}\sum_{i\in I}|Q_i|.\end{aligned}$$

It remains to refer to (1.17). □

### 1.1.4 A weak type inequality for linear operators

We want to illustrate the use of the Calderón–Zygmund decomposition. The arguments that follow are a slight formalization of a procedure in [CZ]. We need a definition.

**Definition 1.4.** Suppose $1 < p \le \infty$, and let $T : L^p \to L^p$ be a bounded linear operator. This operator is said to be *long-range $L^1$-regular* (or often simply long-range regular) if for every function $f \in L^p \cap L^1$ supported on a cube $Q$ and having zero average (i.e., $\int f(s)ds = 0$) we have

$$\int_{\mathbb{R}^n \setminus 2Q} |Tf(s)|\, ds \le \mathrm{lr}(T)\, \|f\|_{L^1} \tag{1.24}$$

with a constant $\mathrm{lr}(T) > 0$ independent of $f$ and $Q$.

Surely, $T$ is long-range regular if it is bounded on $L^1$. However, we shall see that some very important operators not bounded on $L^1$ are long-range regular. At the moment we show (under the guidance of [CZ]) that the long-range regular operators send $L^1$ to $L^{1,\infty}$.

**Theorem 1.5.** *Let $1 < p \le \infty$, and let $T$ be a linear operator that maps the space $L^p = L^p(\mathbb{R}^n)$ into itself boundedly. If $T$ is long-range $L^1$-regular, then for all $f \in L^p \cap L^1$ we have*

$$\|Tf\|_{L^{1,\infty}} \le C\, \|f\|_{L^1} \tag{1.25}$$

*with a constant $C > 0$ independent of $f$.*

We remind the reader that the space $L^{1,\infty}$ (or weak-$L^1$) is determined by the quasinorm $\|f\|_{1,\infty} = \sup_{t>0} t|\{|f| > t\}|$.

*Remark* 1.6. An operator $T$ satisfying (1.25) (equivalently, satisfying $|\{|Tf| > t\}| \le ct^{-1}\|f\|_{L^1}$, $t > 0$) is said to be *of weak type* $(1, 1)$. This definition applies also to sublinear[1] operators that may fail to be linear.

*Remark* 1.7. The proof below yields $C \le d(2^{2n(1-1/p)}\|T\|_{L^p \to L^p} + \mathrm{lr}(T))$, where $d$ is a numerical constant. We leave apart the question about the optimality of this estimate.

*Remark* 1.8. Since $L^1 \cap L^p$ is dense in $L^1$, the theorem shows that $T$ extends by continuity up to a bounded operator from $L^1$ to $L^{1,\infty}$. For this extension, (1.24) remains true if, instead of $f \in L^p \cap L^1$, we demand merely that $f \in L^1$.

*Proof.* We must prove that

$$|\{x : |(Tf)(x)| > t\}| \le Ct^{-1}\, \|f\|_{L^1}$$

[1] An operator $T$ is *sublinear* if $|T(\alpha f + \beta g)| \le |\alpha||Tf| + |\beta||Tg|$.

for all $t > 0$. Fixing $t$, we consider the Calderón–Zygmund decomposition $f = f_{wt} + (f - f_{wt})$ (see (1.11)), for some positive constant $w$. Next, we take two more positive numbers $u$ and $v$ such that $u + v = 1$. Since

$$\{x : |(Tf)(x)| > t\} \subset \{x : |(Tf_{wt})(x)| > ut\} \cup \{x : |T(f - f_{wt})(x)| > vt\}, \quad (1.26)$$

it suffices to estimate the measures of the sets on the right separately.[2]

First, assume that $p < \infty$ (otherwise the proof simplifies; see below). The definition (1.9) of the "good" function $f_{wt}$ implies

$$\|f_{wt}\|_{L^1} \le \|f\|_{L^1};$$

consequently, $f_{wt} \in L^1 \cap L^\infty \subset L^p$. Since $T$ is bounded on $L^p$ and $\|f_{wt}\|_{L^\infty} \le 2^n wt$ (cf. (1.21)), we see that

$$\begin{aligned} |\{x : |(Tf_{wt})(x)| > ut\}| &\le \frac{1}{(ut)^p} \int_{\mathbb{R}^n} |Tf_{wt}(s)|^p \, ds \\ &\le \frac{1}{(ut)^p} \|T\|^p_{L^p \to L^p} \int_{\mathbb{R}^n} |f_{wt}(s)|^p \, ds \le \frac{1}{(ut)^p} \|T\|^p_{L^p \to L^p} (2^n wt)^{p-1} \|f_{wt}\|_{L^1} \\ &\le \|T\|^p_{L^p \to L^p} (2^n w)^{p-1} \frac{1}{u^p} \frac{1}{t} \|f\|_{L^1}. \end{aligned} \quad (1.27)$$

To estimate the second summand in (1.26), we observe that

$$|\{x : |T(f - f_{wt})(x)| > vt\}| \le |\cup 2Q_i| + |\{x \in \mathbb{R}^n \setminus \cup 2Q_i : |T(f - f_{wt})(x)| > vt\}|.$$

We remind the reader that the "bad" function $f - f_{wt}$ is the sum of the blocks $b_i = (f - \frac{1}{|Q_i|} \int_{Q_i} f(s) ds) \chi_{Q_i}$ with properties (1.20). Now, long-range regularity implies

$$\begin{aligned} |\{x \in \mathbb{R}^n \setminus \cup 2Q_i : |T(f - f_{wt})(x)| > vt\}| &\le \frac{1}{vt} \int_{\mathbb{R}^n \setminus \cup 2Q_i} |T(f - f_{wt})(s)| \, ds \\ &\le \frac{1}{vt} \sum_{i \in I} \int_{\mathbb{R}^n \setminus 2Q_i} |(Tb_i)(s)| \, ds \le \operatorname{lr}(T) \frac{1}{vt} \sum_{i \in I} \int_{\mathbb{R}^n} |b_i(s)| \, ds \le \frac{2 \operatorname{lr}(T)}{vt} \int_{\mathbb{R}^n} |f(s)| \, ds. \end{aligned}$$

Therefore, the inequality

$$\left| \bigcup 2Q_i \right| \le 2^n \sum_{i \in I} |Q_i| \quad (1.28)$$

combined with (1.5) yields

$$\begin{aligned} |\{x : |T(f - f_{wt})(x)| > vt\}| &\le 2^n \sum_{i \in I} |Q_i| + \frac{2 \operatorname{lr}(T)}{vt} \int_{\mathbb{R}^n} |f(s)| \, ds \\ &\le \left( \frac{2^n}{w} + \frac{2 \operatorname{lr}(T)}{v} \right) \frac{1}{t} \int_{\mathbb{R}^n} |f(s)| \, ds. \end{aligned} \quad (1.29)$$

[2] If we are not interested in an estimate for $C$ (see Remark 1.7), we can simply take, say, $w = 1$ and $u = v = 1/2$.

Since $\sum_{i\in I}|Q_i| \le t^{-1}\int_{\mathbb{R}^n}|f(s)|ds$ by (1.5), inequalities (1.27) and (1.29) show that, indeed, the measure of the set where $|Tf(x)|>t$ does not exceed $Ct^{-1}\|f\|_{L^1}$, as required. The constant $C$ can easily be written out; the expression for it will involve the norm of $T$ on $L^p$, the quantity $\mathrm{lr}(T)$, and three positive parameters $u$, $v$, and $w$ ($u+v=1$). Minimization over these parameters yields Remark 1.7.

For $p=\infty$, we simply put $u=v=1/2$ and $w=1/(4\cdot 2^n\|T\|_{L^\infty\to L^\infty})$. Then the first summand in (1.26) is 0, and the above estimates imply inequality (1.25) with $C=d(2^{2n}\|T\|_{L^\infty\to L^\infty}+\mathrm{lr}(T))$. □

It should be mentioned that long-range regularity is typical of singular integral operators. The basics of their theory will be presented in the sequel. For the moment we mention that the Hilbert transformation, or the projection $Q$ onto the gradient vector fields (discussed in the Introduction), or the partial sum operators related to wavelet expansions are all long-range regular.

### 1.1.5 Hardy–Littlewood maximal operator

We remind the reader that the Hardy–Littlewood maximal operator is defined on the functions $f\in L^1_{\mathrm{loc}}$ by the formula

$$Mf(x)=\sup_{Q\,:\,x\in Q}\frac{1}{|Q|}\int_Q |f(y)|\,dy. \tag{1.30}$$

Though nonlinear, this operator is subadditive, i.e.,

$$M(\lambda f)=|\lambda|\,Mf, \qquad M(f+g)\le Mf+Mg. \tag{1.31}$$

Also, it is bounded on $L^\infty$, that is

$$\|Mf\|_{L^\infty}\le\|f\|_{L^\infty}.$$

Since also $Mf\ge|f|$ by the definition (1.30) and the Lebesgue theorem (Theorem 0.19), we see that in fact

$$\|Mf\|_{L^\infty}=\|f\|_{L^\infty}. \tag{1.32}$$

The operator $M$ is of weak type (1, 1), i.e., satisfies the inequality

$$\|Mf\|_{L^{1,\infty}}\le c\,\|f\|_{L^1} \tag{1.33}$$

with a constant $c>0$ depending only on the dimension $n$. This fact is very well known, so, rather, the emphasis in the statement below is on an intimate relationship between Lebesgue sets of $Mf$ and the cubes involved in the Calderón–Zygmund decompositions of $f$. (This statement does imply (1.33), though.) The techniques employed will be similar to but somewhat different from those in the proof of Theorem 1.5.[3]

[3] Subsequently, it will be explained that (1.33) can be obtained by direct application of Theorem 1.5 to a certain linear and long-range regular operator bounded on $L^\infty$. Surely, this is also well known.

**Theorem 1.9.** *Suppose that $f \in L^1$ and $s > 0$. Put $t = \frac{s}{10^{n+1}}$ and denote by $Q_i$, $i \in I$, the cubes occurring in the Calderón–Zygmund decomposition of $f$ at the level $t$. Then*

$$\{x : Mf(x) > s\} \subset \bigcup_{i \in I} 2Q_i. \tag{1.34}$$

*Proof.* By the subadditivity of $M$, we can write (as before):

$$\{x : Mf(x) > s\} \subset \left\{x : Mf_t(x) > \frac{s}{2}\right\} \cup \left\{x : M(f - f_t)(x) > \frac{s}{2}\right\}. \tag{1.35}$$

The choice of $t$ shows that $\|f_t\|_{L^\infty} \leq 2^n t \leq \frac{s}{2}$ (see (1.21)), whence $|Mf_t| \leq s/2$ everywhere. Therefore, the first set on the right in (1.35) is empty. We claim that the second set is included in $\bigcup_{i \in I} 2Q_i$.

To establish the claim, take $x \notin \bigcup 2Q_i$. We again use the formula $f - f_t = \sum_{i \in I} b_i$, where the blocks $b_i$ possess the properties (1.20). Therefore, for every cube $Q$ containing $x$ we have

$$\frac{1}{|Q|} \int_Q |f - f_t| \leq \frac{1}{|Q|} \sum_{i : Q_i \cap Q \neq \varnothing} \int_{Q_i} |b_i| \leq \frac{2^{n+1} t}{|Q|} \sum_{i : Q_i \cap Q \neq \varnothing} |Q_i|. \tag{1.36}$$

Since $x$ is in $Q$ but outside $\bigcup_{i \in I} 2Q_i$, geometric considerations show that any $Q_i$ intersecting $Q$ is included in $5Q$. Since the interiors of the $Q_i$ are disjoint, it follows that

$$\sum_{i : Q_i \cap Q \neq \varnothing} |Q_i| \leq |5Q| = 5^n |Q|.$$

So, by (1.36), for $x \notin \cup 2Q_i$ we obtain

$$\begin{aligned} M(f - f_t)(x) &= \sup_{Q \ni x} \frac{1}{|Q|} \int_Q |f - f_t| \leq \sup_{Q \ni x} \frac{2^{n+1} t}{|Q|} \sum_{i : Q_i \cap Q \neq \varnothing} |Q_i| \\ &\leq 2^{n+1} 5^n t \leq \frac{s}{2}, \end{aligned}$$

proving the claim. ☐

As has already been mentioned, the weak type inequality (1.33) is a consequence of the above theorem. Moreover, it implies the following (formally) stronger *Kolmogorov inequality.*

**Theorem 1.10.** *For every $s > 0$, we have*

$$|\{x : Mf(x) > s\}| \leq \frac{c}{s} \operatorname{dist}_{L^1}(f, B_{L^\infty}(\frac{s}{c}))$$

*with a constant $c$ depending only on the dimension $n$.*

*Proof.* Indeed, we put $t = \frac{s}{10^{n+1}}$ as in the preceding theorem and combine it with Lemma 1.3 to obtain

$$|\{x : Mf(x) > s\}| \leq \Big|\bigcup_{i\in I} 2Q_i\Big| \leq 2^n \cdot \frac{2}{t}\,\mathrm{dist}_{L^1}\Big(f, B_{L^\infty}\Big(\frac{t}{2}\Big)\Big)$$
$$\leq 20^{n+1}\frac{1}{s}\,\mathrm{dist}_{L^1}\Big(f, B_{L^\infty}\Big(\frac{s}{20^{n+1}}\Big)\Big). \qquad \square$$

We include a well-known calculation based on the Kolmogorov inequality (i.e., on Theorem 1.10) and showing that the operator $M$ is bounded on $L^p$ for $1 < p < \infty$. Since $\mathrm{dist}_{L^1}(f, B_{L^\infty}(s/c)) \leq \int_{|f|\geq s/c} |f(x)|dx$, we can write

$$\|Mf\|_{L^p}^p = p\int_0^\infty s^{p-1}|\{Mf > s\}|\,ds \leq C\int_0^\infty s^{p-2}\int_{|f|>s/c}|f(x)|\,dx\,ds$$
$$= C\int_{\mathbb{R}^n}|f(x)|\int_0^{c|f(x)|} s^{p-2}\,ds\,dx = C_1\int_{\mathbb{R}^n}|f(x)|^p\,dx.$$

## 1.2 Norms on BMO and Lipschitz spaces

Here we present two more results the proofs of which are spectacular applications of the Calderón–Zygmund procedure. The arguments are similar in nature, but differ at a point, so we give them separately. Both results are about the Morrey–Campanato spaces $\dot{C}_p^{s,k}$ defined in (0.14), (0.15). (The definitions will be reproduced and discussed below.) Now our interest lies in the case where the smoothness parameter $s$ is nonnegative; then, for historical reasons, it is natural to call the spaces in question simply *Campanato spaces*. The first result says that, in the case of zero smoothness, the space does not depend on $p$ for $1 \leq p < \infty$.

**Proposition 1.11.** BMO $= \dot{C}_p^{0,1}$ *for every* $1 \leq p < \infty$.

The second result says that for $s > 0$ the equivalence of norms on $\dot{C}_p^{s,k}$ extends up to $p = \infty$ inclusive.

**Proposition 1.12.** *If* $s > 0$, *the space* $\dot{C}_p^{s,k}$ *does not depend on* $p$, $1 \leq p \leq \infty$. *In other words, the functional*

$$f \mapsto \sup_Q \inf_{q\in\mathcal{P}_k} |Q|^{-s/n}\,\mathrm{ess\,sup}_{x\in Q}|f(x) - q(x)| \tag{1.37}$$

*(see* (0.15)*) is an equivalent seminorm on* $\dot{C}_p^{s,k}$ *for every finite* $p \geq 1$.

### 1.2.1 John–Nirenberg inequality

We remind the reader that BMO is the space of locally integrable functions $f$ on $\mathbb{R}^n$ for which the following seminorm is finite:

$$\|f\|_{\mathrm{BMO}} = \sup_Q \frac{1}{|Q|} \int_Q |f - f_Q|,$$

where $f_Q = |Q|^{-1} \int_Q f$ and the supremum is taken over all cubes $Q$ in $\mathbb{R}^n$ (as usual, with edges parallel to coordinate axes). This seminorm vanishes precisely on the constant functions; so, BMO is sooner a space of classes modulo constants than of functions.

In fact, up to renorming, we deal here with the (Morrey–)Campanato space $\dot{C}_1^{0,1}$. We recall the formula for the seminorm on the latter space:

$$\|f\|_{\dot{C}_1^{0,1}} = \sup_Q \inf_c \frac{1}{|Q|} \int_Q |f(x) - c| dx,$$

where the inner infimum is taken over all constants $c$ (polynomials of degree zero). Comparison of the formulas shows that in the first case we subtract in the integrand the average $f_Q = |Q|^{-1} \int_Q f$, and in the second case we subtract (in fact) a constant of best approximation (the infimum over $c$ is attained), so that, clearly, the second seminorm does not exceed the first. As to the reverse estimate, for every constant $c$ we have $|f_Q - c| \leq |Q|^{-1} \int_Q |f - c|$, so that we make an admissible error when subtracting $f_Q$ in place of the best constant.

It turns out that, in fact, the functions in BMO obey a condition much stronger than local integrability and much stronger than it was claimed in Proposition 1.11.

**Theorem 1.13** (John–Nirenberg inequality). *Suppose $f \in \mathrm{BMO}$. Then for every cube $Q$ and every $t > 0$ we have*

$$|\{x \in Q : |f(x) - f_Q| > t\}| \leq C|Q| \exp\left(\frac{-ct}{\|f\|_{\mathrm{BMO}}}\right), \tag{1.38}$$

*where the constants $C$ and $c$ depend only on the dimension $n$.*

*Proof.* Without loss of generality, we assume that $\|f\|_{\mathrm{BMO}} \leq 1$. We fix a cube $Q$ and put

$$\varphi(t) = \sup_{\|g\|_{\mathrm{BMO}} \leq 1} \frac{|\{x \in Q : |g(x) - g_Q| > t\}|}{\int_Q |g - g_Q|}.$$

Multiplication of the independent variable by a constant generates an isometry of BMO, and the same is true for a shift in $\mathbb{R}^n$. So, the function $\varphi$ does not depend on $Q$ (in particular, we may and do assume that $Q$ is a dyadic cube). Clearly, $\varphi$ is monotone nonincreasing and $\varphi(t) \leq t^{-1}$. We claim that

$$\varphi(d + 2^n A) \leq \varphi(d)/A$$

whenever $d > 0$ and $A > 1$.

To show this, we consider the function $G = |g - g_Q|\chi_Q$, where the BMO-norm of $g$ is at most 1, and apply the Calderón–Zygmund lemma at the level $A$ to it. Since $G_Q \le 1 \le A$ and we have agreed that $Q$ is a dyadic cube, all resulting cubes $Q_i$ will be included in $Q$. Now, we have

$$|g_Q - g_{Q_i}| \le \frac{1}{|Q_i|}\int_{Q_i} |g - g_Q| \le 2^n A, \tag{1.39}$$

whence it follows that, for every $i$, if $x \in Q_i$ and $|g(x) - g_Q| > d + 2^n A$, then $|g(x) - g_{Q_i}| > d$. Since $|g - g_Q| \le A$ outside the union of $Q_i$ on $Q$, we obtain

$$\begin{aligned}\{x \in Q : \ |g(x) - g_Q| > d + 2^n A\} &\subset \bigcup_i \{x \in Q_i : \ |g(x) - g_Q| > d + 2^n A\} \\ &\subset \bigcup_i \{x \in Q_i : \ |g(x) - g_{Q_i}| > d\}.\end{aligned} \tag{1.40}$$

Recalling that $\int_{Q_i} |g - g_{Q_i}| \le |Q_i|$ (because the BMO-norm of $g$ is at most 1), finally we can write

$$\begin{aligned}|\{x \in Q : \ |g(x) - g_Q| > d + 2^n A\}| &\le \sum_i |\{x \in Q_i : \ |g(x) - g_{Q_i}| > d\}| \\ &\le \varphi(d) \sum_i \int_{Q_i} |g - g_{Q_i}| \le \varphi(d) \sum_i |Q_i| \\ &\le \frac{\varphi(d)}{A} \sum_i \int_{Q_i} |g - g_Q| \le \frac{\varphi(d)}{A} \int_Q |g - g_Q|,\end{aligned}$$

which is our claim.

The rest is easy. We put $A = e$, $d = 2^n e$, and recall that $\varphi(d) \le d^{-1}$. Adding consecutively $2^n A$ to the argument, we arrive at the inequality $\varphi(2^n ek) \le \frac{1}{2^n e^k}$, $k \in \mathbb{N}$. Since $\varphi$ is monotone nonincreasing, we obtain $\varphi(t) \le C \exp(-ct)$ if $t \ge 2^n e$. Finally, if $f$ is a function with BMO-norm not exceeding 1, then $|Q| \ge \int_Q |f - f_Q|$ for every cube $Q$, and we arrive at

$$\frac{|\{x \in Q : \ |f - f_Q| > t\}|}{|Q|} \le \varphi(t) \le C \exp(-ct)$$

if $t \ge 2^n e$. But the fraction on the left does not exceed 1 for all $t$. □

The John–Nirenberg inequality allows us to conclude that, whenever $f \in$ BMO is of norm at most 1, for every $p \in (1, \infty)$ and every cube $Q$ we have

$$\frac{1}{|Q|}\int_Q |f - f_Q|^p \le p \int_0^\infty t^{p-1}\varphi(t)dt \le C(p, n)^p,$$

where $C(p,n)$ is a constant depending only on $p$ and $n$. Thus,

$$f \mapsto \sup_Q \left( \frac{1}{|Q|} \int |f - f_Q|^p \right)^{1/p}$$

is an equivalent seminorm on BMO for every $p \geq 1$. This proves Proposition 1.11.

### 1.2.2 Equivalence of Campanato norms

An argument first found in [Me][4] and much similar to the above allows us to prove Proposition 1.12. We reproduce the definition of the seminorm on $\dot{C}_p^{s,k}$:

$$\|f\|_{\dot{C}_p^{s,k}} = \sup_Q \frac{1}{|Q|^{\frac{s}{n}}} \|f - f_Q\|_{L^p(Q, \frac{dx}{|Q|})}, \; 1 \leq p \leq \infty, \; 0 < s \leq k, \tag{1.41}$$

where $f_Q$ is a polynomial of degree strictly smaller than $k$ that provides the best approximation to $f$ on $Q$ in the $L^p$-metric. As usual, $Q$ stands for an arbitrary cube with edges parallel to coordinate axes.

As in the case of BMO, we can change the meaning of $f_Q$ in some "unified" manner without changing the space. Indeed, put $Q_0 = [-1/2, 1/2]^n$, and let $P$ be an arbitrary projection from $L^1(Q_0)$ onto the space $\mathcal{P}_k$ endowed with the metric of $L^1(Q_0)$. Clearly, $P$ remains bounded if we endow its *range* with the metric of $L^\infty(Q_0)$ because the space $\mathcal{P}_k$ is finite-dimensional; *a fortiori*, $P$ is bounded on $L^p(Q_0)$, $1 \leq p \leq \infty$. We transplant $P$ to an arbitrary cube $Q$ by using a shift and a dilation. Then the norm on $L^p(Q; \frac{dx}{|Q|})$ of the resulting projection $P_Q$ does not depend on $Q$.[5] At the same time, if $u$ is an arbitrary polynomial in $\mathcal{P}_k$, then $P_Q(f-u) = P_Q f - u$, whence we see that $\|P_Q f - u\|_{L^p(Q; \frac{dx}{|Q|})} \leq C\|f - u\|_{L^p(Q; \frac{dx}{|Q|})}$. Thus, (1.41) yields an equivalent seminorm on $\dot{C}_p^{s,k}$ if we interpret $f_Q$ as $P_Q f$. Below we shall understand (1.41) this way.

After this modification, the seminorms (1.41) grow with $p$. Therefore, Proposition 1.12 reduces to the following statement.

**Theorem 1.14.** *If $0 < s \leq k$, then the spaces $\dot{C}_1^{s,k}$ and $\dot{C}_\infty^{s,k}$ coincide (with equivalent seminorms).*

Clearly, it suffices to prove the inequality

$$\|f\|_{\dot{C}_\infty^{s,k}} \leq C\|f\|_{\dot{C}_1^{s,k}}. \tag{1.42}$$

Moreover, it suffices to establish the following lemma.

**Lemma 1.15.** *There is a constant $B$ such that, whenever $\|f\|_{\dot{C}_1^{s,k}} \leq 1$, we have $|f(x) - w(x)| \leq B$ for some polynomial $w \in \mathcal{P}_k$ (depending on $f$) and for a.e. $x$ with $\|x\| \leq 1$.*

---

[4] In [Me] only the situation of $k = 1$ and $0 < s \leq 1$ was treated.

[5] Moreover, the inequality $\|P_Q f\|_{L^\infty(Q)} \leq c(|Q|^{-1} \int_Q |f|)$ is fulfilled with $c$ independent of $|Q|$. This will be used below.

Suppose the lemma is proved. Then, taking an arbitrary function $h$ with $\|h\|_{\dot{C}_1^{s,k}} \le 1$ and arbitrary $\lambda > 0$ and $a \in \mathbb{R}^n$, we put $f(u) = \lambda^{-s} h(\lambda u + a)$. It is easily seen that $\|f\|_{\dot{C}_1^{s,k}} \le 1$. Applying the lemma to $f$, we see that there is $w \in \mathcal{P}_k$ such that $|h(\lambda u + a) - w(u)| \le B\lambda^s$ for a.e. $u$ with $\|u\| \le 1$. But then $x = \lambda u + a$ may be an *arbitrary* point with $\|x - a\| \le \lambda$. Since a linear change of variables takes $\mathcal{P}_k$ onto itself, we see that $h$ belongs to $\dot{C}_\infty^{s,k}$.

*Proof of Lemma* 1.15. Let $K$ be a cube. We introduce a function $\varphi_K$ nearly as in the proof of the John–Nirenberg inequality:

$$\varphi_K(t) = \sup_{\|g\|_{\dot{C}_1^{s,1}} \le 1} \frac{|\{x \in K : |g(x) - g_K| > t\}|}{\int_K |g - g_K|}.$$

This function does not change if we translate $K$, but it changes if we dilate it:

$$\varphi_{\lambda K}(t) = \lambda^{-s} \varphi_K(\lambda^{-s} t), \quad \lambda > 0. \tag{1.43}$$

Now, let $Q$ be a cube with edge length 1 (equivalently, with $|Q| = 1$). Take $d > 0$ and $A > 1$, and consider a function $g$ with $\|g\|_{\dot{C}_1^{s,k}} \le 1$. As in the preceding subsection, we apply the Calderón–Zygmund procedure at the level $A$ to $|g - g_Q|\chi_Q$, obtaining some subcubes $Q_i$ of $Q$. By the continuity properties of the operators $P_Q$ discussed above, an analog of formula (1.39) is still valid: if $x \in Q$, then

$$|g_Q(x) - g_{Q_i}(x)| = |(P_{Q_i}(g_Q - g))(x)| \le C\frac{1}{|Q_i|}\int_{Q_i} |g - g_Q| \le 2^n CA.$$

Now, an analog of (1.40) looks like this:

$$\{x \in Q : \ |g(x) - g_Q(x)| > d + C2^n A\} \subset \bigcup_i \{x \in Q_i : \ |g(x) - g_{Q_i}(x)| > d\},$$

whence we obtain

$$\begin{aligned} |\{x \in Q : |g(x) - g_Q(x)| > d + C2^n A\}| &\le \sum_i |\{x \in Q_i : |g(x) - g_{Q_i}(x)| > d\}| \\ \le \sum_i \varphi_{Q_i}(d) \int_{Q_i} |g - g_{Q_i}| &= \sum_i |Q_i|^{-\frac{s}{n}} \varphi_Q(|Q_i|^{-\frac{s}{n}} d) \int_{Q_i} |g - g_{Q_i}|. \end{aligned} \tag{1.44}$$

The properties of the Calderón–Zygmund cubes yield

$$\sum_i |Q_i| \le \frac{1}{A}\int_Q |g - g_Q|. \tag{1.45}$$

In particular, $|Q_i| \le \frac{1}{A}\int_Q |g - g_Q| \le \frac{1}{A}|Q|^{1+\frac{s}{n}} = \frac{1}{A}$ for all $i$. Since $\varphi_Q$ is monotone decreasing, we see that $\varphi_Q(|Q_i|^{-\frac{s}{n}} d) \le \varphi_Q(A^{\frac{s}{n}} d)$. So, the last quantity in (1.44) is dominated by

$$\varphi_Q(A^{\frac{s}{n}}d)\sum_i |Q_i|^{-\frac{s}{n}}|Q_i|^{1+\frac{s}{n}} = \varphi_Q(A^{\frac{s}{n}}d)\sum_i |Q_i| \le \frac{1}{A}\varphi_Q(A^{\frac{s}{n}}d)\int_Q |g-g_Q|,$$

see (1.45). Collecting the estimates, dividing by $\int_Q |g-g_Q|$, and passing to the supremum over $g$, we arrive at

$$\varphi_Q(d + C2^n A) \le \frac{1}{A}\varphi_Q(A^{\frac{s}{n}}d).$$

Since $\varphi_Q$ is monotone decreasing, this cannot be true for large $d$ unless the left-hand side is 0. Specifically, putting, say, $A = 2$ and $d = C\frac{2^{n+1}}{2^{s/n}-1}$, we obtain $\varphi_Q(B) \le \frac{1}{2}\varphi_Q(B)$ with $B = C\frac{2^{n+1+s/n}}{2^{s/n}-1}$, that is, $\varphi_Q(B) = 0$. This means that $|g(x) - g_Q(x)| \le B$ a.e. on $Q$ whenever $\|g\|_{\dot{C}_1^{\alpha,1}} \le 1$, as required. □

We remind the reader of the definition of the (homogeneous) Lipschitz space $\mathrm{Lip}_s$, $0 < s \le 1$: it is determined by the seminorm

$$\|f\|_{\mathrm{Lip}_s} = \sup_{x\ne y}\frac{|f(x)-f(y)|}{\|x-y\|^s},$$

$f$ being a function on $\mathbb{R}^n$. The result of this subsection shows that, for $0 < s \le 1$, the space $\dot{C}_p^{s,1}$ is merely $\mathrm{Lip}_s$ in disguise, whenever $p \in [1,\infty]$. Thus, $\mathrm{Lip}_s$ acquires a family of *Campanato renormings*. They are quite useful (relevant examples will appear subsequently in this book).

**Corollary 1.16.** *If $0 < s \le 1$ and $1 \le p < \infty$, then the spaces $\mathrm{Lip}_s$ and $\dot{C}_p^{s,1}$ coincide (with equivalent norms).*

*Proof.* Indeed, by Theorem 1.14, it suffices to prove the claim for $p = \infty$, but then it is easy. If $f \in \mathrm{Lip}_s$ and $Q$ is a cube centered at $x_0$, then $|f(x) - f(x_0)| \le c\|x - x_0\|^s \le c|Q|^{\frac{s}{n}}$, so that $f \in \dot{C}_\infty^{s,1}$. Conversely, suppose $f \in \dot{C}_\infty^{s,1}$ is of norm at most 1 in this space. An easy approximation argument involving convolution with an approximate identity allows us to assume that $f$ is continuous. For arbitrary $x, y \in \mathbb{R}^n$, let $Q$ be a cube containing $x$ and $y$ whose edge length is the smallest possible, then $|Q|^{\frac{1}{n}} = \|x - y\|$. But by assumption and since now $f$ is continuous, there is a constant $c$ with $|f(z) - c| \le |Q|^{\frac{s}{n}}$ for every $z \in Q$. Consequently,

$$|f(x) - f(y)| \le |f(x) - c| + |f(y) - c| \le 2\|x - y\|^s. \qquad \square$$

A more complete analysis of the same phenomenon for general spaces $\dot{C}_p^{s,k}$ will be given in Chapter 4. Now we only signalize that Corollary 1.16 extends to arbitrary smoothness indices $s$, but this is not easy (though well-known).

## 1.3 Relationship with real interpolation

Much more detail on real interpolation will be presented in Chapter 5. In this section we summarize some information absolutely indispensable already at the present stage. Specifically, we explain how the Calderón–Zygmund decomposition enters interpolatory questions. The *moralité* can be read in Subsections A and B below. In Part 2 of the book, these results will be extended.

Let $X_0$, $X_1$ be Banach spaces that form a Banach couple. As has already been mentioned, the main device for measuring the quality of splitting of a vector $x \in X_0 + X_1$ in two summands (that is, the *principal tool* of real interpolation) is Peetre's $K$-functional defined by the formula

$$K(t, x; X_0, X_1) = \inf_{x_1 \in X_1} (\|x - x_1\|_{X_0} + t\,\|x_1\|_{X_1}), \quad t > 0.$$

An element $x_t \in X_1$ at which the infimum is attained roughly is called a near-minimizer for the $K$-functional of $x$ at the point $t$. It is clear from the definition that the $K$-functional can be expressed in terms of the slightly more geometric $E$-functional[6]

$$E(s, x; X_0, X_1) = \inf_{\|x_1\|_{X_1} \le s} \|x - x_1\|_{X_0}\,.$$

Specifically, we have

$$K(t, x; X_0, X_1) = \inf_{s>0} \; (E(s, x; X_0, X_1) + ts).$$

Conversely, we have the formula

$$E(s, x; X_0, X_1) = \sup_{t>0}(K(t, x; X_0, X_1) - ts).$$

Surely, the $E$-functional is what was called the distance functional in the Preface and Introduction:

$$E(s, x; X_0, X_1) = \operatorname{dist}_{X_0}(x, B_{X_1}(s)).$$

An element $x_{s,E} \in X_1$ is called a near-minimizer for the distance functional (alias, $E$-functional) of a vector $x \in X_0 + X_1$ at a point $s$ if

$$\|x_{s,E}\|_{X_1} \le \gamma s \text{ and } \|x - x_{s,E}\|_{X_0} \le \gamma \operatorname{dist}_{X_0}(x, B_{X_1}(\frac{s}{\gamma})).$$

Now, put

$$\varphi(t) = \frac{2\gamma K(t, x; X_0, X_1)}{t}. \tag{1.46}$$

Then $x_{\varphi(t),E}$ is a near-minimizer for the $K$-functional of $x$ at $t$. Conversely, if $x_t$ is a near-minimizer for the $K$-functional of $x$ at $t$, then $x_t$ is a near-minimizer for the $E$-functional of $x$ roughly at $\varphi(t)$. See Proposition 5.10 for the relevant calculations.

[6]In the literature, the spaces in the notation for the $E$-functional are usually listed in the opposite order. We have taken the liberty to do as we do because, in fact, in this book we write "dist(...)" instead of "$E(\dots)$" whenever possible.

**A) Calderón–Zygmund decomposition and near-minimizers**

In Lemma 1.3 it was shown that the "good" function

$$f_t = \sum_{i\in I}\left(\frac{1}{|Q_i|}\int_{Q_i} f(s)ds\right)\chi_{Q_i} + f\,\chi_{\mathbb{R}^n\setminus\bigcup Q_i}$$

possesses the "approximation property"

$$\|f_t\|_{L^\infty}\le 2^n t \text{ and } \|f-f_t\|_{L^1}\le 2^{n+2}\operatorname{dist}_{L^1}(f,B_{L^\infty}(\tfrac{t}{2})).$$

This means simply that $f_t$ is a near-minimizer, with $\gamma = 2^{n+2}$, for $\operatorname{dist}_{L^1}(f, B_{L^\infty}(t))$ (the $E$-functional of the couple $(L^1, L^\infty)$) at the point $t$. Therefore, putting

$$\varphi(t) = \frac{2^{n+3}K(t,f;L^1,L^\infty)}{t} \tag{1.47}$$

and referring to the above formulas, we see that $f_{\varphi(t)}$ is a near-minimizer for the $K$-functional of the couple $(L^1, L^\infty)$, that is,

$$\left\|f - f_{\varphi(t)}\right\|_{L^1} + t\left\|f_{\varphi(t)}\right\|_{L^\infty}\le cK(t,f;L^1,L^\infty).$$

**B) Volumes of cubes and the distance functional**

In Lemma 1.3 it was shown that

$$\sum_{i\in I}|Q_i|\le\frac{2}{t}\operatorname{dist}_{L^1}(f,B_{L^\infty}(\tfrac{t}{2})).$$

Furthermore, by (1.19)-(1.20) we have

$$\operatorname{dist}_{L^1}(f,B_{L^\infty}(2^n t))\le\|f-f_t\|_{L^1}\le\sum_{i\in I}\|b_i\|_{L^1}\le 2^{n+1}t\sum_{i\in I}|Q_i|\,.$$

We arrive at the two-sided estimate

$$\frac{1}{2^{n+1}}\operatorname{dist}_{L^1}(f,B_{L^\infty}(2^{n+1}t))\le t\sum_{i\in I}|Q_i|\le 2\operatorname{dist}_{L^1}(f,B_{L^\infty}(\tfrac{t}{2})).$$

It is convenient to write the same in the form

$$\operatorname{dist}_{L^1}(f,B_{L^\infty}(t))\approx t\sum_{i\in I}|Q_i|\,. \tag{1.48}$$

Thus, the distance functional can be expressed in terms of the volumes of the cubes $Q_i$ constructed in the Calderón–Zygmund lemma (for the level $t$).

### C) Operator estimates and linearization

Lemma 1.3 says that the linear operators

$$P_{f,t}(h) = \sum_{i \in I} \left( \frac{1}{|Q_i|} \int_{Q_i} h(s) ds \right) \chi_{Q_i} + h \chi_{\mathbb{R}^n \setminus \bigcup Q_i}$$

possess the following properties: $P_{f,t}(f) = f_t$ and

$$\|P_{f,t}\|_{L^1 \to L^1} \leq 1, \|I - P_{f,t}\|_{\mathrm{BMO} \to L^1} \leq \frac{2}{t} \operatorname{dist}_{L^1}(f, B_{L^\infty}(\frac{t}{2})) .$$

Since $\|I - P_{f,t}\|_{L^\infty \to L^1} \leq \|I - P_{f,t}\|_{\mathrm{BMO} \to L^1}$, we see that the operators $V_{f,t} = P_{f,\varphi(t)}$ with $\varphi(t)$ as in (1.47) "linearize" (in a sense) the $K$-functional of the couple $(L^1, L^\infty)$:

$$\|f - V_{f,t}(f)\|_{L^1} + t \|V_{f,t}(f)\|_{L^\infty} \leq cK(t, f; L^1, L^\infty)$$

with a constant $c \geq 1$ independent of $f$ and $t$. We shall see in the sequel that a similar linearization is available for many other couples of spaces.

### D) The couple $(L^1, L^p)$, $1 < p < \infty$

It turns out that, after a parameter change, the *same* Calderón–Zygmund procedure makes it possible to explicitly exhibit a near-minimizer $u^{(s)}$ for the $E$-functional $E(s, f; L^1, L^p)$ with $1 < p < \infty$. Specifically, the following choice fits:

$$u^{(s)} = f_{\varphi(\tau(s)^{1/\theta})}, \tag{1.49}$$

where $\theta = 1 - 1/p$, $\varphi$ is defined by (1.47), and $\tau(s) = s^{-1} \operatorname{dist}_{L^1}(f, B_{L^p}(s))$; as before, the lower index of $f$ signifies the "good" function in the Calderón–Zygmund decomposition. This seems to have passed nearly unnoticed (not quite: for instance, there is a remark in the same spirit in [JRW]), though a minimum of prerequisites is needed for the proof. We present a more detailed statement to be used in the next section.

**Theorem 1.17.** *Suppose that* $\operatorname{dist}_{L^1}(f, B_{L^p}(s)) > 0$. *Then the function defined by* (1.49) *satisfies*

$$\|u^{(s)}\|_{L^p} \leq Cs \text{ and } \|f - u^{(s)}\|_{L^1} \leq C \operatorname{dist}_{L^1}(f, B_{L^p}(s)),$$

*where* $C > 0$ *is independent of* $f$ *and* $s$. *Moreover, the Calderón–Zygmund cubes involved in the definition of* $u^{(s)}$ *satisfy*

$$\sum |Q_i| \leq C \left( \frac{1}{s} \operatorname{dist}(f, B_{L^p}(s)) \right)^{\frac{1}{\theta}} . \tag{1.50}$$

The proof will be given in Chapter 5. It does not require any advanced knowledge, but is surprisingly bulky.

## 1.4 An elementary stability theorem

The material of the preceding subsection allows us to prove the first stability result for near-minimizers. Though rather elementary, it has useful consequences, which will be employed in the treatment of more advanced stability theorems. We feel that these consequences are also interesting in themselves. Our basic couple will be $(L^1(\mathbb{R}^n), L^p(\mathbb{R}^n))$, where $1 < p < \infty$. There also will be some operator $T$ bounded on $L^p$ but only long-range regular on $L^1$. We remind the reader that this sort of behavior is typical of singular integral operators (see the next chapter).

**Theorem 1.18.** *Under the above assumptions, suppose that $f \in L^1$ satisfies $Tf \in L^1$. Then for every $s > 0$ there exists a function $u^{(s)} \in L^1$ such that*

$$\|u^{(s)}\|_{L^p} \le Cs, \quad \|f - u^{(s)}\|_{L^1} \le C \operatorname{dist}_{L^1}(f, B_{L^p}(s)) \tag{1.51}$$

*and*

$$\|Tf - Tu^{(s)}\|_{L^1} \le C(\operatorname{dist}_{L^1}(f, B_{L^p}(s)) + \operatorname{dist}_{L^1}(Tf, B_{L^p}(s)). \tag{1.52}$$

Surely, (1.51) says that $u^{(s)}$ is a near-minimizer for the distance functional for $f$ at $s$, and (1.52) says that $Tu^{(s)}$ behaves much like a near-minimizer for the distance functional of $Tf$ at $s$, in accordance with the discussion in the Introduction. More precisely, it becomes a near-minimizer if the second distance on the right in (1.52) dominates the first, which is similar to condition (0.10). Briefly, Theorem 1.18 says that *the distance functional admits a $T$-stable near-minimizer.* We give two proofs of this theorem.

### 1.4.1 A proof with much interpolation

Maybe, ultimately there will be even *too* much of it, but this will be hidden until Chapter 5, where we shall prove Theorem 1.17. Now we take that statement for granted and simply apply it. It should be emphasized that the arguments in the present subsection will show that the required function $u^{(s)}$ can be obtained as a result of the Calderón–Zygmund decomposition of $f$ at some level (*depending* on $s$). This feature is in full accordance with our general intentions. It is interesting that this $u^{(s)}$ will depend entirely on $f$ and $s$, but not on $T$. Another proof of Theorem 1.18, free of interpolation, will be presented in the next subsection. It gives some supplementary information, but not the specific form of $u^{(s)}$ mentioned above.

Passing to the proof itself, we simply define $u^{(s)}$ by (1.49) and choose a function $v \in L^p$ with $\|v\|_{L^p} \le s$ and $\|Tf - v\|_{L^1} \le 2 \operatorname{dist}_{L^1}(Tf, B_{L^p}(s))$. Then we write

$$\|Tf - Tu^{(s)}\|_{L^1} \le \int_{\mathbb{R}^n \setminus (\cup 2Q_i)} |Tf - Tu^{(s)}| + \int_{\cup 2Q_i} |Tf - v| + \int_{\cup 2Q_i} |Tu^{(s)} - v|, \tag{1.53}$$

where the $Q_i$ are the same as in (1.50). The middle term on the right does not exceed $2\,\mathrm{dist}_{L^1}(Tf, B_{L^p}(s))$ by the choice of $v$. In the first term, by the Calderón–Zygmund construction we write $f - u^{(s)} = \sum(f - c_i)\chi_{Q_i}$, where the summands have zero integrals. Thus, by the long-range regularity of $T$, this term is dominated by

$$C\sum_i \int_{Q_i} |f - c_i| \le C\|f - u^{(s)}\|_{L^1} \le C\,\mathrm{dist}_{L^1}(f, B_{L^p}(s)).$$

The third term on the right in (1.53) is estimated with the help of the Hölder inequality:

$$\int_{\cup 2Q_i} |Tu^{(s)} - v| \le C|\cup 2Q_i|^\theta s,$$

where $\theta = 1 - 1/p$ (recall that both $u^{(s)}$ and $v$ are in a ball of radius roughly $s$ and centered at zero in $L^p$, and $T$ is bounded on $L^p$). It remains to refer to (1.50) and to collect the estimates.

### 1.4.2 Stabilization à la Bourgain

We reproduce Bourgain's argument (see [Bo]; see also [KX] for related material), which converts an arbitrary near-minimizer into a stable one.[7] Specifically, suppose that $f \notin B_{L^p}(s)$ and we are given a function $h$ with $\|h\|_{L^p} \le Cs$ and $0 < \|f - h\|_{L^1} \le C\,\mathrm{dist}_{L^1}(f, B_{L^p}(s))$. The insight due to Bourgain is that, in Theorem 1.18, the following choice of $u^{(s)}$ fits:

$$u^{(s)} = h + (f - h)_t, \tag{1.54}$$

where $t$ is found from the equation

$$t^{p-1}\|f - h\|_{L^1} = s^p.$$

Here $(f - h)_t$ is (again!) the "good" function in the Calderón–Zygmund decomposition of $f - h$ at the level $t$. Thus, *any* near-minimizer for the distance functional becomes $T$-stable (in the sense of Theorem 1.18) after a simple modification that involves the Calderón–Zygmund procedure.

To prove this claim, first we observe that the function $u^{(s)}$ given by (1.54) is still a near-minimizer for the distance functional of $f$ at $s$ (with other estimational constants). Indeed, by the general properties of the Calderón–Zygmund procedure, the $L^1$-norm of $(f - h)_t$ is dominated by that of $f - h$, and

$$\|(f - h)_t\|_{L^p} \le c\left(\int t^{p-1}|(f - h)_t|\right)^{\frac{1}{p}} \le C\left(t^{p-1}\int |f - h|\right)^{\frac{1}{p}} = Cs \tag{1.55}$$

[7] In [Bo], the operator $T$ was a projection, and only the functions $f$ satisfying $f = Tf$ where treated.

by the choice of $t$, so that we make admissible "errors" when passing from $h$ to $u^{(s)}$. To estimate the norm $\|Tf - Tu^{(s)}\|_{L^1}$, we return to formula (1.53), in which $v$ is chosen in the same way as before, but now the $Q_i$ are the Calderón–Zygmund cubes arising in the construction of $(f-h)_t$. We observe that, by the properties of the Calderón–Zygmund procedure and the choice of $t$,

$$\sum |Q_i| \le Ct^{-1}\|f-h\|_{L^1} = C\left(\frac{\|f-h\|_{L^1}}{s}\right)^q \le C_1\left(\frac{\mathrm{dist}_{L^1}(f, B_{L^p}(s))}{s}\right)^q, \tag{1.56}$$

where $q$ is the exponent conjugate to $p$. The first term on the right in (1.53) becomes

$$\int_{\mathbb{R}^n\setminus(\cup 2Q_i)} |T(f-h) - T((f-h)_t)|.$$

Since $T$ is long-range regular, this quantity is dominated by $\int_{\mathbb{R}^n}|f-h|$, which is less than or equal to $C\,\mathrm{dist}_{L^1}(f, B_{L^p}(s))$. The second term on the right does not exceed $C\,\mathrm{dist}_{L^1}(Tf, B_{L^p}(s))$ (we extend integration to $\mathbb{R}^n$ and refer to the choice of $v$). Finally, the third term is controlled with the help of the Hölder inequality: by (1.55) and the $L^p$-continuity of $T$, the integrand has $L^p$-norm not exceeding $Cs$, so the term in question is dominated by $C(\sum|Q_i|)^{1/q}s$. It remains to refer to (1.56) and then collect the estimates.

### 1.4.3 Some consequences

If an operator $T$ is as in Theorem 1.18, then so is also any operator $S$ of the form $Sf = uTf$, where $u$ is a fixed function in $L^\infty$. Thus, that theorem can also be applied to $S$ in place of $T$. If we choose $u$ to be the characteristic function of a measurable set, we immediately obtain the following statement.

**Corollary 1.19.** *If $T$ is as in Theorem 1.18, $f \in L^1$, and $Tf$ is integrable over a set $E \subset \mathbb{R}^n$, then for every $s > 0$ there exists a function $g \in B_{L^p}(cs)$ with*

$$\|f-g\|_{L^1} \le C\,\mathrm{dist}_{L^1}(f, B_{L^p}(s))$$

*and*

$$\int_E |Tf - Tg| \le C(\mathrm{dist}_{L^1}(f, B_{L^p}(s)) + \mathrm{dist}_{L^1(E)}(\chi_E Tf, B_{L^p(E)}(s))).$$

**Corollary 1.20.** *If $T$, $f$, and $E$ are as above, then there exist functions $f_k$ in $L^1 \cap L^p$ convergent to $f$ in $L^1$ and such that the $Tf_k$ are all integrable over $E$ and $\int_E |Tf_k - Tf| \to 0$.*

*Proof.* Since $f$ belongs to the closure of $L^1 \cap L^p$ and $\chi_E Tf$ to the closure of $L^1(E) \cap L^p(E)$, the claim follows easily from the preceding corollary. □

The next technical statement will be employed in subsequent chapters.

**Corollary 1.21.** *Suppose $T$, $f$ are as in Theorem 1.18, and let $E$ be a measurable subset of $\mathbb{R}^n$. Denote by $q$ the exponent conjugate to $p$ ($p^{-1} + q^{-1} = 1$). Let $\psi \in L^\infty \cap L^q$ be a function supported on $E$ and such that $T^*\psi \in L^\infty$. If $Tf$ is integrable over $E$, then*

$$\int_E \psi Tf = \int_{\mathbb{R}^n} (T^*\psi) f.$$

*Proof.* Taking a sequence $f_n$ as in Corollary 1.20, we can write

$$\int_E \psi Tf = \lim_k \int_{\mathbb{R}^n} \psi T f_k = \lim_k \int_{\mathbb{R}^n} (T^*\psi) f_k = \int_{\mathbb{R}^n} (T^*\psi) f.$$

Passage to the adjoint operator in the second identity is justified because $T$ is bounded on $L^p$, $\psi \in L^q$, and $f_k \in L^p$. □

# Notes and remarks

In Remark 1.7, the emphasis is on an at most linear growth of the constant $C$ as a function of $\mathrm{lr}(T)$. Theorem 1.9 about the level sets of the Hardy–Littlewood maximal operator $f \mapsto Mf$ was proved in [CRW]. As has already been said, in the proof of Proposition 1.12 we use the ideas of the paper [Me], in which the proof of the John–Nirenberg inequality given in [JN] was adjusted to Lipschitz spaces. Originally, the phenomenon described by that proposition was discovered by Campanato [C].

In connection with the material of item A) in §1, Subsection 3 (and also with Lemma 1.3), we mention the paper [Pe1], where a formula for the $K$-functional in terms of the Calderón–Zygmund cubes was presented. That formula is not quite convenient, however, because the principal parameter in it was the total volume of the cubes, whose relationship with the parameter "$t$" of the $K$-functional is entangled somewhat.

Theorem 1.17, the elementary stability theorem (Theorem 1.18), and the consequences of the latter presented in Subsection 4.3 of Chapter 1 first appeared in [KK0] and [KK]. The idea of a long-range regular operator has permeated the theory of singular integrals since its foundation, we only took the liberty to suggest a short and self-explanatory term.

# Chapter 2

# Singular integrals

In this chapter we present the first portion of the material concerning singular integral operators. Basically, this will cover all mentions of them made previously in this book. For us it is important that all singular integral operators are long-range regular (though, typically, unbounded) on $L^1$. So, all statements about long-range regular operators proved in the preceding chapter are applicable to them and, typically, are nontrivial indeed.

Precise definitions will be given later, but we signalize that the operators in question will be of the form $f \mapsto \int K(\cdot, y)f(y)dy$ in the case where the kernel $K$ has a nonintegrable singularity. Such operators arise surprisingly often and are treated by an elaborate and still vivid theory.

Of course, the question as to how to understand such integrals is of primary interest. Basically, the answer should be clear, however: this is done with the help of an appropriate limit passage. The most well-known example is when the kernel $K$ in the above formula depends on the difference $x - y$ only: $K(x, y) = k(x - y)$, where $k$ is a (nonintegrable!) function on $\mathbb{R}^n$. Then the operator in question (call it $T$) is convolution with $k$, often understood in the principal value sense. Under certain conditions on $k$, the definition of $T$ can be made consistent; moreover, $T$ will turn out to be bounded on $L^p$ for $1 < p < \infty$ and long-range regular on $L^1$. For example, the most part of standard manuals include the classical statement that this is true if $k$ satisfies the estimates $|k(x)| \leq C|x|^{-n}$ and $|\nabla k(x)| \leq C|x|^{-n-1}$ for $x \neq 0$, and has zero integral over each spherical shell centered at zero. See [S] for the details, which, however, are not critically needed for us.

If $n = 1$ and $k(x) = 1/x$, we obtain the most well-known singular integral operator, called the Hilbert transformation. We start with a detailed treatment of it, which may serve as an excellent illustration of the techniques involved. After that we formulate the general definition of a singular integral and give examples. Homogeneous kernels (of the above form) in many dimensions will be touched upon only briefly. Instead, we discuss Fourier multipliers that can be viewed as singular integrals.

## 2.1 Hilbert transformation

Let $\mathcal{D}(\mathbb{R}^n)$ denote the class of infinitely differentiable compactly supported functions on $\mathbb{R}^n$. The Hilbert transformation is the operator $H$ defined (initially) on the space $\mathcal{D}(\mathbb{R})$ (to be denoted simply by $\mathcal{D}$ in this section) by the formula

$$(Hf)(y) = \lim_{\varepsilon\to 0} \frac{1}{\pi} \int_{|x|>\varepsilon} \frac{f(x-y)}{x}\, dx, \quad f \in \mathcal{D}. \tag{2.1}$$

The limit in (2.1) exists for every $y$. Indeed,

$$(Hf)(y) = \lim_{\varepsilon\to 0} \frac{1}{\pi} \left( \int_{\varepsilon<|x|<1} \frac{f(x-y)-f(y)}{x}\, dx + \int_{|x|\ge 1} \frac{f(x)}{x}\, dx \right)$$

(when subtracting $f(y)$ in the first term, we have used the fact that $\int_{\varepsilon<|x|<1} \frac{dx}{x} = 0$). Now, the first integrand on the right is summable over the interval $|x| < 1$ because $f$ is smooth. The second integral also exists in the usual sense because $f$ is compactly supported.

In other words, the Hilbert transformation is the operator of convolution with the distribution $\frac{1}{\pi}\text{p.v.}\frac{1}{x}$. This distribution is generated by the nonintegrable function $\frac{1}{\pi}\frac{1}{x}$, but passage to a *principal value* (whence the abbreviation p.v.) makes it well-defined. However, here the distributional point of view can be avoided. Specifically, we can rewrite (2.1) in the form

$$(Hf)(y) = \lim_{\substack{\varepsilon\to 0\\ R\to\infty}} \frac{1}{\pi} \int_{\varepsilon<|x|<R} \frac{f(x-y)}{x}\, dx = \lim_{\substack{\varepsilon\to 0\\ R\to\infty}} (f * \Phi_{\varepsilon,R})(y), \quad f \in \mathcal{D}, \tag{2.2}$$

where $\Phi_{\varepsilon,R}(x) = \frac{1}{\pi}\chi_{\{t:\, \varepsilon\le t\le R\}}\frac{1}{x}$ is an integrable function. It turns out that the limit in (2.2) exists not only pointwise, but also in $L^2(\mathbb{R})$. To see this, we pass to Fourier transforms[1]:

$$\widehat{f * \Phi_{\varepsilon,R}} = \hat{\Phi}_{\varepsilon,R}\hat{f}$$

and calculate the function $\hat{\Phi}_{\varepsilon,R}$:

$$(\hat{\Phi}_{\varepsilon,R})(\xi) = \frac{1}{\pi} \int_{\varepsilon<|x|<R} \frac{\exp(-2\pi i x\xi)}{x}\, dx = -\frac{2i}{\pi} \int_{\varepsilon}^{R} \frac{\sin 2\pi x\xi}{x}\, dx$$

$$= -\frac{2i}{\pi} \operatorname{sgn}\xi \int_{2\pi\varepsilon|\xi|}^{2\pi R|\xi|} \frac{\sin t}{t} dt.$$

[1]We define the Fourier transform of a function $h$ on $\mathbb{R}^n$ by the formula $\hat{h}(\xi) = \int_{\mathbb{R}^n} h(x) \cdot \exp(-2\pi i\langle x, \xi\rangle)\, dx$. Accordingly, the formula for an inverse Fourier transform differs only by the sign under the exponential.

Using the well-known fact that $\lim_{u\to\infty}\int_0^u \frac{\sin t}{t}dt = \pi/2$, we see that all functions $\{\hat{\Phi}_{\varepsilon,R}\}_{0<\varepsilon<R<\infty}$ are uniformly bounded and tend to $-i\operatorname{sgn}\xi$ as $\varepsilon\to 0$ and $R\to\infty$. By the Lebesgue dominated convergence theorem, $\hat{\Phi}_{\varepsilon,R}\hat{f}\to -i(\operatorname{sgn}\xi)\hat{f}(\xi)$ in $L^2$. Now, the Plancherel theorem (saying that the Fourier transformation is a unitary operator on $L^2$) shows that

$$(Hf) = \lim_{\substack{\varepsilon\to 0\\ R\to\infty}} (f*\Phi_{\varepsilon,R}) \text{ in } L^2 \text{ for } f\in\mathcal{D} \tag{2.3}$$

and

$$\widehat{Hf}(\xi) = -i\operatorname{sgn}\xi\,\hat{f}(\xi), f\in\mathcal{D}. \tag{2.4}$$

It is most natural to allow $f$ in (2.4) to be an arbitrary function in $L^2(\mathbb{R}^n)$. Then, again by the Plancherel theorem, $H$ becomes a linear isometry of $L^2(\mathbb{R})$ onto itself. This extension of $H$ to $L^2(\mathbb{R})$ is uniquely determined by its values on the compactly supported smooth functions because the class $\mathcal{D}(\mathbb{R})$ is dense in $L^2(\mathbb{R})$. Next, the $L^2$-convergence in (2.3) holds for all $f\in L^2$. Indeed, we have convergence on the dense subset $\mathcal{D}$, and the operators of convolution with $\Phi_{\varepsilon,R}$ are bounded on $L^2$ uniformly in $\varepsilon$ and $R$ (again by the Plancherel theorem: recall that the functions $\widehat{\Phi_{\varepsilon,R}}$ are uniformly bounded).

Now, an important observation is in order: if two functions $f$ and $g$ in $L^2$ are supported on two bounded sets lying at a positive distance from each other, then

$$\int gHf = \lim_{\substack{\varepsilon\to 0\\ R\to\infty}} \int g(\Phi_{\varepsilon,R}*f) = \frac{1}{\pi}\iint \frac{f(x)g(y)}{x-y}\,dx\,dy \tag{2.5}$$

(singularity disappears, and the double integral exists in the usual sense). Formula (2.5) can be rewritten as follows:

$$(Hf)(y) = \frac{1}{\pi}\int \frac{f(x)}{y-x}\,dx \tag{2.6}$$

if $f\in L^2$ has bounded support and $y$ is at a positive distance from the support of $f$.

It is well known that the Hilbert transformation arises naturally in the theory of analytic functions. In vague terms, the relationship is as follows: let $F$ be an analytic function on the upper half-plane, suppose it has boundary values on the real line in a reasonable sense, and let $u$ and $v$ be the real and the imaginary part of the boundary function. Then $v = Hu$. See, e.g., [GR], [Koo] for detailed explanations. Here we explain this feature under the following assumption, which is very far from the weakest possible: $F$ is analytic in the domain $\{z = x+iy : y > -\delta\}$ for some positive $\delta$ and decays rapidly in this domain as $z\to\infty$. For $\xi\le 0$, the function $F(z)\exp(-2\pi i\xi z)$ shares the above properties of $F$, so that (if the decay is sufficiently fast), its integral over the real line is zero by the Cauchy theorem. This means that $\hat{u}(\xi)+i\hat{v}(\xi) = 0$ for $\xi\le 0$. Since the values $\hat{w}(\xi)$ and $\hat{w}(-\xi)$ are complex conjugate if $w$ is a real function, we deduce that also $\hat{u}(\xi) - i\hat{v}(\xi) = 0$ for $\xi \geq 0$. Recalling (2.4), we see that this proves the claim.

It should also be mentioned that for $f \in L^2$ the following holds true almost everywhere:

$$Hf(y) = \lim_{\varepsilon \to 0} \frac{1}{\pi} \int_{|x| \geq \varepsilon} \frac{f(x-y)}{x}\, dx,$$

see, e.g., [S], [GR] (for fixed $\epsilon$, the integral exists as the convolution of two $L^2$-functions). We shall not need this fact.

From (2.4), it is easy to deduce that

$$H^2 = -\,\mathrm{id} \quad \text{and} \quad \int gHf = -\int fHg \tag{2.7}$$

on $L^2(\mathbb{R})$.

### 2.1.1 Hilbert transformation on $L^1$

For $f = \chi_{[0,1]}$, it is easy to calculate $Hf$:

$$Hf = \frac{1}{\pi} \log \left| \frac{x}{x-1} \right|.$$

This function is not integrable (because it behaves like $x^{-1}$ near infinity), so $H$ cannot be defined on $L^1$ as a bounded operator. Nevertheless, Theorem 1.5 shows that $H$ can be extended to a weak type $(1,1)$ operator on $L^1$. To see that, it suffices to prove that $H$ *is long-range $L^1$-regular*. We remind the reader that this means the following: if a function $f \in L^2 \cap L^1$ is supported on an interval $I$ and $\int f = 0$, then

$$\int_{\mathbb{R}\setminus 2I} |Hf| \leq C \int_I |f|. \tag{2.8}$$

To prove (2.8), we denote by $x_0$ the center of $I$. If $y \notin 2I$, formula (2.6) applies, and the relation $\int f = 0$ allows us to rewrite it as follows:

$$(Hf)(y) = \frac{1}{\pi} \int_I f(x) \left( \frac{1}{y-x} - \frac{1}{y-x_0} \right) dx.$$

Then

$$\int_{\mathbb{R}\setminus 2I} |(Hf)(y)|\, dy \leq \frac{1}{\pi} \int_{\mathbb{R}\setminus 2I} \int_I f(x) \frac{|x-x_0|}{|x-y||x_0-y|}\, dx\, dy.$$

Now, the distances $|x-y|$ and $|x_0-y|$ are equivalent, and $|x-x_0| \leq |I|$. So, we have

$$\int_{\mathbb{R}\setminus 2I} \frac{|x-x_0|}{|x-y||x_0-y|}\, dy \leq C \int_{|x_0-y| \geq |I|} \frac{|x-x_0|}{|x_0-y|^2}\, dy \leq C_1, \tag{2.9}$$

which implies (2.8).

Thus, we have proved the estimate $|\{|Hf| > \lambda\}| \le C\lambda^{-1}\|f\|_{L^1}$, $f \in L^1 \cap L^2$, $\lambda > 0$. In other words, $H$ acts from the space $L^1 \cap L^2$ endowed with the $L^1$-norm to the space weak-$L^1$ (the Lorentz space $L^{1,\infty}$). Clearly, $H$ extends to the entire space $L^1$ by continuity.

### 2.1.2 The operator $H$ on $L^p$, $1 < p < \infty$

We have seen that the operator $H$ maps $L^2$ to $L^2$ and $L^1$ to $L^{1,\infty}$. The Marcinkiewicz interpolation theorem says that then it acts boundedly from $L^p$ to $L^p$ for $1 < p < 2$. By the way, the full strength of the Marcinkiewicz theorem is not required to prove this and the claim reduces to a rather short calculation, see, e.g., [S]. We reproduce it for completeness.

Suppose $f \in L^1 \cap L^2$. We want to prove that $\|Hf\|_{L^p} \le C\|f\|_{L^p}$. Fixing $t > 0$, we write $f = f_1 + f_2$, where $f_1 = f\chi_{\{|f|>t\}}$. Since

$$\{|Hf| > t\} \subset \{|Hf_1| > t/2\} \cup \{|Hf_2| > t/2\},$$

we see that

$$\begin{aligned}\lambda(t) = |\{|Hf| > t\}| &\le c\left(t^{-1}\int |f_1| + t^{-2}\int |Hf_2|^2\right)\\ &\le C\left(t^{-1}\int_{|f|>t} |f(x)|\,dx + t^{-2}\int_{|f|\le t} |f(x)|^2\,dx\right).\end{aligned}$$

Thus,

$$\begin{aligned}\|Hf\|_{L^p}^p &= p\int_0^\infty t^{p-1}\lambda(t)dt\\ &\le C\left(\int_0^\infty t^{p-2}\int_{|f|>t} |f(x)|\,dx\,dt + \int_0^\infty t^{p-3}\int_{|f|\le t} |f(x)|^2\,dx\,dt\right)\\ &= C\left(\int_{\mathbb{R}^n} |f(x)| \int_0^{|f(x)|} t^{p-2}\,dt\,dx + \int_{\mathbb{R}^n} |f(x)|^2 \int_{|f(x)|}^\infty t^{p-3}\,dt\,dx\right)\\ &\le C_1\int_{\mathbb{R}^n} |f(x)|^p\,dx,\end{aligned}$$

and we are done.

On $L^q$ with $q \in (2, \infty)$, the operator $H$ is also bounded. Indeed, assume first that $f \in L^2 \cap L^q$. Then, putting $p = \frac{q}{q-1}$ (the exponent conjugate to $q$) and using (2.7), we can write:

$$\begin{aligned}\|Hf\|_q &= \sup\{|\int (Hf)g| : g \in L^p \cap L^2, \|g\|_p \le 1\} \\ &= \sup\{|\int f(Hg)| : g \in L^p \cap L^2, \|g\|_p \le 1\} \\ &\le \|f\|_q \sup\{\|Hg\|_p : g \in L^p, \|g\|_p \le 1\} \le C\|f\|_q.\end{aligned}$$

With this estimate at hand, we again extend $H$ to the entire space $L^q$ by continuity.

Despite numerous limit passages, we deal in fact with one and the same operator: on the set $\mathcal{D}$ (which is dense in all spaces $L^p$, $1 \le p < \infty$; moreover, for any $f \in L^r \cap L^s$ with finite $r$ and $s$ there is a sequence in $\mathcal{D}$ tending to $f$ in $L^r$ and $L^s$ simultaneously) the values of $H$ are given by (2.1) in any case.

Next, $H$ can be defined also on $L^\infty$: it takes this space to BMO. We shall discuss this in Chapter 4 in a more general situation.

## 2.2 General definition

The theory of the Hilbert transformation sketched above splits naturally into two parts. The first part explains why this operator is bounded on $L^2(\mathbb{R})$. The second shows that, thanks to the properties of the kernel $\frac{1}{x-y}$, $L^2$-boundedness allows us to deduce weak type $(1,1)$ and $L^p$-boundedness for $1 < p < \infty$.

The next definition (compare with [M1]) is aimed at general properties that make it possible to elaborate this second part of the theory. Neither reasons for $L^2$-boundedness (in specific cases the Fourier transformation may be useless), nor even the initial definition of the operator on "very good" functions (a principal value integral may not be appropriate sometimes) is touched upon in it. From this point on, everything happens on $\mathbb{R}^n$.

**Definition 2.1.** A bounded linear operator $T : L^2(\mathbb{R}^n) \to L^2(\mathbb{R}^n)$ is said to be a Calderón–Zygmund (singular integral) operator if there is a function $K$ defined almost everywhere on $\mathbb{R}^n \times \mathbb{R}^n$, locally bounded outside the diagonal $\{(x,x) : x \in \mathbb{R}^n\}$, and such that

(I) if $f \in L^2(\mathbb{R}^n)$ has bounded support, then

$$(Tf)(x) = \int_{\mathbb{R}^n} K(x,y)f(y)\,dy$$

for a.e. $x$ at a positive distance from the support of $f$;

(II) $K$ satisfies the Hörmander condition: for every cube $Q$ in $\mathbb{R}^n$ and every $y_0, y_1 \in Q$ we have

$$\int_{\mathbb{R}^n \setminus 2Q} |K(x,y_0) - K(x,y_1)|\,dx \le C,$$

where the constant $C$ is independent of $Q$, $y_0$, and $y_1$.

The function $K$ is called the *kernel* of $T$. It is easily seen that the identity operator (more generally, an operator of multiplication by a bounded function) satisfies the above conditions with *zero* kernel (simply, $(\mathrm{id}\, f)(x) = 0$ outside the support of $f$). Thus, it is not always possible to recover $T$ by its kernel. The function $\frac{1}{\pi}\frac{1}{x-y}$ is the kernel of the Hilbert transformation, see (2.6); moreover, it satisfies condition (II), see (2.9). Surely, (I) can be rewritten as (2.5): if two functions $f, g \in L^2$ have bounded supports that lie at a positive distance from each other, then

$$\int_{\mathbb{R}^n} gTf = \iint_{\mathbb{R}^n \times \mathbb{R}^n} K(x,y)f(x)g(y)\,dx\,dy.$$

It follows that, up to permutation of the variables, the function $K$ provides the kernel also for the adjoint $T^*$ (throughout, if otherwise is not claimed explicitly, we employ the *bilinear* duality form $\langle \varphi, \psi \rangle = \int \varphi\psi$):

$$(T^*g)(y) = \int_{\mathbb{R}^n} K(x,y)g(x)\,dx$$

for a.e. $y$ at a positive distance from the support of $g$. If condition (II) is fulfilled also in the second variable:

(II dual)

$$\int_{\mathbb{R}^n \setminus 2Q} |K(x_0,y) - K(x_1,y)|\,dy \le C,$$

where $Q$ ia an arbitrary cube and $x_0, x_1 \in Q$, we say that $T^*$ is a Calderón-Zygmund operator.

If $K$ depends only on the difference of the arguments: $K(x,y) = k(x-y)$ (as it happens for the Hilbert transformation), then conditions (II) and (II dual) are equivalent.

Every Calderón–Zygmund operator is long-range $L^1$-regular. Indeed, if $f \in L^2$ is supported on a cube $Q$ and has zero integral, then for $x \notin 2Q$ we have

$$(Tf)(x) = \int_Q K(x,y)f(y)\,dy = \int_Q (K(x,y) - K(x,y_0)f(y)\,dy,$$

where $y_0$ is the center of $Q$. Therefore,

$$\int_{\mathbb{R}^n \setminus 2Q} |(Tf)(x)|\,dx \le \int_Q \int_{\mathbb{R}^n \setminus 2Q} |K(x,y) - K(x,y_0)|\,dx\,|f(y)|\,dy \le C\|f\|_{L^1}$$

by (II).

Now, Theorem 1.5 shows that $T$ has weak type $(1,1)$; by interpolation, it acts from $L^p(\mathbb{R}^n)$ to $L^p(\mathbb{R}^n)$ for $1 < p \le 2$. If $T^*$ is also a Calderón–Zygmund operator, then $T^*$ is also bounded on $L^p(\mathbb{R}^n)$ for $1 < p \le 2$, which implies that $T$ is bounded on $L^r(\mathbb{R}^n)$ for $2 \le r < \infty$.

*Remark* 2.2. In the definition of a Calderon–Zygmund operator, we could require that, instead of $L^2$, it should be bounded initially on $L^s$ with some $s \in (1, \infty]$. Then procedures similar to the above yield, via Theorem 1.5, that it is bounded on $L^p$ for $1 < p \le s$.

Condition (II) is fulfilled whenever $K$ satisfies the estimate

$$|\nabla_y K(x,y)| \le C\|x-y\|^{-n-1} \tag{2.10}$$

(in fact, we used precisely this estimate when we analyzed the Hilbert transformation). Indeed, if $Q$ is a cube and $y_0, y_1 \in Q$ but $x \notin 2Q$, then all distances $\|x-y\|$, $y \in Q$, are comparable to $\|x-y_0\|$, whence we obtain

$$|K(x,y_0) - K(x,y_1)| \le C\frac{\|y_0-y_1\|}{\|x-y_0\|^{n+1}} \le C\frac{\operatorname{diam} Q}{\|x-y_0\|^{n+1}}.$$

Integrating over $\mathbb{R}^n \setminus 2Q$, we obtain

$$\begin{aligned}
&\int_{\mathbb{R}^n\setminus 2Q} |K(x,y_0) - K(x,y_1)|\,dx \\
&\qquad \le C(\operatorname{diam} Q)\sum_{k\ge 1}\int_{2^{k+1}Q\setminus 2^k Q} \frac{1}{\|x-y_0\|^{n+1}}\,dx \\
&\qquad \le C_1(\operatorname{diam} Q)\sum_{k\ge 1}\int_{2^{k+1}Q} \frac{1}{2^{k(n+1)}(\operatorname{diam} Q)^{n+1}}\,dx \le \text{const}\,.
\end{aligned}$$

A similar calculation shows that even the estimate

$$|K(x,y_0) - K(x,y_1)| \le C\frac{\|y_0-y_1\|^u}{\|x-y_0\|^{n+u}} \tag{2.11}$$

for some $u > 0$ suffices.

*Remark* 2.3. Under conditions (2.10) or (2.11), the operator $T$ possesses another form of long-range regularity. Specifically, if $f \in L^1 \cap L^2$ is supported on a cube $Q$ and has zero integral, then for every $\lambda > 1$ we have

$$\int_{\mathbb{R}^n\setminus\lambda Q} |Tf| \le C\|f\|_{L^1}.$$

Surely, here $C$ depends on $\lambda$. The proof is much as above. Such operators will be called *strongly long-range regular*. This property will be employed in subsequent chapters.

### 2.2.1 Examples

#### Homogeneous kernels and principal value integrals

Historically, this is the first class of singular integral operators that had once been studied thoroughly. However, for us it lies somewhat aside — it is more convenient to use the multiplier pattern explained in the next subsection. So, the reader may skip the calculation of Fourier transforms that will follow shortly (we present it for completeness only).

In this subsection and in the next one, the symbol $|x|$ denotes the Euclidean length of the vector $x$ (that is, $|x| = (|x_1|^2 + \cdots + |x_n|^2)^{1/2}$).

Let $\Omega$ be a function on $\mathbb{R}^n \setminus \{0\}$ homogeneous of degree 0 (i.e., $\Omega(\lambda x) = \Omega(x)$, $\lambda > 0$) and sufficiently smooth on the unit sphere. (In many examples, $\Omega$ is infinitely differentiable.) Suppose that $\Omega$ has zero average over the sphere. Then the operator of convolution with $\frac{\Omega(x)}{|x|^n}$ is well defined on $\mathcal{D}(\mathbb{R}^n)$ in the principal value sense:

$$Tf(x) = \lim_{\varepsilon\to 0} \int_{|y|>\varepsilon} \frac{\Omega(y)}{|y|^n} f(x-y)\, dy. \tag{2.12}$$

As in the case of the Hilbert transformation, the limit exists because

$$Tf(x) = \lim_{\varepsilon\to 0} \int_{\varepsilon<|y|\le 1} \frac{\Omega(y)}{|y|^n} (f(x-y) - f(x))\, dy + \int_{|y|>1} \frac{\Omega(y)}{|y|^n} f(x-y)\, dy,$$

and the first integrand is summable on the unit ball of $\mathbb{R}^n$. (Note that $\int\limits_{\varepsilon<|y|\le 1} \frac{\Omega(y)}{|y|^n}\, dy$ $= 0$ because $\Omega$ has zero average over the sphere.)

In order to prove that $T$ is bounded on $L^2(\mathbb{R}^n)$, we note that in (2.12) we deal with the convolution of two $L^2$-functions: $f$ and $\Phi_\varepsilon(y) = \chi_{\{t:\, \varepsilon<|t|\}}(y)\frac{\Omega(y)}{|y|^n}$. By the Plancherel theorem, it suffices to prove that the Fourier transforms $\widehat{\Phi_\varepsilon}$ (in the $L^2$-sense) are $L^\infty$-functions bounded uniformly in $\varepsilon$. For $R > \varepsilon$, we put $\Phi_{\varepsilon,R}(y) = \chi_{\varepsilon<|y|<R}(y)\frac{\Omega(y)}{|y|^n}$ and observe that these functions belong to $L^1 \cap L^2$ and (again by the Plancherel theorem) their Fourier transforms tend in $L^2$ to $\widehat{\Phi_\varepsilon}$ as $R \to \infty$.

Now,

$$\widehat{\Phi_{\varepsilon,R}}(\xi) = \int_{\varepsilon<|y|\le R} \frac{\Omega(y)}{|y|^n} e^{-2\pi i\langle y,\xi\rangle}\, dy$$

(because $\Omega$ has zero average over the sphere)

$$= \int_{\varepsilon<|y|\le R} \frac{\Omega(y)}{|y|^n} ((\cos(2\pi\langle y,\xi\rangle) - 1) - i\sin(2\pi\langle y,\xi\rangle)\, dy$$

$$= \int_{\mathbb{S}^{n-1}} \Omega(w) \left[\int_\varepsilon^R (\cos(2\pi r\langle w,\xi\rangle) - 1)\frac{dr}{r} - i\int_\varepsilon^R \sin(2\pi r\langle w,\xi\rangle)\frac{dr}{r}\right] d\sigma(w)$$

($\sigma$ is the normalized Lebesgue measure on the sphere)

$$= \int_{\mathbb{S}^{n-1}} \Omega(w) \left[\int_{2\pi\varepsilon|\langle w,\xi\rangle|}^{2\pi R|\langle w,\xi\rangle|} (\cos\rho - 1)\frac{d\rho}{\rho} - i\,\mathrm{sgn}\langle w,\xi\rangle \int_{2\pi\varepsilon|\langle w,\xi\rangle|}^{2\pi R|\langle w,\xi\rangle|} \sin\rho\frac{d\rho}{\rho}\right] d\sigma(w).$$

We rewrite the first integral in the square brackets as follows, irrespectively of whether $2\pi R|\langle w,\xi\rangle|$ is smaller or greater than 1:

$$\int_{2\pi\varepsilon|\langle w,\xi\rangle|}^{1}\frac{\cos\rho-1}{\rho}\,d\rho+\int_{1}^{2\pi R|\langle w,\xi\rangle|}\frac{\cos\rho}{\rho}\,d\rho-\log(2\pi R)+\log\frac{1}{|\langle w,\xi\rangle|}.$$

The growing term $-\log(2\pi R)$ disappears after integration with $\Omega$ over the sphere, so finally we obtain

$$\begin{aligned}\widehat{\Phi_{\varepsilon,R}}(\xi)=\int_{\mathbb{S}^{n-1}}\Omega(w)\Bigg[&\int_{2\pi\varepsilon|\langle w,\xi\rangle|}^{1}\frac{\cos\rho-1}{\rho}\,d\rho+\int_{1}^{2\pi R|\langle w,\xi\rangle|}\frac{\cos\rho}{\rho}\,d\rho\\&-i\,\mathrm{sgn}\langle w,\xi\rangle\int_{2\pi\varepsilon|\langle w,\xi\rangle|}^{2\pi R|\langle w,\xi\rangle|}\frac{\sin\rho}{\rho}\,d\rho+\log\frac{1}{|\langle w,\xi\rangle|}\Bigg]\,d\sigma(w).\end{aligned}$$

Both terms in the square brackets that involve $R$ tend to some limits as $R\to\infty$. Next, both (as functions of $w$) have majorants integrable over the sphere and independent of $R\geq\varepsilon$. Indeed, the term involving $\frac{\sin\rho}{\rho}$ is dominated in the absolute value by a numerical constant. The same is true for the other term provided $R$ satisfies $2\pi R|\langle w,\xi\rangle|\geq 1$. But for other values of $R$ this term is dominated by $\log\frac{1}{|\langle w,\xi\rangle|}+\log\frac{1}{2\pi\varepsilon|\langle w,\xi\rangle|}$ (we put the modulus under the integral sign reversing the order of the limits of integration, replace $|\cos\rho|$ by 1, and then calculate the integral and use the fact that $R\geq\varepsilon$). This function is integrable over the sphere.

So, the Lebesgue dominated convergence theorem allows us to pass to the limit as $R\to\infty$ in the integral over the sphere. The second term in the square brackets tends to a constant, which disappears after integration with $\Omega$ over the sphere. We arrive at

$$\begin{aligned}\lim_{R\to\infty}\widehat{\Phi_{\varepsilon,R}}(\xi)&=\int_{\mathbb{S}^{n-1}}\Omega(w)\\&\times\left[\int_{2\pi\varepsilon|\langle w,\xi\rangle|}^{1}\frac{\cos\rho-1}{\rho}\,d\rho-i\,\mathrm{sgn}\langle w,\xi\rangle\int_{2\pi\varepsilon|\langle w,\xi\rangle|}^{\infty}\frac{\sin\rho}{\rho}\,d\rho+\log\frac{1}{|\langle w,\xi\rangle|}\right]d\sigma(w).\end{aligned}$$

This limit can be none other than $\widehat{\Phi_\varepsilon}(\xi)$, and the formula readily shows that these functions are bounded uniformly in $\xi$ and $\varepsilon$, as required.

Furthermore, we may pass to the limit as $\varepsilon\to 0$, obtaining

$$\lim_{\varepsilon\to 0}\widehat{\Phi_\varepsilon}(\xi)=\int_{\mathbb{S}^{n-1}}\Omega(w)\left(\log\frac{1}{|\langle w,\xi\rangle|}-i\frac{\pi}{2}\,\mathrm{sgn}\langle w,\xi\rangle\right)dw.\tag{2.13}$$

Denoting the right-hand side in (2.13) by $m(\xi)$, we see that it is a bounded function, and formula (2.12) can be rewritten as follows:

$$\widehat{Tf} = m\hat{f}, \quad f \in \mathcal{D}. \tag{2.14}$$

Surely, we can allow here arbitrary functions $f \in L^2$, and (2.14) describes a unique extension to $L^2$ of the operator $T$ given by (2.12).

Clearly, the kernel of $T$ in the sense of formula (I) is the function $K(x,y) = \frac{\Omega(x-y)}{|x-y|^n}$, which satisfies (2.10) if $\Omega$ is continuously differentiable on the sphere.

Let $\mathcal{A}$ denote the set of functions on $\mathbb{R}^n$ that are infinitely differentiable on the sphere, homogeneous of degree zero, and have zero average over the sphere. It can be seen from (2.13) that $m \in \mathcal{A}$ provided $\Omega \in \mathcal{A}$. In fact, the mapping $\Omega \mapsto m$ is a *bijection* of $\mathcal{A}$ onto itself. A very short proof of this can be found in [Dou].

**Multipliers**

Quite often, an operator is defined initially by (2.14), where $m$ is some bounded measurable function on $\mathbb{R}^n$. Then the operator is called a (Fourier) *multiplier*, and $m$ is referred to as its *symbol*. In the sequel, we shall denote by $T_m$ the multiplier determined by the symbol $m$. It is interesting then to know when $T_m$ is a Calderón–Zygmund operator. A partial answer to this question was given by the last paragraph of the preceding subsection. We shall discuss more general results. We denote by $D^\alpha$ the operation of taking the partial derivative with multiindex $\alpha = (\alpha_1, \dots, \alpha_n)$ (all $\alpha_j$ are nonnegative integers). As usual, we write $|\alpha| = \alpha_1 + \dots + \alpha_n$. Also, we put $\xi^\alpha = \xi_1^{\alpha_1} \dots \xi_n^{\alpha_n}$ for $\xi \in \mathbb{R}^n$. We recall that in this subsection the symbol $|a|$ is employed also to denote the Euclidean length of a vector $a$ (the context will allow us to avoid the confusion with the order of a multiindex, as always in similar situations).

**Theorem 2.4.** *Let $m$ be a bounded function on $\mathbb{R}^n$ infinitely differentiable on $\mathbb{R}^n \setminus \{0\}$. Suppose that*

$$|D^\alpha m(\xi)| \le c_\alpha |\xi|^{-|\alpha|}, \ \xi \in \mathbb{R}^n, \tag{2.15}$$

*for all multiindices $\alpha$. Then the operator $T$ defined by $\widehat{Tf} = m\hat{f}$, $f \in L^2$, is a singular integral operator of Calderón–Zygmund type.*

The assumptions of this theorem are highly excessive. However, they make it possible to refine the result, as stated below. This refinement will be useful in subsequent chapters. We remind the reader that, among other things, Theorem 2.4 claims the existence of a kernel for $T$, i.e., of a function $K(x,y)$ defined for $x \ne y$ and such that $Tf(x) = \int K(x,y)f(y)dy$ whenever $f \in L^2$ is a function with bounded support and $x$ is at a positive distance from the support of $f$.

**Theorem 2.5.** *Under the assumptions of Theorem 2.4, the kernel of $T$ has the form $K(x,y) = k(x-y)$ for some function $k$ infinitely differentiable everywhere except the origin and satisfying*

$$|(D^\alpha k)(x)| \le C_\alpha |x|^{-n-|\alpha|}, \ x \ne 0, \tag{2.16}$$

*for all multiindices* $\alpha$.

*Remark* 2.6. 1. To verify the claim of Theorem 2.4, it suffices to know (2.16) only for all $\alpha$ with $|\alpha| = 1$, see (2.10). The arguments below show that this is ensured if (2.15) is only fulfilled for the derivatives of $m$ up to a certain finite order (accordingly, the infinite differentiability itself of $m$ is not required either). We shall comment on this later.

2. It should be noted that estimates (2.15) are true *a fortiori* if $m$ is homogeneous of degree zero and infinitely differentiable on the sphere. It is not surprising that the condition $\int_{\mathbb{S}^{n-1}} m = 0$ (see the end of the preceding subsection) has disappeared: we can easily recapture it. Indeed, consider the constant function $m_0 = \int_{\mathbb{S}^{n-1}} m$, then the corresponding operator $T_{m_0} = m_0 \,\mathrm{id}$ is a multiple of the identity, hence a Calderón–Zygmund operator with zero kernel. Next, $T_m = T_{m-m_0} + m_0 \,\mathrm{id}$, etc.

We pass to the proof of Theorems 2.4 and 2.5. Take a function $\psi \in \mathcal{D}(\mathbb{R})$ with $\psi(\tau) = 0$ for $|\tau| > 2$ and $\psi(\tau) = 1$ for $|\tau| \le 1$, and put $\varphi(\xi) = \psi(|\xi|) - \psi(2|\xi|)$ for $\xi \in \mathbb{R}^n$. Then, clearly, $\varphi$ is infinitely differentiable on $\mathbb{R}^n$, $\operatorname{supp}\varphi \subset \{1/2 < |\xi| < 2\}$, and $\sum_{j\in\mathbb{Z}} \varphi(2^{-j}\xi) = 1$ for $\xi \ne 0$ (in the last series, at most two terms are nonzero at every point). We split the function $m$ accordingly: $m = \sum_{j\in\mathbb{Z}} m_j$, where $m_j(\xi) = m(\xi)\varphi(2^{-j}\xi)$. Next, let $k_j$ be the inverse Fourier transform of $m_j$, $j \in \mathbb{Z}$.

Our immediate goal is to estimate partial derivatives of the functions $k_j$. For this, we deduce certain estimates for $m_j$. Differentiating $m_j$, by the Leibnitz formula we obtain

$$D^\alpha m_j(\xi) = D^\alpha[m(\xi)\varphi(2^{-j}\xi)] = \sum_{\beta+\gamma=\alpha} c_{\beta\gamma} 2^{-j|\beta|}[D^\gamma m(\xi)][(D^\beta\varphi)(2^{-j}\xi)].$$

Since any derivative of $\varphi$ is bounded on $\mathbb{R}^n$ by a constant depending on the index of the derivative and vanishes outside the spherical shell $\{2^{j-1} \le |\xi| \le 2^{j+1}\}$, we employ (2.15) to deduce the inequality

$$|D^\alpha m_j(\xi)| \le d_\alpha 2^{-|\alpha|j}, \ \xi \in \mathbb{R}^n,$$

with some constant $d_\alpha$ depending on $\alpha$.

Now, we consider the function $\xi^\beta m_j(\xi)$ ($\beta$ is a multiindex) and again calculate the $\alpha$th partial derivative of it by the Leibnitz formula. Since $m_j$ and all its derivatives vanish outside the spherical shell $\{2^{j-1} \le |\xi| \le 2^{j+1}\}$, the preceding estimate implies

$$|D^\alpha(\xi^\beta m_j(\xi))| \le d_{\alpha\beta} 2^{(|\beta|-|\alpha|)j}, \ \xi \in \mathbb{R}^n.$$

Hence,

$$\int |D^\alpha(\xi^\beta m_j(\xi))| d\xi \le d_{\alpha\beta} 2^{(|\beta|-|\alpha|)j} |\{2^{j-1} \le |\xi| \le 2^{j+1}\}| \le d'_{\alpha\beta} 2^{(n+|\beta|-|\alpha|)j}.$$

Taking the inverse Fourier transform, we arrive at $|x^\alpha D^\beta k_j(x)| \le c_{\alpha\beta} 2^{j(n+|\beta|-|\alpha|)}$. Next, we fix a nonnegative integer $l$ and add these inequalities for all multiindices $\alpha$ with $|\alpha| = l$. This will lead to

$$|D^\beta k_j(x)| \le C_{l,\beta} |x|^{-l} 2^{j(n+|\beta|-l)}. \tag{2.17}$$

This is the estimate we were striving for.

The required function $k$ is given by $k(x) = \sum_{j\in\mathbb{Z}} k_j(x)$. We claim that this series converges absolutely and uniformly on compact sets not containing the origin, and remains convergent in the same sense after differentiation of any order. This can be shown as follows. Fixing a multiindex $\beta$, we use (2.17) with $l = 0$ and $l = |\beta| + n + 1$ to obtain

$$|D^\beta k_j(x)| \le c_\beta \min\{2^{j(n+|\beta|)}, |x|^{-|\beta|-n-1} 2^{-j}\}.$$

The minimum is attained at the first term precisely when $2^j \le |x|^{-1}$ (we denote the set of these $j$ by $A$ and that of all other $j$ by $B$). So

$$\sum_j |D^\beta k_j(x)| \le c_\beta \left( \sum_{j\in A} 2^{j(n+\beta)} + \sum_{j\in B} |x|^{-|\beta|-n-1} 2^{-j} \right) \approx |x|^{-n-|\beta|}.$$

It follows that the function $k$ is well defined and infinitely differentiable everywhere except the origin, and it satisfies (2.16).

It only remains to show that $k(x-y)$ is the kernel of $T$. Let $f$ be a function in $L^2$, then

$$\widehat{Tf} = \lim_{N\to\infty} \sum_{|j|\le N} \hat{f} m_j$$

in $L^2$. So,

$$Tf = \lim_{N\to\infty} \sum_{|j|\le N} f * k_j$$

in $L^2$ by the Plancherel theorem. On the other hand, if, moreover, $f$ is supported on a bounded set $B$ and $x$ is at a positive distance from $B$, then the dominated convergence of the series $\sum k_j$ proved above implies the following (note that the set $x - B$ is bounded and bounded away from the origin):

$$Tf(x) = \lim_{N\to\infty} \int_{x-B} f(x-y) \sum_{|j|\le N} k_j(y) dy = \int_{x-B} f(x-y) k(y) dy,$$

which is equivalent to our claim. This finishes the proof.

At a point in Chapter 4, we shall need some information obtained in the course of the above proof. We collect it in the following proposition. Recall the standard notation $\mathcal{S} = \mathcal{S}(\mathbb{R}^n)$ for the *Schwartz class*, i.e., the space of all infinitely differentiable functions $\varphi$ on $\mathbb{R}^n$ such that $\lim_{x\to\infty} (D^\alpha \varphi)(x) p(x) = 0$ for every polynomial $p$ on $\mathbb{R}^n$ and every multiindex $\alpha$.

**Proposition 2.7.** *Under the assumptions of Theorem* 2.4, *there exist functions* $k^{(l)} \in \mathcal{S}$ *such that* $Tf = \lim_l f * k^{(l)}$ *(in* $L^2$*) for every* $f \in L^2$. *Moreover, the functions* $k^{(l)}$ *satisfy* (2.16) *uniformly in* $l$ *and* $D^s k^{(l)} \to D^s k$ *(as* $l \to \infty$*) for every multiindex* $s$, *uniformly on each compact set not containing the origin.*

*Proof.* Set $k^{(l)} = \sum_{|j| \le l} k_j$. □

It has already been mentioned that any smooth function $m$ homogeneous of degree zero satisfies the assumptions of Theorems 2.4 and 2.5. Such symbols $m$ often arise in the study of elliptic differential operators. Here is a typical question: "What can be said about a second-order derivative of a function if we know its Laplacian?" The solution is as follows. Suppose that $g = \Delta f$ (where, say, $f$ is a tempered distribution; or, as another extreme option, a function in $\mathcal{D}(\mathbb{R}^n)$), and let $h = \frac{\partial^2 f}{\partial x_i \partial x_j}$ ($i$ and $j$ may or may not coincide). Then $\hat{h} = m\hat{g}$, where

$$m(\xi) = \frac{\xi_i \xi_j}{\xi_1^2 + \cdots + \xi_n^2}.$$

Thus, $g$ is taken to $h$ by a Calderón–Zygmund operator. In particular, $\|h\|_{L^p} \le C_p \|g\|_{L^p}$, $1 < p < \infty$.

Calderón–Zygmund operators (the same as above, by the way) arise also in the study of gradient vector fields discussed in the Introduction as an example. (This was already mentioned repeatedly.) The reader will easily invent many similar settings (see, e.g., §2 in Chapter 10). We denote by $\mathcal{H}$ the space of all square integrable $n$-component vector fields on $\mathbb{R}^n$, and by $\mathcal{G}$ the subspace of $\mathcal{H}$ consisting of all gradient vector fields:

$$\mathcal{G} = \{(f_1, \dots, f_n) \in \mathcal{H} : \text{ there exists a tempered distribution } F \\ \text{such that } f_j = \frac{\partial F}{\partial x_j},\ 1 \le j \le n\}.$$

The orthogonal projection $P$ of $\mathcal{H}$ onto $\mathcal{G}$ can be written out explicitly in terms of Fourier transforms. This is done as follows. By the Plancherel theorem, the space $\widehat{\mathcal{H}}$ again consists of arbitrary collections $(\varphi_1, \dots, \varphi_n)$ of square-integrable functions, and $\widehat{\mathcal{G}}$ consists of the collections of the form

$$(-2\pi i \xi_1 u(\xi), \dots, -2\pi i \xi_n u(\xi)),$$

where $u$ is some function such that all components of the above vector-function are square integrable. The orthogonal complement to $\widehat{\mathcal{G}}$ is formed by the collections $(v_1, \dots, v_n) \in \widehat{\mathcal{H}}$ satisfying

$$\xi_1 v_1(\xi) + \cdots + \xi_n v_n(\xi) = 0. \tag{2.18}$$

We denote the orthogonal projection of $\widehat{\mathcal{H}}$ onto $\widehat{\mathcal{G}}$ by $Q$. Let $(\varphi_1, \dots, \varphi_n) \in \widehat{\mathcal{H}}$, then $Q(\varphi_1, \dots, \varphi_n) = -2\pi i(\xi_1 u(\xi), \dots, \xi_n u(\xi))$ for some function $u$. Now,

$(\varphi_1, \ldots, \varphi_n) - Q(\varphi_1, \ldots, \varphi_n)$ is orthogonal to $\widehat{\mathcal{G}}$; applying (2.18)), we obtain

$$\xi_1(\varphi_1(\xi) + 2\pi i \xi_1 u(\xi)) + \cdots + \xi_n(\varphi_n(\xi) + 2\pi i \xi_n u(\xi)) = 0.$$

From this equation, we find $u(\xi) = (-2\pi i)^{-1}|\xi|^{-2} \sum_k \varphi_k(\xi)\xi_k$, yielding finally

$$Q(\varphi_1, \ldots, \varphi_n) = \left( \sum_k \varphi_k \frac{\xi_1 \xi_k}{|\xi|^2}, \ldots, \sum_k \varphi_k \frac{\xi_n \xi_k}{|\xi|^2} \right).$$

Passage to inverse Fourier transforms gives a formula for the projection $P$:

$$P(f_1, \ldots, f_n) = (\sum_k T_{m_{1k}} f_1, \ldots, \sum_k T_{m_{nk}} f_n),$$

where the symbols $m_{jk}$ are defined by $m_{jk} = (\xi_j \xi_k)/|\xi|^2$. So, by Theorem 2.4, the orthogonal projection $P$ from $\mathcal{H}$ onto $\mathcal{G}$ is given by an operator matrix all entries of which are Calderón-Zygmund operators.

### 2.2.2 Additional information

In Remark 2.6 it was indicated how to relax the assumptions of Theorem 2.4. We want to formulate an even sharper result in this direction, called the Mikhlin–Hörmander theorem. This statement will not be used in this book; we dwell on it because it gives the opportunity to slightly extend the notion of a Calderón–Zygmund operator.

**Definition 2.8.** By a Calderón–Zygmund operator we shall also understand any bounded linear operator $S : L^2 \to L^2$ with the following properties. There is a sequence $\{T_j\}$ of operators satisfying the requirements of Definition 2.1 (their kernels will be denoted by $K_j$) and such that

(a) $$\sup_j \|T_j\|_{L^2 \to L^2} < \infty \text{ and } T_j f \to Sf \text{ in } L^2 \text{ for all } f \in L^2;$$

(b) $$\text{the Hörmander conditions } \int_{\mathbb{R}^n \setminus 2Q} |K_j(x, y_0) - K_(x, y_1)|\, dx \le C,$$

where $Q$ is a cube, $y_0, y_1 \in Q$, are fulfilled with one and the same constant $C$ for all $j$.

Such operators $S$ are still of weak type $(1, 1)$ and all $L^p$-boundedness results remain valid for them (to include the case of $p > 2$, a uniform Hörmander condition should be imposed also in the second variable). Indeed, we know well that long-range $L^1$-regularity suffices for that. But this property is satisfied indeed because if $f \in L^1 \cap L^2$ has zero integral and $\operatorname{supp} f \subset Q$ ($Q$ is some cube), then $\int_{\mathbb{R}^n \setminus 2Q} |T_j f| \le C \int_Q |f|$ with a constant $C$ *independent* of $j$. Now, the required inequality follows from the Fatou theorem, because $T_j f \to Sf$ in $L^2$.

We are ready to state the Mikhlin-Hörmander theorem.

**Theorem 2.9.** *Let $\varkappa$ be an integer greater than $n/2$, and let $m$ be a bounded function on $\mathbb{R}^n$ satisfying*

$$\frac{1}{R^n}\int_{\frac{1}{2}R\le|\xi|\le 2R}\left|R^{|\alpha|}D^{\alpha}m(\xi)\right|^2 d\xi \le B^2, 0<R<\infty, |\alpha|<\varkappa. \tag{2.19}$$

*Then the operator $S$ defined by $\widehat{Sf}=\hat{f}m$, $f\in L^2$, is a Calderón–Zygmund operator in the sense of the last definition.*

Note that condition (2.19) is true whenever the uniform condition (2.15) is fulfilled for $|\alpha|\le\varkappa$.

We refer the reader, e.g., to [H] for the proof. It should only be noted that the existence of a kernel for $T$ under the mere assumption (2.19) is questionable, but this is not important because the operators $T_j=\sum_{|l|\le j}T_{m_j}$ (the functions $m_j$ were introduced in the proof of Theorems 2.4 and 2.5) do possess kernels that satisfy the Hörmander condition uniformly. The relevant calculations are in the same spirit as those in that proof, but they are more involved.

## 2.3 Vector-valued analogs

When dealing with the projection onto gradient vector fields, we in fact encountered a singular integral operator on functions that take values in an $n$-dimensional space. Vector patterns will arise in this book repeatedly. In our setting, they will not be much harder than the scalar case.

The definition of a singular integral operator extends to spaces of vector-valued functions as follows. Let $1<r\le\infty$, and let $X$ and $Y$ be two Banach spaces. A bounded linear operator $T: L^r(\mathbb{R}^n,X)\to L^r(\mathbb{R}^n,Y)$ is said to be *singular integral of Calderón–Zygmund type* if it possesses a kernel $K(s,t)$ with properties similar to those listed in Definition 2.1 (with Remark 2.2 taken into account). Specifically, now the values of $K$ are bounded linear operators from $X$ to $Y$. Condition (I) in Definition 2.1 rewrites in the same form (but it should be kept in mind that now $f\in L^r(\mathbb{R}^n,X)$ and the integrand is in $L^r(\mathbb{R}^n,Y)$). In condition (II) we simply replace the absolute value under the integral sign by the norm in the space of operators from $X$ to $Y$. Presently, we may content ourselves with a condition of the form (2.11), specifically

$$\|K(s,t_0)-K(s,t_1)\|_{\mathcal{L}(X,Y)}\le C\frac{\|t_0-t_1\|^u}{\|s-t_0\|^{n+u}} \tag{2.20}$$

for some positive $u$ if, say, $\|t_0-t_1\|\le 2\|s-t_0\|$. As in the scalar case, it is easy to show that then $T$ is long-range regular on $L^1(X)$: if an integrable $X$-valued function $f$ is supported on a cube $K$ and has zero integral, then

$$\int_{\mathbb{R}^n\setminus 2K}\|Tf(s)\|_Y ds\le C\|f\|_{L^1(X)}.$$

Precisely as in the scalar case, this implies the weak type (1,1) inequality

$$|\{s : \|Tf(s)\|_Y > \alpha\}| \leq C\frac{\|f\|_{L^1(\mathbb{R}^n;X)}}{\alpha}.$$

By interpolation, it follows that $T$ acts boundedly from $L^p(\mathbb{R}^n;X)$ to $L^p(\mathbb{R}^n;Y)$ if $1 < p \leq r$.

By way of illustration, we explain that this vector setting allows us to obtain for free various estimates of maximal operators.

Consider a positive function $\varphi \in \mathcal{D}(\mathbb{R}^n)$ with unit integral, and for $w > 0$ put $\varphi_w(x) = w^{-n}\varphi(x/w)$. We introduce the maximal operator

$$\mathcal{M}f(x) = \sup_w (f * \varphi_w)(x), \tag{2.21}$$

defined on all locally integrable functions $f$. It is easy to realize that this operator is closely related to the Hardy–Littlewood maximal operator $M$ (see (1.30)):

$$c\mathcal{M}f \leq Mf \leq C\mathcal{M}f$$

almost everywhere, with constants $c$, $C$ depending only on $\psi$ and $n$. So, in many cases we may estimate $\mathcal{M}$ instead of $M$. We show that $\mathcal{M}$ can be treated as a singular integral operator whose values are vector-valued functions. Let $(\mu f)(s,w)$ denote the expression under the supremum sign in the definition of $\mathcal{M}$:

$$(\mu f)(s,w) = (f * \varphi_w)(s).$$

For every $s$, we treat $(\mu f)(s,\cdot)$ as an $L^\infty$-function in the second variable. We check that the operator $\mu$ enjoys the above definition with $r = \infty$, $X = \mathbb{R}$, and $Y = L^\infty$. Indeed, it is a bounded operator from the space of essentially bounded scalar functions to the space of essentially bounded $L^\infty$-valued functions, and it has a kernel, namely, the function $(s,t) \mapsto w^{-n}\varphi((s-t)/w)$. (The value of this function at $(s,t)$ is an operator from $\mathbb{R}$ to $L^\infty$, i.e., an essentially bounded function of the variable $w$.) We show that this function satisfies condition (2.20). First,

$$\frac{|\varphi((s-t_0)/w) - \varphi((s-t_1)/w|}{w^n} \leq \frac{\|t_0 - t_1\|}{w^{n+1}} \max|\nabla\varphi(\xi)|,$$

where the maximum is taken over the segment joining $\frac{s-t_0}{w}$ and $\frac{s-t_1}{w}$. Since $\varphi$ is compactly supported, this maximum is 0 if $w < c\|s - t_0\|$, where $c$ depends only on $\varphi$. Thus,

$$\sup_{w>0} \frac{|\varphi((s-t_0)/w) - \varphi((s-t_1)/w|}{w^n} \leq \sup_{w \geq c\|s-t_0\|} \frac{\|t_0 - t_1\|}{w^{n+1}} \leq C\frac{\|t_0 - t_1\|}{\|s - t_0\|^{n+1}},$$

as required.

Thus, the theory of the Hardy–Littlewood maximal function embeds indeed in that of singular integral operators. It should be noted that the Fefferman–Stein maximal theorem (Theorem 0.23, see the Introduction) also fits into this pattern. Hint: write the counterpart of that theorem for $\mu$ instead of $M$ and observe that the statement is clear for $p = q$; then write out the kernel of the corresponding operator (it takes $l^q$-valued functions to $l^q(L^\infty)$-valued functions) and of its adjoint and verify condition (2.20) for them (the estimates will be quite similar to each other; in fact, they reduce to the above calculation because the two operators are "diagonal" in a sense).

# Notes and remarks

The entire material is standard. We intentionally restrict ourselves to the facts indispensable for the stability theorems in Chapters 1, 3, and 10. Much more information can be found, e.g., in the books [GR] and [M]. To get some idea of the quite spectacular more recent development of the theory, the reader may consult [Vol] and the references therein.

The Mikhlin–Hörmander theorem (Theorem 2.9) was proved by Hörmander [H0] (see also the exposition in [H]), but the phenomenon was discovered by Mikhlin [M]. Mikhlin's conditions on the symbol of the multiplier were rather in the spirit of Theorem 2.4; the difference is that (2.15) was required only for the derivatives up to some finite order.

# Chapter 3

# Classical covering theorems

In this chapter we prove classical covering theorems. The arguments will be not quite standard, and these new thoughts will be used in Part 2 in the proofs of controlled covering theorems.

With the help of the Whitney lemma, a smooth analog of the Calderón–Zygmund decomposition will be constructed. It will be used to prove a stability theorem for near-minimizers in the case of the couple $(L^1, L^\infty)$.

## 3.1 Classical covering theorems and partitions of unity

We shall need three well-known covering theorems. We remind the reader that, throughout, $r(Q)$ stands for the radius of a cube $Q \subset \mathbb{R}^n$, that is, half the edge length of $Q$. Next, the distance between two points $x, y \in \mathbb{R}^n$ is defined to be

$$\|y - x\| = \max_{i=1,\dots,n} |y_i - x_i|, \qquad x = (x_1, \dots, x_n), \qquad y = (y_1, \dots, y_n).$$

We write $Q_x$ to denote a cube centered at $x$, and $\{Q_x\}_{x\in\Omega}$ to denote a family of cubes with centers at the points of a set $\Omega \subset \mathbb{R}^n$. Various modifications of the algorithm described below will play a central role in Chapter 8. It will be called the *Besicovitch q-process* in honor of Besicovitch.

### 3.1.1 The Besicovitch $q$-process

Suppose $\Omega \subset \mathbb{R}^n$, and we are given a family $\{Q_x\}_{x\in\Omega}$ of cubes with the following property:

$$\sup_{x\in\Omega} r(Q_x) = \sigma < \infty.$$

**Algorithm 3.1** (Besicovitch $q$-process). We fix $q > 0$ and consider the half-open intervals $\left(\frac{\sigma}{(1+q)^k}, \frac{\sigma}{(1+q)^{k-1}}\right]$ with $k = 1, 2, \dots$. Take the first of them $\left(\frac{\sigma}{1+q}, \sigma\right]$, corresponding to $k = 1$, and consider the set $\Omega_1$ of all cubes $Q_x$ such that $r(Q_x) \in \left(\frac{\sigma}{1+q}, \sigma\right]$. It is convenient to place the origin at the center of some cube $Q_{x_1}$ in $\Omega_1$; thus, we assume (without loss of generality) that $x_1 = 0$. Then we choose a cube $Q_{x_2} \in \Omega_1$ whose center $x_2$ lies outside $Q_{x_1}$ and is "almost" the closest to the origin, that is,

$$\|x_2\| \leq 2 \inf \left\{\|x\| : Q_x \in \Omega_1; x \notin Q_{x_1}\right\}.$$

Next, we choose a cube $Q_{x_3} \in \Omega_1$ whose center $x_3$ lies outside the cubes chosen previously, i.e., outside $Q_{x_1} \cup Q_{x_2}$, and is "almost" the closest to the origin. Continuing in this manner, we obtain some family $\{Q_{x_i}\}_{i \in I_1}$ of cubes in $\Omega_1$. The index set $I_1$ is finite if the process stops at some moment, i.e., the centers of all cubes in $\Omega_1$ lie in the union of the cubes chosen previously; otherwise, it is countably infinite. At the second step we consider the set $\Omega_2$ of all cubes $Q_x$ with $r(Q_x) \in \left(\frac{\sigma}{(1+q)^2}, \frac{\sigma}{1+q}\right]$ (this interval corresponds to $k = 2$). To $\Omega_2$, we again apply the above procedure. The only difference is that the center of any new cube must lie outside *all* the cubes chosen previously, including those chosen at the first step. As a result, we arrive at a family $\{Q_{x_i}\}_{i \in I_2}$ of cubes. Continuing in the same manner, finally we arrive at a family $\{Q_{x_i}\}_{i \in I}$ (the index set $I$ is the union of the sets $I_k$ arising at all steps).

**Definition 3.2.** The resulting subfamily $\{Q_{x_i}\}_{i \in I}$ of $\{Q_x\}_{x \in \Omega}$ will be called a family of Besicovitch cubes corresponding to the parameter $q$.

**Lemma 3.3.** *The cubes of the family $\{Q_{x_i}\}_{i \in I}$ possess the following properties:*

- *the cubes $Q_{x_i}$ are almost disjoint, i.e.,*

$$\min\left(r(Q_{x_i}), r(Q_{x_j})\right) < \|x_i - x_j\| \tag{3.1}$$

*for every $i, j \in I$, $i \neq j$;*

- *if $i, j \in I$ and $i \neq j$, then*

$$\frac{1}{2+q} Q_{x_i} \cap \frac{1}{2+q} Q_{x_j} = \varnothing; \tag{3.2}$$

- *for every $Q_x$ in the initial family $\{Q_x\}_{x \in \Omega}$, there is an index $i \in I$ such that*

$$x \in Q_{x_i} \text{ and } Q_x \subset (2+q) Q_{x_i}. \tag{3.3}$$

*Proof.* Property (3.1) follows from the fact that if $Q_{x_j}$ was chosen after $Q_{x_i}$, then $x_j$ lies outside $Q_{x_i}$, i.e.,

$$r(Q_{x_i}) < \|x_i - x_j\|. \tag{3.4}$$

To prove (3.2), we observe that, by construction, if $Q_{x_j}$ was chosen after $Q_{x_i}$, then

$$(1+q) r(Q_{x_i}) > r(Q_{x_j}). \tag{3.5}$$

Thus, if there exists $z \in \frac{1}{2+q}Q_{x_i} \cap \frac{1}{2+q}Q_{x_j}$, by (3.4) and (3.5) we obtain

$$r(Q_{x_i}) < \|x_i - x_j\| \le \|x_i - z\| + \|z - x_j\| \le \frac{1}{2+q}r(Q_{x_i}) + \frac{1}{2+q}r(Q_{x_j})$$
$$< \frac{1}{2+q}r(Q_{x_i}) + \frac{1}{2+q}(1+q)r(Q_{x_i}) = r(Q_{x_i}),$$

which is impossible, and (3.2) follows.

To verify (3.3), first we prove that if the family $\{Q_{x_i}\}_{i\in I_k}$ of the cubes chosen at the $k$th step is infinite, then

$$\lim_{i\in I_k} \|x_i\| = \infty. \tag{3.6}$$

Indeed, the radii of all cubes among $\{Q_{x_i}\}_{i\in I_k}$ exceed $\frac{\sigma}{(1+q)^k}$. Therefore, by (3.2) and volume considerations, any cube centered at the origin may include only finitely many cubes of the family $\{Q_{x_i}\}_{i\in I_k}$, and we are done.

Now, we verify (3.3). Take the natural number $k$ with

$$r(Q_x) \in \left(\frac{\sigma}{(1+q)^k}, \frac{\sigma}{(1+q)^{k-1}}\right].$$

If the cube $Q_x$ was chosen at the $k$th step, the claim is clear. If it was not, we observe that $x \in Q_i$ for some $i \in I_l$ with $l \le k$. (Indeed, otherwise, by construction, the set $I_k$ is infinite and $\|x_j\| \le 2\|x\|$ for all $j \in I_k$, which contradicts (3.6).) Taking such $i$, we observe that also

$$r(Q_x) < (1+q)r(Q_{x_i}).$$

Thus, if $z \in Q_x$, then

$$\|z - x_i\| \le \|z - x\| + \|x - x_i\| \le r(Q_x) + r(Q_{x_i}) < (2+q)r(Q_{x_i}).$$

This is precisely the required inclusion $Q_x \subset (2+q)Q_{x_i}$. □

### 3.1.2 Besicovitch theorem

To illustrate the use of the above construction, we prove the classical Besicovitch covering theorem (see [B]).

**Theorem 3.4.** *Suppose $\Omega \subset \mathbb{R}^n$, and we are given a family $\{Q_x\}_{x\in\Omega}$ of cubes satisfying*

$$\sup_{x\in\Omega} r(Q_x) < \infty. \tag{3.7}$$

*Then this family has an at most countable subfamily $\{Q_{x_i}\}_{i\in I}$ that covers the set $\Omega$ of centers,*

$$\Omega \subset \bigcup_{i\in I} Q_{x_i}, \tag{3.8}$$

*and is of finite multiplicity with a constant $M$ depending on the dimension $n$ only:*

$$\sum_{i\in I} \chi_{Q_{x_i}} \le M. \tag{3.9}$$

*Proof.* As $\{Q_{x_i}\}_{i\in I}$, we take the collection of Besicovitch cubes with some parameter $q$, as described above. It is at most countable by construction. Clearly, (3.8) is a consequence of (3.3). Thus, only finite multiplicity needs a proof. For arbitrary $x \in \Omega$, we shift the origin of the Cartesian coordinate system to $x$ and consider the $2^n$ "octants" in which $\mathbb{R}^n$ is split by the shifted coordinate hyperplanes. Next, we distribute all cubes $Q_{x_i}$, $i \in I$, that contain $x$ among $2^n$ subfamilies, according to the "octant" in which the center $x_i$ happens to fall. Take one of these subfamilies and consider the cube $Q_{x_{i_0}}$ in it that was chosen first. Then, by construction, we have $r(Q_{x_i}) \le (1+q)r(Q_{x_{i_0}})$ for all other cubes $Q_{x_i}$ in this subfamily and, therefore (see Figure 3.1; there an "octant" is a quadrant), their centers lie in one "octant" of the cube $K$ centered at $x$ and of radius $(1+q)r(Q_{x_{i_0}})$ (the volume of this "cubic octant" is $((1+q)r(Q_{x_{i_0}}))^n$). Next, all $Q_{x_i}$ in question contain $x$ but their centers lie outside $Q_{x_{i_0}}$. Consequently, $r(Q_{x_i}) > r(Q_{x_{i_0}})$. So, any of the (mutually nonintersecting) cubes $\frac{1}{2+q}Q_{x_i}$ cuts a volume of at least $(\frac{1}{2+q}r(Q_{x_{i_0}}))^n$ out of the "cubic octant" in question.

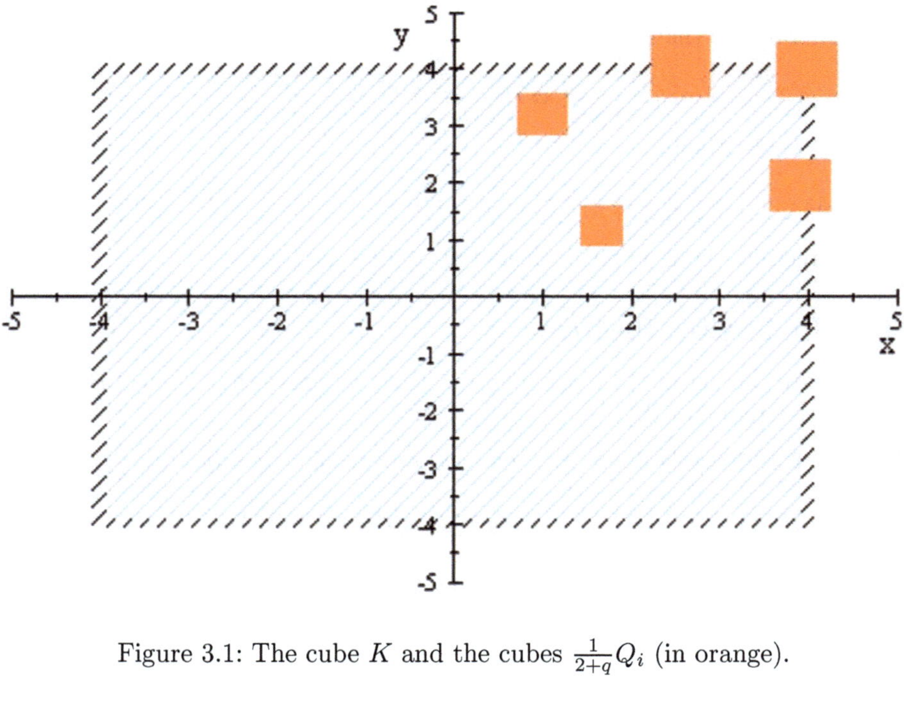


Figure 3.1: The cube $K$ and the cubes $\frac{1}{2+q}Q_i$ (in orange).

Therefore, there are at most

$$\frac{((1+q)r(Q_{x_{i_0}}))^n}{(\frac{1}{2+q}r(Q_{x_{i_0}}))^n} = (1+q)^n(2+q)^n$$

such cubes. There are $2^n$ "octants", so we arrive at the following multiplicity estimate:

$$M \leq 2^n(1+q)^n(2+q)^n. \qquad \square$$

A family of mutually nonintersecting cubes will be called a *packing.* This meaning of the word is not quite standard but convenient. We shall often employ the following fact.

**Proposition 3.5.** *A family of cubes in* $\mathbb{R}^n$ *that is of finite multiplicity with constant* $M$ *is a union of at most* $2^n(M-1)+1$ *packings.*

The question seems to be first addressed in [BrK], though the proof of this particular statement is only implicit there.

*Proof.* First, we verify the claim for a *finite* family $\{Q_i\}_{i\in I}$ of cubes. We proceed by induction on the cardinality of $I$. If $I$ is a singleton, there is nothing to prove. Otherwise, consider a cube $Q_d$ of minimal radius among the $\{Q_i\}_{i\in I}$ and observe that, by minimality, each cube $Q_i$ that intersects $Q_d$ must contain a vertex of $Q_d$. The vertices of $Q_d$ total $2^n$, and each of them may belong to at most $M-1$ cubes $Q_j$ with $j \neq d$. Thus, there are at most $2^n(M-1)$ cubes $Q_j$ with $j \neq d$ that intersect $Q_d$.

The inductive hypothesis allows us to split the family $\{Q_i\}_{i\in I\setminus\{d\}}$ into at most $2^n(M-1)+1$ packings. Adding empty packings if necessary, we may assume that the number of packings is precisely $2^n(M-1)+1$. Now, $Q_d$ must have empty intersection with all cubes of at least one of these packings (because otherwise more than $2^n(M-1)$ cubes with indices different from $d$ intersect $Q_d$, which is impossible). We adjoin $Q_d$ to this particular packing, which finishes a step of induction.

The case of an infinite family $\{Q_i\}_{i\in I}$ reduces to that of a finite family by a compactness argument. First, note that the index set $I$ must be *countably* infinite (because every cube contains a point with rational coordinates and no such point can belong to more than $M$ cubes of the family). So, we may assume that $I$ is the set $\mathbb{N}$ of positive integers. Let $\mathcal{S}_k$ denote the (nonempty!) set of all splittings of the family $\{Q_i\}_{i\leq k}$ in at most $2^n(M-1)+1$ packings. For each $j \leq k$, each packing in $\{Q_i\}_{i\leq k}$ induces a packing in $\{Q_i\}_{i\leq j}$ in a natural way (we simply eliminate all cubes with indices greater than $j$); so, each $P \in \mathcal{S}_k$ also induces an element $R \in \mathcal{S}_j$ (we call it the *j-print* of $P$). Thus, we have defined a natural mapping $\beta_{j,k}$ of $\mathcal{S}_k$ to $\mathcal{S}_j$, which takes any element of $\mathcal{S}_k$ to its $j$-print.

Clearly, $\beta_{l,k}\beta_{k,j} = \beta_{l,j}$ whenever $l \geq k \geq j$. Now, the quickest way to finish the proof is to use the Tikhonov theorem saying that an arbitrary product of compact spaces is compact. Indeed, the compact (finite discrete) spaces $\mathcal{S}_s$ and the above (continuous) mappings form an inverse spectrum. It is a standard consequence of the Tikhonov theorem that its projective limit is nonempty. This means that there is a sequence $\{P_k\}_{k\geq 1}$, $P_k \in \mathcal{S}_k$, with $P_j = \beta_{k,j} P_k$ for arbitrary $j$ and $k$ satisfying $j \leq k$. A moment's reflection shows that this implies the claim. □

*Remark* 3.6. It should be noted that the full strength of the Tikhonov theorem is not required here, moreover, the existence of the above sequence $\{P_k\}_{k\geq 1}$ can be proved by an easy diagonal procedure. Indeed, since the set $\mathcal{S}_j$ is finite, in any sequence $P_s \in \mathcal{P}_{k_s}$ (the $k_s$ being strictly monotone increasing), for any $j$ there exist a subsequence such that the $j$-prints of its members coincide. Now, we pick some elements $R_i \in \mathcal{S}_i$ arbitrarily. By the above, the sequence $R_i$ possesses a subsequence whose members have one and the same 1-print; from this subsequence, we extract a subsequence whose members have one and the same 2-print; and so on. This leads to a countable nesting family of subsequences, and it is clear that the corresponding diagonal sequence has the required properties.

### 3.1.3 Wiener lemma

A slight modification of the Besicovitch $q$-process yields the lemma established by Wiener in 1939 for the proof of an ergodic theorem.

**Lemma 3.7** (Wiener). *Let $\Omega \subset \mathbb{R}^n$. Suppose we are given a family $\{Q_x\}_{x\in\Omega}$ of cubes with uniformly bounded radii (i.e., condition (3.7) of the Besicovitch theorem is satisfied). Fix a parameter $q > 0$. Then there exists a packing $\{Q_{x_i}\}_{i\in I}$ in $\{Q_x\}_{x\in\Omega}$ such that for every cube $Q_x$ of the initial family there exists an index $i \in I$ with*

$$Q_x \subset (2+q)Q_{x_i}. \tag{3.10}$$

*Remark* 3.8. In distinction to the Besicovitch theorem, the cubes chosen may fail to cover $\Omega$. In recompense, in place of the mere finite multiplicity property, they do not intersect.

*Proof.* The construction differs from the Besicovitch $q$-process at one point: at each stage, the *entire* cube newly chosen (and not merely its center) must lie outside the cubes selected at the previous stages. We give the details. Putting $\sup_{x\in\Omega} r(Q_x) = \sigma < \infty$, we introduce the half-open intervals $\left(\frac{\sigma}{(1+q)^k}, \frac{\sigma}{(1+q)^{k-1}}\right]$, $k = 1, \dots$. Take the interval $\left(\frac{\sigma}{1+q}, \sigma\right]$, which corresponds to $k = 1$, and consider the set $\Omega_1$ of all cubes $Q_x$ whose radius $r(Q_x)$ falls within this interval. In $\Omega_1$, we choose a cube $Q_{x_1}$ arbitrarily and place the origin at its center (i.e., we assume without loss of generality that $x_1 = 0$). Then we choose a cube $Q_{x_2} \in \Omega_1$ that lies entirely outside $Q_{x_1}$ and with center $x_2$ "almost" the nearest to the origin, that is

$$\|x_2\| \leq 2 \inf\left\{\|x\| : Q_x \in \Omega_1; Q_x \cap Q_{x_1} = \varnothing\right\}.$$

Then we choose a cube $Q_{x_3} \in \Omega_1$ that lies entirely outside $Q_{x_1} \cup Q_{x_2}$ and whose center $x_3$ is almost the nearest to the origin. Continuing in this way, we obtain a family $\{Q_{x_i}\}_{i \in I_1}$ of cubes in $\Omega_1$.

At the second step, we consider the set $\Omega_2$ of all cubes in $\{Q_x\}_{x\in\Omega}$ whose radius falls within the interval $\left(\frac{\sigma}{(1+q)^2}, \frac{\sigma}{1+q}\right]$, corresponding to $k = 2$. With $\Omega_2$, we do the same as before, with the difference that all cubes newly chosen must lie outside *all* cubes selected previously, including those selected at the first step. This will result in a certain family $\{Q_{x_i}\}_{i\in I_2}$ of cubes. Continuing in a similar manner, we finally arrive at a family $\{Q_{x_i}\}_{i\in I}$ (the index set $I$ is the union of the sets $I_k$ obtained at all steps).

Now, if the family $\{Q_{x_i}\}_{i\in I_k}$ of cubes selected at the $k$th step is infinite, then, as in the Besicovitch process,

$$\lim_{i\in I_k} \|x_i\| = \infty. \tag{3.11}$$

Indeed, the radii of all cubes in the family $\{Q_{x_i}\}_{i\in I_k}$ exceed $\frac{\sigma}{(1+q)^k}$. Therefore, their volumes are at least $\left(\frac{2\sigma}{(1+q)^k}\right)^n$. Since the cubes $Q_{x_i}$ do not intersect, volume considerations show that only finitely many of them may have centers within a given distance from the origin, and (3.11) follows.

Now, we prove (3.10). Take the natural number $k$ such that

$$r(Q_x) \in \left(\frac{\sigma}{(1+q)^k}, \frac{\sigma}{(1+q)^{k-1}}\right].$$

If the cube $Q_x$ in question was chosen at the $k$th step, it coincides with some $Q_{x_i}$, and (3.10) is obvious. If $Q_x$ was not chosen when we dealt with the cubes with radii within the interval $\left(\frac{\sigma}{(1+q)^k}, \frac{\sigma}{(1+q)^{k-1}}\right]$, we employ (3.11) and the fact that each time we choose a cube "almost" the closest to the origin. This implies the existence of a cube $Q_{x_i}$ such that

$$Q_x \cap Q_{x_i} \neq \varnothing \text{ and } (1+q)r(Q_{x_i}) > r(Q_x).$$

Therefore, if $z \in Q_x$, then

$$\|z - x_i\| \leq \|z - x\| + \|x - x_i\| \leq r(Q_x) + r(Q_{x_i}) < (2+q)r(Q_{x_i}).$$

This means that $Q_x \subset (2+q)Q_{x_i}$. □

### 3.1.4 Whitney lemma, WB-coverings, and partitions of unity

A covering of quite a different sort is provided by a lemma proved in 1934 by H. Whitney (see [Wh]), in connection with smoothness-preserving extension of functions from a closed set to $\mathbb{R}^n$. We state the Whitney lemma in a form convenient to us.

**Lemma 3.9.** *Let $\Omega$ be an open subset of $\mathbb{R}^n$ with nonempty complement $F = \mathbb{R}^n \setminus \Omega$. Then there exists a family $\{Q_i\}_{i\in I}$ of dyadic cubes with mutually disjoint interiors such that*

$$\bigcup_{i\in I} Q_i = \bigcup_{i\in I} 2Q_i = \Omega \tag{3.12}$$

*and for all $i \in I$ we have*

$$5Q_i \cap F \neq \varnothing. \tag{3.13}$$

*Proof.* Consider the following family of dyadic cubes:

$$D' = \{Q \in D : 2Q \subset \Omega\}.$$

As $\{Q_i\}_{i\in I}$, we take the collection of maximal cubes in $D'$, i.e., those not included in any other cube in $D'$. Since $\Omega$ is open and $F = \mathbb{R}^n \setminus \Omega$ is nonempty, (3.12) follows immediately, and we only need to show that $5Q_i \cap F \neq \varnothing$. Take the dyadic cube $\tilde{Q}_i$ such that $Q_i$ is obtained from it by splitting into $2^n$ congruent cubes. Since $Q_i$ is maximal, we see that $2\tilde{Q}_i \cap F \neq \varnothing$. It remains to observe that $2\tilde{Q}_i \subset 5Q_i$. □

*Remark* 3.10. The following example shows that the constant 5 in (3.13) cannot be reduced with preservation of (3.12). Consider the set $\Omega = (-1, 10) \subset \mathbb{R}$. Since the dyadic interval $Q(1,1) = [0,2]$ does not satisfy the condition $2Q \subset \Omega$, the family $\{Q_i\}_{i\in I}$ must contain a dyadic interval $Q_{i_0}$ that lies entirely in the dyadic interval $Q(\frac{3}{2}, \frac{1}{2}) = [1,2]$. It remains to observe that $(5-\varepsilon)Q_{i_0}$ does not intersect $F = \mathbb{R} \setminus \Omega$ for every $\varepsilon > 0$.

The cubes $\{Q_i\}_{i\in I}$ will be called the *Whitney cubes.* From (3.12) and (3.13) it follows that the cubes become smaller and smaller as they approach the boundary of $\Omega$. Moreover, (3.12) and (3.13) show that the distance of a Whitney cube to the boundary of $\Omega$ is comparable to its radius.

We shall need the family $K_i = (1+\frac{1}{6})Q_i$, $i \in I$, of slightly modified Whitney cubes. It turns out that this family possesses a number of remarkable properties, which provide a basis for many applications of the Whitney lemma. In the definition below, the capital letters W (for Whitney) and B (for Besicovitch) are used to signalize that the covering in question shares the features of the two constructions.

**Definition 3.11.** A family $\{K_i\}_{i\in I}$ of cubes is called a WB-covering of a set $\Omega$ if it has the following properties:

- the Whitney property

$$\bigcup_{i\in I}\left(\frac{1}{1+\delta}K_i\right) = \bigcup_{i\in I} K_i = \Omega; \tag{3.14}$$

- finite multiplicity

$$\sum_{i\in I} \chi_{K_i} \leq M; \tag{3.15}$$

- strong engagement

$$\text{if } K_i \cap K_j \neq \varnothing, \text{ then } |K_i \cap K_j| \geq \varepsilon \max(|K_i|, |K_j|), \qquad i, j \in I. \quad (3.16)$$

Here $\delta$, $M$, and $\varepsilon$ are positive constants independent of $i, j \in I$. They will be called the Whitney constant, the multiplicity constant, and the engagement constant, respectively.

In many cases a family of cubes with the above properties will emerge in no relationship with any set $\Omega$ prescribed beforehand; then we still use the term "a WB-covering" meaning that the family covers its union.

The term "Whitney property" was used for (3.14) because, by (3.12), the family $K_i = 2Q_i$, $i \in I$ ($Q_i$ being the Whitney cubes) possesses this property with $\delta = 1$.

Next, finite multiplicity (see (3.15)) is a feature typical of the cubes in the Besicovitch Theorem 3.4. In fact, the present form of the definition of a WB-covering has been chosen merely to emphasize this similarity, because the cubes in a WB-covering always have a stronger property.

**Definition 3.12.** A collection of cubes is said to have *finite overlap* if there is a constant $\nu$ such that each cube in the collection intersects at most $\nu$ other cubes. This $\nu$ is called the *overlap number.*

Clearly, finite overlap implies finite multiplicity, but the converse is not true in general. However, we have the following statement.

**Lemma 3.13.** *If $\{K_i\}_{i \in I}$ is a family of cubes that satisfies* (3.15) *and* (3.16), *then this family has finite overlap. The overlap number depends only on $M$, $\varepsilon$, and the dimension.*

*Proof.* By Proposition 3.5, the family in question splits in at most $2^n(M-1)+1$ packings. By (3.16), each packing may contain at most $1/\varepsilon$ cubes intersecting $K_i$ and different from $K_i$. □

It should be noted that, curiously enough, all specific constructions of WB-coverings presented in this book will yield finite overlap directly rather than via Lemma 3.13. We emphasize that, in the setting of the classical Besicovitch theorem, only finite multiplicity is available.

Here is an example of a WB-covering of the entire space $\mathbb{R}^n$.

*Example* 3.14. Split $\mathbb{R}^n$ into cubes $\{Q_i\}_{i \in I}$ of equal volume and put $K_i = (1 + \frac{1}{2})Q_i$. It is easily seen that the cubes $\{K_i\}_{i \in I}$ form a WB-covering of $\mathbb{R}^n$ with $\delta = \frac{1}{2}$, $M = 2^n$, $\nu = 3^n$, and $\varepsilon = \frac{1}{3^n}$. (To calculate the overlap number, we observe that it is equal to the number of neighbors of each $Q_i$, including $Q_i$ itself. The cubes mentioned pave $3Q_i$.)

Now, we show that the Whitney lemma provides a WB-covering.

**Theorem 3.15.** *Let $\{Q_i\}_{i\in I}$ be the family of Whitney cubes for an open set $\Omega$ with $\mathbb{R}^n \setminus \Omega \neq \varnothing$. Then the cubes $K_i = (1+\frac{1}{6})Q_i$, $i \in I$, constitute a* WB*-covering of $\Omega$ with overlap number $\nu$ and engagement constant $\varepsilon$ depending only on the dimension $n$, and with $\delta = \frac{1}{6}$ in the Whitney property* (3.14).

Before proving the theorem, we establish a property of Whitney cubes.

**Lemma 3.16.** *Suppose two Whitney cubes $Q_i$, $Q_j$ are such that $(1+\frac{1}{6})Q_i \cap (1+\frac{1}{6})Q_j \neq \varnothing$. Then these cubes are neighbors, that is,*

$$Q_i \cap Q_j \neq \varnothing. \tag{3.17}$$

*Proof.* Let $x_i$ and $x_j$ denote the centers of $Q_i$ and $Q_j$ and let, for definiteness, $r(Q_i) \geq r(Q_j)$. First, we observe that all neighbors of $Q_i$ have radius at least $\frac{1}{4}r(Q_i)$. Indeed, should a neighbor $Q_k$ of $Q_i$ be of radius strictly smaller than $\frac{1}{4}r(Q_i)$, we would have $r(Q_k) \leq \frac{1}{8}r(Q_i)$ (because all cubes are dyadic), implying $5Q_k \subset 2Q_i \subset \Omega$. But this contradicts (3.13). The inclusion $5Q_k \subset 2Q_i$ is proved as follows:

$$5Q_k \subset Q(x_i, r(Q_i) + r(Q_k) + 5r(Q_k)) \subset Q(x_i, (1+\frac{6}{8})r(Q_i)) \subset 2Q(x_i, r(Q_i)).$$

The fact proved above shows that the cube $(1+\frac{1}{2})Q_i$ is included in the union of $Q_i$ and its neighbors. So, if (3.17) fails, then the cubes $(1+\frac{1}{2}-\varepsilon)Q_i$ and $Q_j$ do not intersect for an arbitrary positive $\varepsilon$, and consequently

$$\|x_i - x_j\| > (1+\frac{1}{2}-\varepsilon)r(Q_i) + r(Q_j).$$

On the other hand, since $(1+\frac{1}{6})Q_i \cap (1+\frac{1}{6})Q_j \neq \varnothing$, we have

$$\|x_i - x_j\| \leq (1+\frac{1}{6})r(Q_i) + (1+\frac{1}{6})r(Q_j),$$

and therefore

$$(1+\frac{1}{2}-\varepsilon)r(Q_i) + r(Q_j) < (1+\frac{1}{6})r(Q_i) + (1+\frac{1}{6})r(Q_j).$$

Since $r(Q_i) \geq r(Q_j)$, this inequality cannot be true for $\varepsilon$ sufficiently small. The lemma follows. □

*Remark* 3.17. In the course of the proof of the above lemma, it was shown that $(1+\frac{1}{2})Q_i$ is included in the union of $Q_i$ and its neighbors, whenever $Q_i$ is a Whitney cube. Also, the radii of all neighbors are at least $\frac{1}{4}r(Q_i)$.

Now, we prove the theorem.

*Proof.* By the Whitney lemma, we have

$$\bigcup_{i\in I} Q_i = \bigcup_{i\in I} 2Q_i = \Omega.$$

Therefore, $\bigcup_{i\in I} K_i = \bigcup_{i\in I}(1+\frac{1}{6})Q_i = \Omega$, and the Whitney condition (3.14) with $\delta = \frac{1}{6}$ follows from the identity $\frac{1}{1+\frac{1}{6}}K_i = Q_i$. We verify the finite overlap property. Take a cube $Q_{i_0}$ arbitrarily. By (3.17), we see that the number of cubes $K_j = (1+\frac{1}{6})Q_j$ intersecting $K_{i_0} = (1+\frac{1}{6})Q_{i_0}$ does not exceed the number of neighbors of $Q_{i_0}$ plus 1 (the cube $K_{i_0}$ itself). We estimate the number of neighbors of $Q_{i_0}$. From Remark 3.17, it follows that the "layer" $(1+\frac{1}{2})Q_{i_0}\setminus Q_{i_0}$ (its volume is $((1+\frac{1}{2})^n - 1)\,|Q_{i_0}|$) cuts out a volume of at least $\frac{1}{4^n}\,|Q_{i_0}|$ from every neighbor of $Q_{i_0}$. Thus, there are at most $((1+\frac{1}{2})^n - 1)\cdot 4^n = 6^n - 4^n$ neighbors. Consequently, the overlap number $\nu$ obeys the inequality

$$\nu \le 6^n - 4^n + 1.$$

To prove (3.16), it suffices to observe that combination of Remark 3.17 and Lemma 3.16 implies the following: if $K_i \cap K_j \neq \varnothing$, then the volume of $K_i \cap K_j$ is at least $\frac{1}{6^n}\max(r(Q_i)^n, r(Q_j)^n)$. Now, since $|K_i| = (1+\frac{1}{6})^n 2^n r(Q_i)^n = \frac{14^n}{6^n} r(Q_i)^n$, we have

$$|K_i \cap K_j| \ge \frac{1}{6^n}\max(r(Q_i)^n, r(Q_j)^n) = \frac{1}{14^n}\max(|K_i|\,, |K_j|).$$

Thus, we have proved (3.16) with $\varepsilon = \frac{1}{14^n}$. □

*Remark* 3.18. It was shown in the course of the proof that, for the WB-covering $\{K_i\}_{i\in I}$ constructed above, the Whitney constant $\delta$ equals $\frac{1}{6}$ (this was even included in the claim originally), and the overlap number and engagement constant can be chosen as follows: $\nu = 6^n - 4^n + 1$, $\varepsilon = \frac{1}{14^n}$. Surely, the multiplicity constant $M$ does not exceed $\nu$.

WB-coverings are important because they allow us to construct a $C^\infty$-partition of unity adjusted to the family $\{K_i\}_{i\in I}$, as described in the following theorem.

**Theorem 3.19.** *Suppose that cubes $\{K_i\}_{i\in I}$ form a* WB-*covering with constants ($\delta$, $M$, and $\varepsilon$). Then there exists a family $\{\psi_i\}_{i\in I}$ of infinitely differentiable functions that forms a partition of unity on $\cup K_i$:*

$$\sum_{i\in I} \psi_i = \chi_{\cup K_i} \tag{3.18}$$

*and has the following properties:*

$$supp\,\psi_i \subset \frac{1+\frac{\delta}{2}}{1+\delta}K_i,\ i\in I; \tag{3.19}$$

$$\psi_i(x) \geq \frac{1}{M} \text{ for } x \in \frac{1}{1+\delta}K_i, \ i \in I; \tag{3.20}$$

$$\left|D^k\psi_i\right| \leq c\frac{1}{|K_i|^{\frac{|k|}{n}}}, \qquad |k| = \sum_{j=1}^n k_j, \qquad D^k = \frac{\partial^k}{\partial^{k_1}\partial^{k_2}\dots\partial^{k_n}}. \tag{3.21}$$

*Property* (3.21) *holds true for every* $k = (k_1, \dots, k_n) \in Z^n$ *with* $k_j \geq 0$, $j = 1, \dots, n$. *The constant* $c = c(k, \delta, M, \varepsilon) > 0$ *depends only on the parameters indicated.*

*Proof.* Consider a $C^\infty$-function $\varphi$ on the real line such that

$$0 \leq \varphi \leq 1, \quad \operatorname{supp}\varphi \subset \left[-\frac{1+\frac{\delta}{2}}{1+\delta}, \frac{1+\frac{\delta}{2}}{1+\delta}\right],$$

and $\varphi$ is equal to 1 on $\left[-\frac{1}{1+\delta}, \frac{1}{1+\delta}\right]$.

Denoting the center of $K_i$ by $x_i = (x_{i,1}, \dots, x_{i,n})$, on the space $\mathbb{R}^n$ we introduce the function

$$\varphi_i(x) = \varphi\left(\frac{x_1 - x_{i,1}}{r(K_i)}\right) \times \dots \times \varphi\left(\frac{x_n - x_{i,n}}{r(K_i)}\right), \qquad x = (x_1, \dots, x_n).$$

Next, we put

$$\psi_i = \frac{\varphi_i}{\sum_{j\in I}\varphi_j}, \qquad i \in I. \tag{3.22}$$

The inclusion (3.19) and the statement that $\{\psi_i\}_{i\in I}$ is a partition of unity for $\bigcup_{i\in I} K_i$ follow from the construction and the Whitney property

$$\bigcup_{i\in I}\frac{1}{1+\delta}K_i = \bigcup_{i\in I}K_i$$

of the cubes. Next, (3.20) is a consequence of (3.22) because in (3.22) the numerator $\varphi_i$ is equal to 1 on $\frac{1}{1+\delta}K_i$ and, by finite multiplicity, after evaluation at a point, the denominator involves at most $M$ summands, each not exceeding 1.

We prove (3.21) by induction. Suppose it is established for $|k| < N$. We prove it for $|k| = N$. For that, we rewrite (3.22) as

$$\psi_i \cdot \sum_{j\in I}\varphi_j = \varphi_i$$

and apply $D^k$ to both sides. By the Leibnitz formula,

$$(D^k\psi_i)\cdot\sum_{j\in I}\varphi_j = D^k\varphi_i - \sum_{|l|<|k|} c_l D^l\psi_i \cdot \left(D^{k-l}(\sum_{j\in I}\varphi_j)\right).$$

Since $\sum_{j\in I}\varphi_j \geq 1$ on $\bigcup_{i\in I}\frac{1}{1+\delta}K_i = \bigcup_{i\in I}K_i$ and $\operatorname{supp}\psi_i \subset K_i$, we obtain

$$\left\|D^k\psi_i\right\|_{L_\infty} \leq \left\|D^k\varphi_i\right\|_{L_\infty} + \sum_{|l|<N} c_l \left\|D^l\psi_i\right\|_{L_\infty} \left\|D^{k-l}\Big(\sum_{j\in I}\varphi_j\Big)\right\|_{L_\infty(K_i)}.$$

Now, $\left\|D^k\varphi_i\right\|_{L_\infty} \leq c\frac{1}{|K_i|^{\frac{|k|}{n}}}$ and $\left\|D^{k-l}\varphi_i\right\|_{L_\infty} \leq c\frac{1}{|K_i|^{\frac{|k-l|}{n}}}$ with a constant $c > 0$ depending only on the initial function $\varphi$, $\delta$, and the order of the derivative. By the inductive hypothesis, the required statement will follow if we show that

$$\left\|D^{k-l}\Big(\sum_{j\in I}\varphi_j\Big)\right\|_{L_\infty(K_i)} \leq c\frac{1}{|K_i|^{\frac{|k-l|}{n}}}. \tag{3.23}$$

To prove (3.23), we recall that the number of summands in $\sum_{j\in I}\varphi_j$ that are nonzero on $K_i$ is controlled in terms of the overlap number $\nu$ (see Lemma 3.13). But by formula (3.16), the volumes of the cubes $K_j$ intersecting $K_i$ are proportional to the volume of $K_i$. Consequently, the right-hand side of (3.23) dominates the $L^\infty(K_i)$-norm of each function $D^{k-l}\varphi_j$, $j \in I$, and we are done. □

In what follows, a family $\{\psi_i\}$ with properties (3.18)–(3.21) is said to be *adjusted* to the WB-covering $\{K_i\}$.

## 3.2 Another Calderón–Zygmund procedure

Suppose we are given a function $f \in L^1(\mathbb{R}^n)$ and a number $t > 0$. In the classical Calderón–Zygmund decomposition, the "good" function was defined by the formula

$$f_t = \sum_{i\in I} c_i\chi_{Q_i} + f\chi_{\mathbb{R}^n\setminus\bigcup Q_i}, \text{ where } c_i = \frac{1}{|Q_i|}\int_{Q_i} f(s)ds,$$

and the family $\{Q_i\}_{i\in I}$ was formed by the maximal dyadic cubes in the family $D_{f,t} = \left\{Q \in D : \frac{1}{|Q|}\int_Q |f(s)|\,ds \geq t\right\}$ (see the proof of the Calderón–Zygmund lemma). This $f_t$ is not smooth and even is discontinuous. As we shall see below, this presents an obstruction in many situations. We describe a procedure that eliminates this drawback.

Consider the set

$$\tilde{\Omega}_{f,t} = \left\{Q : \frac{1}{|Q|}\int_Q |f(s)|\,ds > t\right\}; \tag{3.24}$$

this time we take all (rather than only dyadic) cubes satisfying the above condition. Put

$$\Omega_{f,t} = \bigcup_{Q\in\tilde{\Omega}_{f,t}} Q. \tag{3.25}$$

This set is open by the countable additivity (hence, continuity) of the Lebesgue integral. For all $Q \in \tilde{\Omega}_{f,t}$ we have

$$|Q| \le \frac{1}{t} \int_Q |f(s)|\, ds \le \frac{1}{t} \|f\|_{L^1}, \tag{3.26}$$

so the Wiener lemma is applicable. By the first inequality in (3.26), the sum of the volumes of the cubes in the packing given by that lemma is controlled by $(1/t)\|f\|_{L^1}$. So, $\Omega_{f,t}$ has finite measure and, therefore, nonempty complement. Thus, we can use the Whitney lemma and Theorem 3.15 to obtain a WB-covering $\{K_i\}_{i\in I}$ of $\Omega_{f,t}$ that possesses the property

$$5K_i \cap (\mathbb{R}^n \setminus \Omega_{f,t}) \neq \varnothing, \qquad i \in I. \tag{3.27}$$

Now, we take a smooth partition of unity $\{\psi_i\}_{i\in I}$ adjusted to the family $\{K_i\}_{i\in I}$, and define the "good" function $f_t$ by

$$f_t = \sum_{i\in I} c_i\psi_i + f\chi_{\mathbb{R}^n\setminus\cup K_i}, \text{ where } c_i = \frac{1}{\int \psi_i} \int f(s)\psi_i(s)\, ds. \tag{3.28}$$

Clearly, for the "bad" part $f - f_t$ we have

$$f - f_t = \sum_{i\in I} (f - c_i)\psi_i, \qquad c_i = \frac{1}{\int \psi_i} \int f(s)\psi_i(s)\, ds \tag{3.29}$$

(because the family $\{\psi_i\}_{i\in I}$ is a partition of unity).

To explain the choice of the formula for the coefficients $c_i$, we consider the "blocks"

$$b_i = (f - c_i)\psi_i = \left(f - \frac{1}{\int \psi_i} \int f(s)\psi_i(s)\, ds\right)\psi_i \tag{3.30}$$

that constitute the bad function. The $c_i$ are chosen so the blocks will have zero average:

$$\int b_i(s)\, ds = \int (f(s) - c_i)\psi_i(s)\, ds = 0. \tag{3.31}$$

The above decomposition

$$f = f_t + (f - f_t)$$

will be referred to as a *smooth* Calderón–Zygmund decomposition. We show that it enjoys certain properties similar to properties (CZ1)-(CZ4) of the classical procedure.

We define an operator $P_{f,t}$ by the formula

$$P_{f,t}(h) = \sum_{i\in I} \left(\frac{1}{\int \psi_i} \int h(s)\psi_i(s)\, ds\right)\psi_i + h\chi_{\mathbb{R}^n\setminus\cup K_i}. \tag{3.32}$$

Then

$$(I - P_{f,t})(h) = \sum_{i\in I} \left( h - \frac{1}{\int \psi_i} \int h(s)\psi_i(s)\, ds \right) \psi_i. \tag{3.33}$$

In particular,

$$P_{f,t}(f) = f_t \text{ and } (I - P_{f,t})f = f - f_t. \tag{3.34}$$

(Compare with (1.12), (1.13), and (1.14).)

**Lemma 3.20.** *The objects defined above in connection with a smooth Calderón–Zygmund decomposition have the following properties.*

(CZS1) (*capacity estimate*)

$$\sum_{i\in I} |K_i| \le \frac{c}{t} \operatorname{dist}_{L^1}(f, B_{L^\infty}(\frac{t}{2})); \tag{3.35}$$

(CZS2) (*splitting the "bad" function in blocks*)

$$f - f_t = \sum_{i\in I} b_i, \tag{3.36}$$

*and*

$$\operatorname{supp} b_i \subset \frac{1+\frac{\delta}{2}}{1+\delta} K_i, \qquad \int_{\mathbb{R}^n} b_i(s)\, ds = 0, \qquad \|b_i\|_{L^1} \le ct\, |K_i| \tag{3.37}$$

*for all $i \in I$;*

(CZS3) (*approximation property*)

$$\|f_t\|_{L^\infty} \le ct, \tag{3.38}$$

$$\|f - f_t\|_{L^1} \le c \operatorname{dist}_{L^1}(f, B_{L^\infty}(\frac{t}{2})); \tag{3.39}$$

(CZS4) (*operator estimates*)

$$\|P_{f,t}\|_{L^1\to L^1} \le c, \quad \|I - P_{f,t}\|_{\mathrm{BMO}\to L^1} \le \frac{c}{t} \operatorname{dist}_{L^1}(f, B_{L^\infty}(\frac{t}{2})). \tag{3.40}$$

*Throughout, $c \ge 1$ is a constant depending only on $n$.*

*Proof.* (CZS1) Since $K_i = (1+\frac{1}{6})Q_i$ (see Theorem 3.15), it follows that

$$\sum_{i\in I} |K_i| = (1+\frac{1}{6})^n \sum_{i\in I} |Q_i| = (1+\frac{1}{6})^n \, |\Omega_{f,t}|, \tag{3.41}$$

and we need to estimate $|\Omega_{f,t}|$. For this, we consider the family (3.24) of cubes. Since $f \in L^1(\mathbb{R}^n)$, we have

$$|Q| < \frac{1}{t} \|f\|_{L^1}$$

for every $Q \in \tilde{\Omega}_{f,t}$, so the Wiener lemma can be applied to $\tilde{\Omega}_{f,t}$. By that lemma (with $q = 3$), there is a family $\{\tilde{Q}_j\}_{j\in J}$ of mutually disjoint cubes in $\tilde{\Omega}_{f,t}$ such that

$$\bigcup_{Q\in\tilde{\Omega}_{f,t}} Q \subset \cup 5\tilde{Q}_j.$$

Since $\Omega_{f,t} = \bigcup_{Q\in\tilde{\Omega}_{f,t}} Q$, we obtain

$$|\Omega_{f,t}| \le 5^n \sum_{j\in J} \left|\tilde{Q}_j\right|.$$

Now, we take an arbitrary function $g \in L^\infty$ with $\|g\|_{L^\infty} \le \frac{1}{2}t$. We have

$$\begin{aligned}\|f-g\|_{L^1} &\ge \sum_{j\in J}\int_{\tilde{Q}_j} |f(s)-g(s)|\,ds \ge \sum_{j\in J}\int_{\tilde{Q}_j} |f(s)|\,ds - \sum_{j\in J}\int_{\tilde{Q}_j} |g(s)|\,ds \\ &\ge t\sum_{j\in J}\left|\tilde{Q}_j\right| - \frac{1}{2}t\sum_{j\in J}\left|\tilde{Q}_j\right| = \frac{1}{2}t\sum_{j\in J}\left|\tilde{Q}_j\right| \ge \frac{1}{2\cdot 5^n}t\,|\Omega_{f,t}|.\end{aligned}$$

Since $g$ is arbitrary, (3.41) implies

$$\sum_{i\in I}|K_i| = (1+\frac{1}{6})^n\,|\Omega_{f,t}| \le (1+\frac{1}{6})^n\cdot 2\cdot 5^n\frac{1}{t}\,\mathrm{dist}_{L_1}(f, B_{L_\infty}(\frac{t}{2})).$$

(CZS2) Property (3.36) readily follows from the definitions. The first relation in (3.37) is a consequence of the fact that the support of $\psi_i$ is included in $\frac{1+\frac{\delta}{2}}{1+\delta}K_i$ (see (3.19)). The second relation has already been explained. To prove the third relation in (3.37), we note that $5K_i \cap (\mathbb{R}^n \setminus \Omega_{f,t}) \ne \varnothing$ (see (3.27)) and, therefore,

$$\frac{1}{|5K_i|}\int_{5K_i} |f(y)|\,dy \le t.$$

Now, the required inequality stems from the following estimates:

$$\begin{aligned}\|b_i\|_{L_1} &= \left|\int\left(f(y) - \frac{1}{\int\psi_i}\int f(s)\psi_i(s)\,ds\right)\psi_i(y)dy\right| \le 2\int_{K_i}|f(y)|\,dy \\ &\le 2\int_{5K_i}|f(y)|\,dy \le 2\,|5K_i|\,t = 2\cdot 5^n\,|K_i|\,t.\end{aligned}$$

(CZS3) First, note that, if $x \in \mathbb{R}^n \setminus \cup K_i = \mathbb{R}^n \setminus \Omega_{f,t}$, then for every cube $Q$ containing $x$ we have

$$\frac{1}{|Q|}\int_Q |f(s)|\,ds \le t.$$

Hence, the Lebesgue theorem implies

$$\left\|f\chi_{\mathbb{R}^n\setminus\cup K_i}\right\|_{L^\infty} \le t. \tag{3.42}$$

We estimate $|f_t(x)|$ for $x \in \cup K_i$. On $\frac{1}{1+\delta}K_i$, we have $\psi_i \geq \frac{1}{M}$ (see (3.20)). Consequently,

$$\int \psi_i \geq \frac{|K_i|}{M(1+\delta)^n}. \tag{3.43}$$

Since $0 \leq \psi_i \leq 1$ and $\operatorname{supp}\psi_i \subset K_i$, we obtain

$$\left|\frac{1}{\int \psi_i} \int f(s)\psi_i(s)\,ds\right| \leq \frac{M(1+\delta)^n}{|K_i|} \int_{K_i} |f(s)|\,ds \leq \frac{M(1+\delta)^n 5^n}{|5K_i|} \int_{5K_i} |f(s)|\,ds.$$

Since $5K_i$ is not included in $\tilde{\Omega}_{f,t}$, we have $\frac{1}{|5K_i|}\int_{5K_i} |f(s)|\,ds \leq t$ and, therefore,

$$\left|\frac{1}{\int \psi_i} \int f(s)\psi_i(s)\,ds\right| \leq M(1+\delta)^n 5^n \cdot t.$$

Recalling that the family $\{K_i\}_{i\in I}$ has finite multiplicity, from $\operatorname{supp}\psi_i \subset K_i$, $i \in I$, we deduce that

$$|f_t(x)| = \left|\sum_{i\in I} c_i\psi_i(x)\right| \leq \sum_{i:x\in K_i} \left|\frac{1}{\int \psi_i} \int f(s)\psi_i(s)\,ds\right| \leq M^2(1+\delta)^n 5^n \cdot t$$

for $x \in \cup K_i$. Together with (3.42), this yields (3.38).

Now, we prove formula (3.39). By (3.36) and (3.37), we have $\|f - f_t\|_{L^1} \leq ct\sum_{i\in I}|K_i|$, and the required inequality follows from (3.35).

(CZS4) The first inequality is a consequence of the definitions and the finite multiplicity of the family $\{K_i\}_{i\in I}$:

$$\begin{aligned}\|P_{f,t}(h)\|_{L_1} &\leq \sum_{i\in I} \int \left|\frac{1}{\int \psi_i} \int h(s)\psi_i(s)\,ds\right| \psi_i(y)\,dy + \int \left|h\chi_{\mathbb{R}^n\setminus\cup K_i}(y)\right|\,dy \\ &= \sum_{i\in I} \int |h(s)|\,\psi_i(s)\,ds + \int \left|h\chi_{\mathbb{R}^n\setminus\cup K_i}(y)\right|\,dy \leq M\,\|h\|_{L_1}.\end{aligned}$$

The second inequality follows from (3.33) and (3.43): first, we have

$$\begin{aligned}\|(I - P_{f,t})h\|_{L_1} &\leq \sum_{i\in I} \int \left|\left(h(y) - \frac{1}{\int \psi_i} \int h(s))\psi_i(s)\,ds\right)\psi_i(y)\right|\,dy \\ &\leq \sum_{i\in I} \frac{1}{\int \psi_i} \iint |(h(y) - h(s))\psi_i(s)\psi_i(y)|\,ds\,dy \\ &\leq \sum_{i\in I} \frac{M(1+\delta)^n}{|K_i|} \iint_{K_i\times K_i} |(h(y) - h(s))|\,ds\,dy.\end{aligned}$$

Next, letting $c = \frac{1}{|K_i|}\int_{K_i} h(u)du$, we write

$$\iint_{K_i\times K_i} |(h(y)-h(s))|\,ds\,dy \le \iint_{K_i\times K_i} |(h(y)-c+c-h(s))|\,ds\,dy$$
$$\le \iint_{K_i\times K_i} |(h(y)-c)|\,ds\,dy + \iint_{K_i\times K_i} |(c-h(s))|\,ds\,dy \le 2\,|K_i|^2\,\|h\|_{\mathrm{BMO}}\,.$$

Therefore,

$$\|(I-P_{f,t})h\|_{L_1} \le 2M(1+\delta)^n\,\|h\|_{\mathrm{BMO}}\sum_{i\in I}|K_i|\,,$$

and the required inequality follows by (3.35). □

## 3.3 Stability of near-minimizers for the couple $(L^1, L^\infty)$

### 3.3.1 Statement and proof

In a previous chapter, we proved an elementary stability result for the couple $(L^1, L^p)$ with $p < \infty$ (see Theorem 1.18). Here we are going to say something *nontrivial* about the couple $(L^1, L^\infty)$. For that, we shall need the smooth Calderón-Zygmund decomposition, and a new estimate of long-range regular (in practice, singular integral) operators. A similar pattern will be employed in Chapter 10. We signalize immediately that the space $L^\infty$ presents some fundamental difficulties for the theory discussed, which will result in "incompletely" stable near-minimizers.

We saw (cf. (3.38) and (3.39)) that the "good" function $f_t$ belongs to the ball of radius *ct* and centered at zero in $L^\infty$ and approximates $f$ in a nearly optimal way, i.e.,

$$\|f-f_t\|_{L^1} \le c\ \mathrm{dist}_{L^1}(f, B_{L^\infty}(\frac{t}{2})),$$

where the constant $c \ge 1$ depends only on $n$. Suppose, as before, that we are given a linear operator $T$ sending elements of $L^1$ to measurable functions but, generally speaking, unbounded on $L^1$. We want to find some conditions on $T$ under which $Tf_t$ yields a good approximation to $Tf$ in the norm of $L^1$. Recall that, should $T$ be $L^1$-bounded, we would have

$$\|Tf-Tf_t\|_{L^1} \le \|T\|_{L^1\to L^1}\,\|f-f_t\|_{L^1} \le c\,\|T\|_{L^1\to L^1}\,\mathrm{dist}_{L^1}\left(f, B_{L^\infty}\left(\frac{t}{2}\right)\right). \tag{3.44}$$

As has already been repeated several times, in general (for $T$ nonbounded) it cannot be expected without additional assumptions that $\|Tf-Tf_t\|_{L^1}$ is estimated in terms of the distance on the right in (3.44). At least, much as we did before, it is natural to assume that the function $Tf$ (which we are going to approximate) belongs to $L^1$. Some requirements about $T$ are also indispensable.

We shall assume that $T$ is a linear operator that acts continuously from $L^1$ to the space of measurable functions (with convergence in measure) and has the following properties:

(S1) $T$ is strongly long-range regular in the sense of Remark 2.3: if $f \in L^1$ is supported on a cube $Q$ and has zero integral, then for every $\lambda > 1$ we have

$$\left\|T(f) \cdot \chi_{\mathbb{R}^n \setminus \lambda Q}\right\|_{L^1} \le c(\lambda) \|f\|_{L^1}.$$

(We emphasize that this differs somewhat from a similar condition employed in Theorems 1.5 and 1.18.)

(S2) $T$ acts boundedly from $L^\infty$ to BMO.

(S3) $T$ is weakly $L^1$-bounded: there is a natural number $k$ such that for every $C^\infty$-function $\varphi$ supported on a cube $Q$ and every $f \in L^1$ such that $Tf$ is integrable over $Q$ we have

$$\left|\int T(f)(s)\varphi(s)ds\right| \le C\|f\|_{L^1}\left(\|\varphi\|_{L^\infty} + \sum_{j\le k}(\operatorname{diam} Q)^j \left\|\nabla^j\varphi\right\|_{L^\infty}\right) \tag{3.45}$$

with a constant $C > 0$ independent of $Q$, $f$, and $\varphi$.

Here by $\nabla^l\varphi$ we mean the collection of all derivatives of order $l$ of $\varphi$; the $L^\infty$-norm of this collection can be defined as (say) the sum of the $L^\infty$-norms of its components. Note that we shall apply condition (S3) only to positive functions $\varphi$.

We already know very well that condition (S1) is typical of singular integral operators. In the next chapter, we shall see the same about condition (S3). As to (S2), it is pertinent to *adjoints* to singular integrals. More precisely, it will be shown in the next chapter that $T = R^*$ satisfies (S2) if $R$ is long-range regular and (say) $L^2$-bounded. So, in practice the stability theorem below applies when both $T$ and its adjoint are singular integral operators.

Now, we formulate the main result of this section.

**Theorem 3.21.** *Suppose that a linear operator $T$ satisfies conditions* (S1)–(S3). *Consider a function $f \in L^1$ such that $Tf \in L^1$ and denote by $f_t$ the "good" part of the smooth Calderón–Zygmund decomposition of $f$. Then*

$$\|Tf - Tf_t\|_{L^1} \le c(\operatorname{dist}_{L^1}(f, B_{L^\infty}(\tfrac{t}{2})) + \operatorname{dist}_{L^1}\left(Tf, B_{\mathrm{BMO}}\left(\frac{t}{2}\right)\right), \tag{3.46}$$

*with a constant $c \ge 1$ independent of $f$ and $t > 0$.*

*Remark* 3.22. If $Tf = f$, the theorem says that

$$\|Tf - T(f_t)\|_{L^1} \le c \operatorname{dist}_{L^1}\left(f, B_{L^\infty}\left(\frac{t}{2}\right)\right). \tag{3.47}$$

*Remark* 3.23. However, $Tf_t$ may fall out of $L^\infty$ in any case, and we meant precisely this when talking earlier about "imperfect" stability: by (S2), we can only guarantee that $Tf_t$ lies in the ball of radius roughly $t$ in BMO. Formally, we could eliminate this feature by including the requirement that $T$ should be bounded on $L^\infty$, but then the content of the statement would become nearly empty. Our principal examples are singular integral operators "in both variables", and normally such an operator does not take $L^\infty$ to itself unless it is degenerate. In future, we shall consider stability theorems in which the second space in the couple is a class of functions smooth in a sense. Typically, the adjoint to a singular integral operator takes such spaces to themselves, so the "imperfection" will disappear.

*Proof.* We choose a function $u$ in such a way that

$$\|u\|_{\mathrm{BMO}} \le \frac{t}{2} \text{ and } \|Tf - u\|_{L^1} \le 2\,\mathrm{dist}_{L^1}\left(Tf, B_{\mathrm{BMO}}\left(\frac{t}{2}\right)\right). \tag{3.48}$$

Note that $T(f_t)$ belongs to BMO and, therefore, is locally integrable. Thus, the following calculation is meaningful:

$$\begin{aligned}\|Tf - Tf_t\|_{L^1} &= \|T(I-P_{f,t})f\|_{L^1}\\ &\le \|P_{f,t}T(I-P_{f,t})f\|_{L^1} + \|(I-P_{f,t})T(I-P_{f,t})f\|_{L^1}\\ &\le \|P_{f,t}T(I-P_{f,t})f\|_{L^1} + \|(I-P_{f,t})(Tf-u)\|_{L^1}\\ &\quad + \|(I-P_{f,t})TP_{f,t}f\|_{L^1} + \|(I-P_{f,t})u\|_{L^1}\,.\end{aligned}$$

We need to estimate the four summands on the right.

All summands except the first are handled easily. Indeed, (3.48) and the first inequality in (3.40) imply

$$\|(I-P_{f,t})(Tf-u)\|_{L^1} \le c\,\|Tf-u\|_{L^1} \le 2c\,\mathrm{dist}_{L^1}\left(Tf, B_{\mathrm{BMO}}\left(\frac{t}{2}\right)\right). \tag{3.49}$$

Next, the second inequality in (3.40) yields

$$\begin{aligned}\|(I-P_{f,t})u\|_{L^1} &\le \frac{c}{t}\,\mathrm{dist}_{L^1}(f, B_{L^\infty}(\tfrac{t}{2}))\,\|u\|_{\mathrm{BMO}}\\ &\le c\,\mathrm{dist}_{L^1}\left(f, B_{L^\infty}\left(\frac{t}{2}\right)\right)\end{aligned} \tag{3.50}$$

and

$$\begin{aligned}\|(I-P_{f,t})TP_{f,t}f\|_{L^1} &\le \frac{c}{t}\,\mathrm{dist}_{L^1}\left(f, B_{L^\infty}\left(\frac{t}{c}\right)\right)\|Tf_t\|_{\mathrm{BMO}}\\ &\le c\,\|T\|_{L^\infty\to\mathrm{BMO}}\,\mathrm{dist}_{L^1}\left(f, B_{L^\infty}\left(\frac{t}{c}\right)\right).\end{aligned} \tag{3.51}$$

Thus, we are left with the first summand $\|P_{f,t}T(I-P_{f,t})f\|_{L^1}$. By (3.32), we have

$$P_{f,t}(T(f-P_{f,t}f)) = \sum_{i\in I}\left(\frac{1}{\int\psi_i}\int\psi_i T(f-P_{f,t}f)\right)\psi_i + (T(f-P_{f,t}f))\chi_{\mathbb{R}^n\setminus\cup K_i}. \tag{3.52}$$

The second term on the right is treated easily. We remind the reader that $f - P_{f,t}f = \sum_j b_j$, where the blocks $b_j$ are given by

$$b_j(y) = \left(f(y) - \frac{1}{\int\psi_j}\int f(s)\psi_j(s)\,ds\right)\psi_j(y).$$

So, we obtain

$$\int_{\mathbb{R}^n\setminus\cup K_i}|T(f-P_{f,t}f)| \le \sum_i \int_{\mathbb{R}^n\setminus K_i}|Tb_i|.$$

Since the support of $b_i$ is included in the cube $\frac{1+\frac{\delta}{2}}{1+\delta}K_i$ and $\int b_i = 0$, by strong long-range regularity (see (S1)) we obtain

$$\int_{\mathbb{R}^n\setminus K_i}|(Tb_i)(y)|\,dy \le c\,\|b_i\|_{L_1} \le ct\,|K_i|\,.$$

Now, (3.35) implies

$$\sum_{i\in I}\int_{\mathbb{R}^n\setminus K_i}|(Tb_i)(y)|\,dy \le ct\sum_{i\in I}|K_i| \le c\,\mathrm{dist}_{L^1}\left(f, B_{L^\infty}\left(\frac{t}{2}\right)\right). \tag{3.53}$$

To estimate the first sum in the first term on the right in (3.52), we fix $i$ and consider the $i$th summand separately. In this summand, we split the function $f - P_tf$ as follows:

$$f - P_tf = \sum_{j:K_j\cap K_i\neq\varnothing} b_j + \sum_{j:K_j\cap K_i=\varnothing} b_j \stackrel{\text{def}}{=} u_i + v_i.$$

The sum defining $u_i$ is finite by the finite overlap property (see Theorem 3.15), and the series defining $v_i$ converges to $v_i$ in $L^1$ (see the third formula in (3.37)); therefore,

$$Tv_i = \sum_{j:K_j\cap K_i=\varnothing} T(b_j)$$

in the sense of convergence in measure.

Consider the measurable function

$$V_i = \sum_{j:K_j\cap K_i=\varnothing} |T(b_j)|$$

(it may happen to be infinite on a set of positive measure). Then, again by strong long-range regularity,

$$\sum_i \int_{K_i} V_i = \sum_j \sum_{i:K_j\cap K_i=\varnothing} \int_{K_i} |T(b_j)| \leq C \sum_j \int_{\mathbb{R}^n\setminus K_j} |T(b_j)|,$$

which is dominated by $\operatorname{dist}_{L^1}(f, B_{L^\infty}(\frac{t}{2}))$ (see (3.53)). Since $|Tv_i| \leq V_i$, we see that $Tv_i$ is integrable over $K_i$ and

$$\sum_i \left| \int_{K_i} \psi_i T v_i \right| \leq C \operatorname{dist}_{L^1}\left(f, B_{L^\infty}\left(\frac{t}{2}\right)\right).$$

Now, $Tu_i = T(f - P_{f,t}f) - Tv_i$ is also integrable over $K_i$, and we employ condition (S3) for the functions $\varphi = \psi_i$. Taking (3.21) into account, we obtain

$$\sum_i \left| \int (Tu_i)(s)\psi_i(s)ds \right| \leq C \sum_i \sum_{j:K_i\cap K_j\neq\varnothing} \|b_j\|_{L_1}.$$

Next, again by the finite overlap property (Theorem 3.15), the number of occurrences of every summand $\|b_j\|_{L_1}$ on the right is finite and depends only on the dimension $n$, and again the entire sum is dominated by $\operatorname{dist}_{L^1}(f, B_{L^\infty}(\frac{t}{2}))$. It remains to collect the estimates. □

### 3.3.2 Vector form of the stability theorem

Let $\mathcal{E}$ be a finite-dimensional normed space. The norm on it will be denoted by $|\cdot|_{\mathcal{E}}$. A particular choice of the norm will not be important, because all norms on $\mathcal{E}$ are equivalent.

If the dimension of $\mathcal{E}$ is $N$, we may identify $\mathcal{E}$, up to renorming, with $\mathbb{R}^N$ or, in the case of complex scalars, with $\mathbb{C}^N$. Accordingly, $\mathcal{E}$-valued functions $F$ on $\mathbb{R}^n$ can be viewed as $N$-tuples $F = (F_1, \dots, F_N)$ of scalar-valued functions on $\mathbb{R}^n$. The spaces $L^p(\mathcal{E})$ consist of the $\mathcal{E}$-valued measurable functions on $\mathbb{R}^n$ for which the following norms are finite:

$$\|F\|_{L^p(\mathcal{E})} = \left( \int_{\mathbb{R}^n} |F(t)|_{\mathcal{E}}^p \, dt \right)^{\frac{1}{p}}, \qquad p < \infty,$$
$$\|F\|_{L^\infty(\mathcal{E})} = \operatorname{ess\,sup}_{t\in\mathbb{R}^n} |F(t)|.$$

If $A$ is a measurable subset of $\mathbb{R}^n$, the spaces $L^p(A;\mathcal{E})$ of $\mathcal{E}$-valued functions defined on $A$ are introduced similarly. Clearly, in a coordinate representation (see above), $F \in L^p(\mathcal{E})$ if and only if all coordinate functions $F_l$, $l = 1, \dots, N$, belong to $L^p$. The space $\mathrm{BMO}(\mathcal{E})$ also can be defined by demanding that all components $F_l$,

$l = 1, \ldots, N$, lie in the scalar space BMO, but, surely, it is convenient to use the following norm on it:

$$\|F\|_{\mathrm{BMO}(\mathcal{E})} = \sup_Q \frac{1}{|Q|} \|F - F_Q\|_{L^1(Q;\mathcal{E})}.$$

Here the supremum is taken over all cubes $Q$ in $\mathbb{R}^n$ and $F_Q$ is the average of $F$ over $Q$.

Now, suppose we are given a function $F \in L^1(\mathcal{E})$ and a number $t > 0$. The smooth Calderón–Zygmund decomposition for $F$ is defined in the following way. On $\mathbb{R}^n$, consider the function

$$f(t) = |F(t)|_{\mathcal{E}},$$

and, starting with $f$ and $t$, define the set $\Omega_{f,t}$ (see (3.25)), the family of cubes $\{K_i\}_{i\in I}$, and the corresponding smooth partition of unity $\{\psi_i\}_{i\in I}$. Next, we split $F$ into a "good" and a "bad" part. We define

$$F_t = \sum_{i\in I} \left( \frac{1}{\int \psi_i} \int F(s)\psi_i(s)\, ds \right) \psi_i + F\chi_{\mathbb{R}^n \setminus \cup K_i}$$

(the formula is the same as in the scalar case, but now the integrand is $\mathcal{E}$-valued). Then

$$F - F_t = \sum_{i\in I} \left( F - \frac{1}{\int \psi_i} \int F(s)\psi_i(s)\, ds \right) \psi_i,$$

and we can introduce the vector blocs $B_i$, $i \in I$, by

$$B_i = \left( F - \frac{1}{\int \psi_i} \int F(s)\psi_i(s)\, ds \right) \psi_i.$$

As is in the scalar case, the support of a $\mathcal{E}$-valued measurable function $F$ is the set of all points $x$ at which $F(x) \neq 0$. The support is well defined up to a set of measure zero because, as usual, we identify measurable functions coinciding a.e. In a coordinate representation, the support of $F$ is the union of the supports of its scalar components $F_j$.

In order to state the counterpart of Lemma 3.20, we define an operator $P_{F,t}$ on $\mathcal{E}$-valued functions $H$ by

$$P_{F,t}(H) = \sum_{i\in I} \left( \frac{1}{\int \psi_i} \int H(s)\psi_i(s)\, ds \right) \psi_i + H\chi_{\mathbb{R}^n \setminus \cup K_i}.$$

The next statement is proved precisely as Lemma 3.20.

**Lemma 3.24.** *The objects discussed above have the following properties:*

- $\sum_{i\in I} |K_i| \leq \frac{c}{t} \operatorname{dist}_{L^1}(F, B_{L^\infty(\mathcal{E})}(\frac{t}{2}))$;

- $F - F_t = \sum_{i \in I} B_i$ *and* $\operatorname{supp} B_i \subset \frac{1+\frac{\delta}{2}}{1+\delta} K_i$, $\int B_i(s)ds = 0, \|B_i(s)\|_{L^1(\mathcal{E})} < ct\,|K_i|$ *for all* $i \in I$;
- $\|F_t\|_{L^\infty(\mathcal{E})} \le ct$, $\|F - F_t\|_{L^1(\mathcal{E})} \le c \operatorname{dist}_{L^1(\mathcal{E})}(F, B_{L^\infty(\mathcal{E})}(\frac{t}{2}))$;
- $\|P_{F,t}\|_{L^1(\mathcal{E}) \to L^1(\mathcal{E})} \le c$, $\|I - P_{F,t}\|_{\mathrm{BMO}(\mathcal{E}) \to L^1(\mathcal{E})} \le \frac{c}{t} \operatorname{dist}_{L^1(\mathcal{E})}(F, B_{L^\infty(\mathcal{E})}(\frac{t}{2}))$.

*Here* $c \ge 1$ *depends only on the dimension* $n$.

We are ready to formulate the result. Let $T$ be a continuous linear operator from $L^1(\mathcal{E})$ to the space of measurable $\mathcal{G}$-valued functions ($\mathcal{G}$ is another finite-dimensional normed space) that obeys the vector analogs of conditions (S1), (S2), and (S3), as given below.

(SV1) $T$ is strongly long-range regular: if $F \in L^1(\mathcal{E})$ has zero average and is supported on a cube $Q$, then for every $\lambda > 1$ we have

$$\left\|T(F) \cdot \chi_{\mathbb{R}^n \setminus \lambda Q}\right\|_{L^1(\mathcal{G})} \le c(\lambda) \left\|F\right\|_{L^1(\mathcal{E})}.$$

(SV2) $T$ maps boundedly $L^\infty(\mathcal{E})$ to $\mathrm{BMO}(\mathcal{G})$.

(SV3) $T$ is weakly $L^1$-bounded: there is a natural number $k$ such that for every scalar-valued $C^\infty$-function $\varphi$ supported on a cube $Q$ and every integrable $\mathcal{E}$-valued function $F$ with $TF$ integrable over $Q$, we have

$$\left|\int T(F)\varphi\right| \le C \left\|F\right\|_{L^1(\mathcal{E})} \left( \left\|\varphi\right\|_{L^\infty} + \sum_{j \le k} (\operatorname{diam} Q)^j \left\|\nabla^j \varphi\right\|_{L^\infty} \right),$$

with a constant $C > 0$ independent of $Q$ and $\varphi$.

**Theorem 3.25.** *Suppose* $T$ *obeys* (SV1)–(SV3). *Consider a function* $F \in L^1(\mathcal{E})$ *with* $T(F) \in L^1(\mathcal{G})$. *Then*

$$\begin{aligned} &\|T(F) - T((F)_t)\|_{L^1(\mathcal{G})} \\ &\le c \left( \operatorname{dist}_{L^1(\mathcal{E})} \left( F, B_{L^\infty(\mathcal{E})} \left( \frac{t}{2} \right) \right) + \operatorname{dist}_{L^1(\mathcal{E})} \left( T(F), B_{\mathrm{BMO}(\mathcal{G})} \left( \frac{t}{2} \right) \right) \right), \end{aligned}$$

*with a constant* $c \ge 1$ *independent of* $F$ *and* $t > 0$.

The proof is quite similar to that of Theorem 3.21, and we omit it. Stability theorems of §1.3 and of Chapter 10 also admit vector counterparts, and the same is true concerning the entire theory of near-minimizers discussed in Part 2 of this book. Much like the above discussion, in general the vector-valued theory can be built in accordance with the following recipe: perceive $|\cdot|$ as the norm in a finite-dimensional space rather than the absolute value, and act as with scalars.

It only remains to say that conditions (SV1)-(SV3) can be verified componentwise. Specifically, let $N$ and $M$ be the dimensions of $\mathcal{E}$ and $\mathcal{G}$. Introducing coordinates in these spaces, we may identify $T$ with a matrix with $N$ columns and $M$ rows whose entries $T_{ij}$ are operators taking scalar-valued functions to scalar-valued functions. Clearly, *if the* $T_{ij}$ *satisfy* (S1)-(S3), *then conditions* (SV1)-(SV3) *are fulfilled for* $T$.

We have already seen that (S1) is typical of singular integral operators, and in the next chapter we shall see the same about (S2) and (S3) (more precisely, (S2) is typical of the adjoints to singular integral operators). So, matrices composed of singular integral operators "in both variables" may play the role of "$T$" in the above theorem. For instance, the projection onto the space of gradient vector fields, analyzed in §2 of Chapter 2, is such.

# Notes and remarks

The idea of employing the Whitney lemma in order to get a smooth analog of the Calderón–Zygmund decomposition is due to Stein (see [S1]); in the paper [FS1], the same idea was used to prove the boundedness of singular integral operators on the Hardy spaces $H^p$. The stability Theorem 3.21 is published here for the first time, though, basically, the pattern can be found in [KK0] and [KK].

# Chapter 4

# Spaces of smooth functions and operators on them

In this chapter we present some very basic facts about spaces of smooth functions (specifically, about Sobolev and Morrey–Campanato spaces). Next, we give some additional information about singular integrals (note that in the stability theorem in the last chapter we employed some facts not covered before). We prove that the adjoints to singular integral operators take $L^\infty$ to BMO and discuss the weak $L^1$-boundedness condition (3.45). More generally, we shall see that, under some additional assumptions, the adjoints to singular integral operators take some Morrey-Campanato spaces to themselves. Also, we analyze partial sum operators for wavelet expansions from a singular integral theory viewpoint. All this will be of much importance in Part 2 of the book.

Proofs included in this chapter will be more technical than previously. For this reason, first we summarize the content.

## 4.1 Summary

### 4.1.1 Homogeneous spaces of smooth functions

#### Sobolev spaces

Let $\Omega$ be a domain in $\mathbb{R}^n$. The (*homogeneous*) Sobolev space $\dot{W}_q^k(\Omega)$ is defined to be the set of all distributions $g$ on $\Omega$ whose $k$th derivatives belong to $L^q(\Omega)$. This space is endowed with the natural seminorm

$$\|g\|_{\dot{W}_q^k} = \sup_{k_1+\cdots+k_n=k} \left\| \frac{\partial^k}{\partial^{k_1} x_1 \partial^{k_2} x_2 \dots \partial^{k_n} x_n} g(x) \right\|_{L^q}. \tag{4.1}$$

It is well known that any $g \in \dot{W}_q^k(\Omega)$ lies in $L^q_{\mathrm{loc}}(\Omega)$, see, e.g., [Maz], Subsection 1.1.13. The following statement is an adaptation of the Sobolev embedding theorem to the case of homogeneous spaces; see, e.g., [Maz], the end of Subsection 1.4.5 (especially the nonlabeled display there); or else, see Lemmas 3.1 and 3.2 in [BS]. Not to enter in the discussion of the properties of $\Omega$ that ensure this type of behavior, we state the result for the cube $Q_0 = [-1/2, 1/2]^n$.

**Theorem 4.1** (Sobolev embedding). *Suppose $0 < p < \infty$ and $1 \le q < \infty$ satisfy $\frac{k}{n} + \frac{1}{p} - \frac{1}{q} \ge 0$. Then for every function $u \in \dot{W}_q^k(Q_0)$ there exists a polynomial $r \in \mathcal{P}_k$ such that*

$$\|u - r\|_{L^p(Q_0)} \le C\|u\|_{\dot{W}_q^k(Q_0)}, \tag{4.2}$$

*where $C$ depends only on $p$, $q$, the smoothness order $k$, and the dimension $n$.*

In what follows, we shall only need the Sobolev spaces on the entire $\mathbb{R}^n$. For them, we use the notation $\dot{W}_q^k$. The above result leads to the following.

**Theorem 4.2** (Sobolev–Poincaré inequality). *Assume that condition (0.2) is fulfilled. Let $g \in \dot{W}_q^k$, and let $Q$ be a cube. Then there exists a constant $c = c(n,k,p,q)$ and a polynomial $g_Q$ of degree strictly smaller than $k$ such that*

$$\begin{aligned}\left(\frac{1}{|Q|}\int_Q |g(x) - g_Q|^p\,dx\right)^{\frac{1}{p}} \\ \le c\,|Q|^{\frac{k}{n}} \sup_{k_1+\cdots+k_n=k} \left(\frac{1}{|Q|}\int_Q \left|\frac{\partial^k}{\partial^{k_1}x_1\ldots\partial^{k_n}x_n} g(x)\right|^q dx\right)^{\frac{1}{q}}.\end{aligned} \tag{4.3}$$

*Proof.* Indeed, this is reduced to (4.2) by a shift and dilation that take $Q$ to $Q_0$. □

### Morrey–Campanato spaces

Now, we want to discuss in more detail the Morrey–Campanato spaces $\dot{C}_p^{s,k}$, already mentioned several times in this book. Certain maximal operators adjusted to measure smoothness enter implicitly their definition. These maximal functions were introduced by Calderón [Ca] in a somewhat different form. Subsequently, they were investigated thoroughly in the monograph [DeVSh]. Suppose $\alpha \in \mathbb{R}$, $1 \le p \le \infty$, and $k$ is a nonnegative integer. For $f \in L^1_{\mathrm{loc}}(\mathbb{R}^n)$, we define the maximal function $M_{k,p,\alpha}(f)$ by

$$M_{k,p,\alpha}(f)(x) = \sup_{x\in Q}\inf_{g\in\mathcal{P}_k} \frac{1}{|Q|^{\alpha/n}}\left(\frac{1}{|Q|}\int_Q |f-g|^p\right)^{\frac{1}{p}}. \tag{4.4}$$

We remind the reader that $\mathcal{P}_k$ is the space of polynomials of degree strictly smaller than $k$. If $k = 0$, we agree that this class contains only the zero function. So, if $k = \alpha = 0$ and $p = 1$, we obtain the classical maximal function of Hardy and Littlewood (for it, we retain the notation $Mf$ used previously).

Formula (0.7) for the seminorm shows that $\dot{C}_p^{s,k} = \{f : M_{k,p,s}f \in L^\infty\}$. Recall that here the "smoothness" parameter $s$ is subject to the condition $s \in (-n/p, k]$. In fact, only for $s \geq 0$ should it really be interpreted as a smoothness exponent. Also, formally taking $s$ outside the interval $(-n/p, k]$ leads to degeneration (we do not dwell on this).

It has already been mentioned that for $s > 0$ the spaces $\dot{C}_p^{s,k}$ are a disguise of certain more familiar classes defined in terms of smoothness moduli. Moreover, one and the same function class may arise as $\dot{C}_p^{s,k}$ for different triples $(p, s, k)$. Consequently, the choice of $p$, $s$, and $k$ in the results of §9.3 boils down to merely fixing a convenient equivalent seminorm on some "standard" space of smooth functions. We want to give a detailed description of the situation. Proofs will be included for completeness. However, before that we want to present a characterization (due to Calderón) of the Sobolev spaces in terms of the maximal functions $M_{k,p,s}$ (see the corollary to the next theorem). This is instructive, though will not be used below. We write $|\nabla^k f|(x) = \sup_{|\beta|=k} |D^\beta f(x)|$ (we recall that $D^\beta$ stands for the partial derivative with multiindex $\beta$).

**Theorem 4.3.** *Suppose $k/n + p^{-1} - r^{-1} > 0$ and $1 \leq r \leq q$, then $c|\nabla^k f|(x) \leq M_{k,p,k}f(x) \leq C(M(|\nabla^k f|^r)(x))^{1/r}$ for every $f$ that belongs to $\dot{W}_q^k$ locally and every $x \in \mathbb{R}^n$.*

We give a hint to the proof. The upper estimate is a direct consequence of the Sobolev–Poincaré inequality because, *a fortiori*, $f$ is locally in $\dot{W}_r^k$. The lower estimate becomes stronger as $p$ becomes smaller, so it suffices to prove it for $p = 1$. For that, see [DeVSh], Theorem 5.6.

**Corollary 4.4.** *If* (0.2) *is fulfilled and $q > 1$, then $f \mapsto \|M_{k,p,k}f\|_{L^q}$ is an equivalent seminorm on $\dot{W}_q^k$.*

### Reasonable values of $k$ in the definition of Morrey–Campanato spaces

Basically, the *reasonable* value is defined to be the smallest nonnegative integer strictly greater then $s$. However, if $s$ is a positive integer, both $k = s$ and $k = s+1$ are reasonable (by definition), and the corresponding spaces differ indeed. Taking greater (so, "nonreasonable") $k$ does not add much to the essence (yet can be useful technically).

**Proposition 4.5.** *Let $l$ be a nonnegative integer satisfying $s < l < k$. Then for every $f \in \dot{C}_p^{s,k}$ there exists a polynomial $p \in \mathcal{P}_k$ such that $\|f - p\|_{\dot{C}_p^{s,l}} \leq C\|f\|_{\dot{C}_p^{s,k}}$, where $C$ does not depend on $f$.*

The proof will be given later. Now we continue our commentary.

### Morrey spaces

This term pertains to the case of $s < 0$. Here the reasonable value of $k$ is 0, so the norm on a Morrey space $\dot{C}_p^{s,0}$ is given by the expression

$$\sup_Q |Q|^{-s/n} \left( \frac{1}{|Q|} \int_Q |f|^p \right)^{1/p}, \tag{4.5}$$

the supremum being taken over all cubes. The following statement will be used later in this chapter.

**Corollary 4.6.** *The functional*

$$f \mapsto \sup_Q \inf_c |Q|^{-s/n} \left( \frac{1}{|Q|} \int_Q |f-c|^p \right)^{1/p} \tag{4.6}$$

*(the inner infimum is taken over all constants* $c$*) is an equivalent norm on* $\dot{C}_p^{s,0}$*.*

*Proof.* The functional (4.6) is the standard (semi)norm on $\dot{C}_p^{s,1}$. Clearly, if $f \in \dot{C}_p^{s,0}$, this seminorm is dominated by the norm of $f$ in $\dot{C}_p^{s,0}$. But by Proposition 4.5, there is also a constant $a$ such that $\|f-a\|_{\dot{C}_p^{s,0}}$ is dominated by $\|f\|_{\dot{C}_p^{s,1}}$. Now, $a$ must be 0, because no nonzero constant lies in $\dot{C}_p^{s,0}$ (consider large cubes in the definition of the corresponding norm). Therefore, the norms (4.5) and (4.6) are equivalent on $\dot{C}_p^{s,0}$. □

### Zero smoothness

Suppose $s = 0$ and $k = 1$ (a reasonable value). Then for $p = 1$ we obtain the space BMO. In Chapter 1 it was already explained that the seminorms given by (0.13) and by (0.7) with $k = 0$, $p = 1$ are equivalent. Moreover, we remind the reader that the John–Nirenberg inequality (Theorem 1.13 proved in the same chapter) implies the following statement.

**Proposition 4.7.** BMO $= \dot{C}_p^{0,1}$ *for every* $1 \le p < \infty$.

The other reasonable value $k = 0$ for $s = 0$ leads to the space $L^\infty$.

### Positive smoothness: Campanato spaces

It remains to consider the case of $s > 0$. Then the spaces $\dot{C}_p^{s,k}$ are called Campanato spaces. We already proved in Chapter 1 that the equivalence of norms mentioned in the preceding item extends now up to $p = \infty$ inclusive.

**Proposition 4.8.** *If* $s > 0$*, the space* $\dot{C}_p^{s,k}$ *does not depend on* $p$*, and moreover, the functional*

$$f \mapsto \sup_Q \inf_{q \in \mathcal{P}_k} |Q|^{-s/n} \operatorname{ess\,sup}_{x \in Q} |f(x) - q(x)| \tag{4.7}$$

*is an equivalent seminorm on* $\dot{C}_p^{s,k}$*.*

In a natural way, we write $\dot{C}^{s,k}_\infty$ if we mean the seminorm (4.7) on the space in question. Up to the end of the present item, we shall work precisely with $\dot{C}^{s,k}_\infty$. Moreover, we assume that $k$ is reasonable, i.e., it is the smallest positive integer greater than $s$ if $s$ is not an integer, and it may be either $s$ or $s+1$ otherwise. To finish the description of Campanato spaces, we recall the definition of the $k$th difference $\Delta^k_h$. If $f$ is a function on $\mathbb{R}^n$ and $h \in \mathbb{R}^n$, we put by induction

$$\Delta^1_h f(x) = f(x+h) - f(x), \; \Delta^{k+1}_h f = \Delta^1_h \Delta^k_h f, \; k = 1, 2, \dots .$$

**Proposition 4.9.** *Under the above assumption about $k$ and $s$, the space $\dot{C}^{s,k}_\infty$ coincides with the class of all locally integrable functions $f$ satisfying*

$$|\Delta^k_h f(x)| \le C|h|^s \tag{4.8}$$

*with $C$ independent of $x$ and $h$. The infimum of such $c$ provides an equivalent seminorm on $\dot{C}^{s,k}_\infty$.*

As a particular case, the proposition includes Corollary 1.16 proved in Chapter 1: if $0 < s \le 1$ and $k = 1$, then $\dot{C}^{s,k}_p$ coincides with the Lipschitz space $\mathrm{Lip}_s$ for every $p \in [1, \infty]$. Next, the proposition also shows that $\dot{C}^{1,2}_p$ is the so-called Zygmund class. This class is different from $\mathrm{Lip}_1$. It is widely known that, say, if $s > 1$ is not an integer, then the condition $|\Delta^k_h f(x)| \le c|h|^s$ determines the space of all functions whose derivatives of order $[s]$ satisfy the Lipschitz condition of order $s - [s]$ ($[s]$ stands for the greatest integer strictly smaller than $s$). We shall not dwell on this, but we shall prove Proposition 4.9 in what follows.

### 4.1.2 Singular integral operators

Not to enter immediately in complicated details, we now restrict ourselves to some partial cases of what will be proved. We signalize that the statements to be discussed provide "raw material" for the stability theorems in Chapters 3 and 10: it will turn out that an arbitrary operator $T$ covered by Theorems 4.10 and 4.11 below possesses $T$-stable near-minimizers.

**Theorem 4.10.** *Let $m$ be a function infinitely differentiable on $\mathbb{R}^n \setminus \{0\}$ and satisfying* (2.15), *i.e.,*

$$|D^\alpha m(\xi)| \le c_\alpha |\xi|^{-|\alpha|}, \; \xi \in \mathbb{R}^n$$

*for all multiindices $\alpha$. Then the multiplier $T_m$ can be defined consistently as a bounded operator from $L^\infty$ to* BMO, *from* BMO *to itself, and from the Morrey space $\dot{C}^{s,0}_p$ ($s < 0$) to itself. Also, $T_m$ can be defined consistently as a bounded operator from the Campanato space $\dot{C}^{s,k}_\infty$ ($s > 0$) to itself whenever $k$ is the smallest integer strictly greater than $s$ (then $k$ is reasonable in the above terminology). Next, it is a bounded operator on the Sobolev space $\dot{W}^k_p$ whenever $1 < p < \infty$.*

We remind the reader that in Chapter 2 we proved (see Theorem 2.4) that $T_m$ is a singular integral operator under the above assumptions; also, we established some finer estimates on the kernel of $T_m$ in Theorem 2.5 — precisely they will be important in the proof of the above statement. We note immediately that the assertion about Sobolev spaces is clear: by considering Fourier transforms, we see that $T_m$ commutes with differentiation, so it is bounded on $\dot{W}_p^k$ simply because it is bounded on $L^p$.

By Proposition 4.9, it follows that, in particular, $T_m$ operates on $\mathrm{Lip}_s$ for $0 < s < 1$.

The exposition, however, will not be confined to singular integral operators of the form $T_m$: we shall discuss also operators that do not necessarily commute with translations (so, their kernels $K(x, y)$ do not necessarily depend only on the difference $x - y$). This will be done to cover more examples, in particular, the partial sum operators related to wavelet expansions. Let us explain what this means.

For a function $\psi \in L^2(\mathbb{R})$, we write $\psi_{j,m}(t) = 2^{j/2}\psi(2^j t + l)$, $j, m \in \mathbb{Z}$. We say that $\{\psi_{j,m}\}$ is a *system of wavelets* with *generating wavelet* $\psi$ if it is an orthogonal basis of $L^2(\mathbb{R})$. Such systems do exist and have become quite "popular" since the middle 1980s, when certain unified methods to construct them were discovered. See, e.g., [M], [NPS] for more information.

A feature of wavelet systems very important by itself but also of primary interest to us is that, usually, for every set $A$ of pairs $(j, m)$ of integers, the orthogonal projection $\pi_A$ onto the closed linear span of $\{\psi_{j,m}\}_{(j,m)\in A}$ is a singular integral operator. More precisely, this is true for *sufficiently smooth and well-localized* wavelets. The assumptions of the next theorem are excessive and are imposed in this form to lighten the statement.

**Theorem 4.11.** *Suppose we are given a wavelet system $\{\psi_{j,m}\}$ such that the Fourier transform of the generating wavelet $\psi$ belongs to $\mathcal{D}(\mathbb{R})$. Fix a numerical sequence $t_{i,m}$ uniformly bounded by* 1 *(say) in modulus, and define an operator $T$ on $L^2(\mathbb{R})$ by the formula*

$$Tf = \sum_{j,m} t_{j,m} \langle f, \psi_{j,m} \rangle \psi_{j,m}.$$

*Then $T$ is a singular integral operator having all properties listed in Theorem* 4.10.

It should be noted that we work throughout with *homogeneous* spaces of smooth functions. The elements of these spaces are classes of functions modulo polynomials. For this reason, the question of defining a singular integral operator consistently as a mapping from one such space to another (see the claim of Theorem 4.10) is meaningful and nontrivial. This question will be considered in detail in this chapter. Now we want to mention that the so-called *nonhomogeneous* spaces of smooth functions may fail to be natural domains for singular integrals. We illustrate this by a simplest example. For $0 < \alpha < 1$, consider the nonhomogeneous Lipschitz space $\mathrm{Lipschitz}_\alpha(\mathbb{R})$ determined by the norm $|f|_\alpha = \|f\|_{\mathrm{Lip}_\alpha(\mathbb{R})} + \|f\|_{L^\infty(\mathbb{R})}$.

Already the Hilbert transformation is unbounded on this space. Moreover, *there is no constant $c$ such that $|Hf|_\alpha \le c|f|_\alpha$ for every $f \in \mathcal{D}(\mathbb{R})$.*

Indeed, we argue by contradiction. Suppose such $c$ exists. Taking $f \in \mathcal{D}(\mathbb{R})$, we write the estimate in question for the function $f_v$, $f_v(x) = f(vx)$, $v > 0$. Since $H$ commutes with dilations, this results in

$$\|Hf\|_{L^\infty} + v^\alpha \|Hf\|_{\mathrm{Lip}_\alpha} \le c\|f\|_{L^\infty} + v^\alpha \|f\|_{\mathrm{Lip}_\alpha}.$$

Letting $v \to 0$, we arrive at the boundedness of $H$ in the $L^\infty$-norm, which is not true (recall that the Hilbert transform of the characteristic function of a bounded interval can be calculated explicitly and is not a bounded function).

## 4.2 Morrey–Campanato spaces: proofs

In this section, we prove Propositions 4.5 and 4.9. To treat the first of them, we need a lemma.

**Lemma 4.12.** *Suppose $l < k$ are two nonnegative integers, $V$ is a cube, $x \in V$, and $p \in \mathcal{P}_k$. There is a polynomial $q \in \mathcal{P}_l$ such that for every cube $U$ centered at $x$ and included in $V$ we have*

$$\|p - q\|_{L^\infty(U)} \le C\left(\frac{|U|}{|V|}\right)^{l/n} \|p\|_{L^\infty(V)}. \tag{4.9}$$

*Here $C$ depends only on $k$ and the dimension $n$.*

*Proof.* We write $p$ in the form

$$p(y) = \sum_{|j|<k} c_j (y-x)^j$$

($(y-x)^j$ is the product $(y_1 - x_1)^j \dots (y_n - x_n)^j$) and put $q(y) = \sum_{|j|<l} c_j (y-x)^j$. Then the claim is an easy consequence of Lemma 0.25. □

We prove a slightly stronger version of Proposition 4.5.

**Proposition 4.13.** *Let $l$ be a nonnegative integer satisfying $s < l < k$. Then for every $f$ such that $M_{k,p,s}(f)$ is finite a.e. there exists a polynomial $p \in \mathcal{P}_k$ such that $M_{l,p,s}(f-p) \le CM_{k,p,s}(f)$, where $C$ does not depend on $f$.*

The next lemma establishes a main partial case of Proposition 4.13.

**Lemma 4.14.** *Let $Q$ be a cube, and consider a function $f$ as in the proposition. Then there exists $u^Q \in \mathcal{P}_k$ with the following property: for every cube of the form $2^jQ$ with $j = 0, 1, 2 \dots$, there is a polynomial $w_j \in \mathcal{P}_l$ such that*

$$\left(\frac{1}{|2^jQ|}\int_{2^jQ} |f - u^Q - w_j|^p\right)^{\frac{1}{p}} \le c|2^jQ|^{\frac{s}{n}} \inf_{x\in 2^jQ} M_{k,p,s}f(x). \tag{4.10}$$

*Here $c$ does not depend on $Q$, $j$, and $f$. Moreover, $w_0 = 0$.*

*Proof.* For any cube $K$, we denote by $A(K)$ the distance from $f$ to $\mathcal{P}^k$ in the metric of $L^p(K; dt/|K|)$.

To define the required polynomial $u^Q$, consider all cubes $2^m Q$, $m \in \mathbb{Z}_+$. For each $m$, there is a polynomial $u_m$ in $\mathcal{P}_k$ for which the quantity $(|2^m Q|^{-1} \int_{2^m Q} |f - u_m|^p)^{1/p}$ is equal to $A(2^m Q)$. Since $u_{m+1} - u_m \in \mathcal{P}_k$, we have

$$\|u_{m+1} - u_m\|_{L^\infty(2^m Q)} \le c \left( \frac{1}{|2^m Q|} \int_{2^m Q} |u_{m+1} - u_m|^p \right)^{\frac{1}{p}}$$

with $c$ independent of $m$ and $Q$. Now, it easily follows that

$$\|u_{m+1} - u_m\|_{L^\infty(2^m Q)} \le C(A(2^{m+1}Q) + A(2^m Q)) \le C|2^m Q|^{s/n} \inf_{x \in 2^m Q} M_{k,p,s} f(x),$$

where $C$ is independent of $m$, $Q$, and $f$. By Lemma 4.12, there exists a polynomial $v_m \in \mathcal{P}_l$ such that for every $d \in \mathbb{Z}_+$ with $d \le m$ we have

$$\begin{aligned} \|u_{m+1} - u_m - v_m\|_{L^\infty(2^d Q)} &\le C \left( \frac{|2^d Q|}{|2^m Q|} \right)^{\frac{l}{n}} |2^m Q|^{s/n} \inf_{x \in 2^m Q} M_{k,p,s} f(x) \\ &= C 2^{-(m-d)(l-s)} |2^d Q|^{\frac{s}{n}} \inf_{x \in 2^m Q} M_{k,p,s} f(x). \end{aligned} \tag{4.11}$$

This estimate shows that the polynomial

$$u^Q = u_0 + \sum_{m \ge 0} (u_{m+1} - u_m - v_m)$$

is well defined. Indeed, by (4.11), the remainder $\sum_{m \ge d} (u_{m+1} - u_m - v_m)$ of the series converges normally on the cube $2^d Q$, $d = 0, 1, 2, \ldots$.

We prove that this polynomial has the desired property. Indeed, for every $j \in \mathbb{Z}_+$ we can write

$$u^Q = u_j + \sum_{m \ge j} (u_{m+1} - u_m - v_m) - w_j,$$

where $w_j = \sum_{m<j} v_m \in \mathcal{P}_l$. (Note that then $w_0 = 0$, as claimed at the end of the lemma.) Now, denote the middle term (the sum over $m$) in the above formula by $g_j$, then $\|g_j\|_{L^\infty(2^j Q)} \le c|2^j Q|^{s/n} \inf_{x \in 2^j Q} M_{k,p,s} f(x)$ by (4.11). It follows that

$$\begin{aligned} \left( \frac{1}{|2^j Q|^{-1}} \int_{2^j Q} |f - u^Q - w_j|^p \right)^{\frac{1}{p}} &= \left( \frac{1}{|2^j Q|^{-1}} \int_{2^j Q} |f - u^j - g_j|^p \right)^{\frac{1}{p}} \\ &\le c|2^j Q|^{s/n} \inf_{x \in 2^j Q} M_{k,p,s} f(x) \end{aligned}$$

with some other $c$, and we are done. □

Now, we prove Proposition 4.13. Let again $f$ satisfy the assumptions of the proposition. We fix some $Q$ (say, $Q = [-1/2, 1/2]^n$), take another cube $K$, and consider the polynomials $u^Q$ and $u^K$ given by the above lemma. Next, we fix some $j \in \mathbb{Z}_+$ with $K \subset 2^j Q$. Then it is easy to realize that there exists $i \in \mathbb{Z}_+$ such that $2^j Q \subset 2^i K$ and the volumes of $2^j Q$ and $2^i K$ are roughly the same (that is, their ratio is bounded and bounded away from zero by some constant depending only on the dimension). Therefore, by (4.10), there are two polynomials $w, \tilde{w} \in \mathcal{P}_l$ such that *the two quantities*

$$\left(|2^j Q|^{-1} \int\limits_{2^j Q} |f - u^Q - w|^p\right)^{\frac{1}{p}} \quad \text{and} \quad \left(|2^j Q|^{-1} \int\limits_{2^j Q} |f - u^K - \tilde{w}|^p\right)^{\frac{1}{p}}$$

are dominated by $C|2^j Q|^{\frac{s}{n}} \inf_{x\in K} M_{k,p,s} f(x)$ (when treating the second quantity, we first write (4.10) for the cubes $K$ and $2^i K$, and then use the rough comparability of the volumes mentioned above).

It follows that $\|u^Q - u^K + w - \tilde{w}\|_{L^\infty(2^j Q)} \le C|2^j Q|^{s/n} \inf_{x\in K} M_{k,p,s} f(x)$. We apply Lemma 4.12 with $2^j Q$ in the role of "$V$" and the center of $K$ in the role of "$x$", obtaining a polynomial $h \in \mathbb{P}_l$ with

$$\begin{aligned} &\|u^Q - u^K + w - \tilde{w} - h\|_{L^\infty(K)} \\ &\quad \le C\left(\frac{|K|}{|2^j Q|}\right)^{l/n} |2^j Q|^{s/n} \inf_{x\in K} M_{k,p,s} f(x) \le C|K|^{s/n} \inf_{x\in K} M_{k,p,s} f(x) \end{aligned}$$

(because $l > s$). Now, put $\pi = w - \tilde{w} - h$; then $\pi \in \mathbb{P}_l$. We can write

$$\left(\frac{1}{|K|}\int_K |f - u^Q + \pi|^p\right)^{\frac{1}{p}} \le C\left(\frac{1}{|K|}\int_K |f - u^K|^p\right)^{\frac{1}{p}} + C\|u^Q - u^K + \pi\|_{L^\infty(K)}.$$

The second summand on the right is dominated by $C|K|^{s/n} \inf_{x\in K} M_{k,p,s} f(x)$ by the preceding inequality, and the first by Lemma 4.14 (see the last statement there).

This being done, the proof can be finished as follows. Take $y \in \mathbb{R}^n$, and let $K$ be a cube with $y \in K$. The above shows that there is $\pi \in \mathcal{P}_l$ with

$$|K|^{-\frac{s}{n}} \left(\frac{1}{|K|}\int_K |f - u^Q + \pi|^p\right)^{\frac{1}{p}} \le C M_{k,p,s}(f)(y).$$

Passing to the supremum over all $K$ containing $y$, we prove Proposition 4.13 (and, with it, Proposition 4.5).

We proceed to the proof of Proposition 4.9. If $h \in \dot{C}_\infty^{s,k}$, then (4.8) is fulfilled because the $k$th difference of every polynomial of degree strictly smaller than $k$ is identically zero. For the other direction, we show that (4.8) implies $h \in \dot{C}_1^{s,k}$, which suffices because of Proposition 1.12.

Denote temporarily by $X_{k,s}$ the space of locally integrable functions $f$ satisfying (4.8), and endow it with the seminorm $|\cdot|_{k,s}$ equal to the smallest possible $C$ in (4.8). The following lemma is quite easy; we formulate it as a separate statement for further references.

**Lemma 4.15.** *If $t > 0$, then the transformation $f(v) \mapsto t^{-s} f(tv)$ is an isometry of $X_{k,s}$ and of $\dot{C}_p^{s,k}$ for every $p \in [1, \infty]$.*

The lemma and the fact that any shift is also an isometry of the spaces in question imply that it suffices to show the following.

**Proposition 4.16.** *Let $f$ be locally integrable with $|f|_{k,s} \le 1$, and let $K = [-\frac{1}{2}, \frac{1}{2}]^n$; then there exists $q \in \mathcal{P}^k$ such that*

$$\int_K |f - q| \le C, \tag{4.12}$$

*where $C$ is independent of $f$.*

*Proof.* We approximate $f$ by its convolution $g$ with an appropriate infinitely differentiable and compactly supported function with unit integral. The discrepancy $\int_K |f - g|$ can be made arbitrarily small; on the other hand, it is easily seen that still $\|g\|_{k,s} \le 1$. So, it suffices to prove Proposition 4.16 for $g$ in place of $f$. In other words, we may assume that $f \in C^\infty$.

We follow an argument due to Brudnyi. It suffices to find a function $u \in C^\infty(K)$ such that $\int_K |h - u| \le C$ and $\max_{|j|=k} \|D^j u\|_{L^\infty} \le C$. Indeed, if this is done, we simply apply Theorem 4.1 to $u$ (say, with $p = q = 1$), which will yield the required polynomial.

To construct $u$, we recall two well-known simple formulas for the $k$th difference. The first can easily be proved by induction on $k$, much as the binomial formula:

$$\Delta_h^k f(x) = \sum_{j=0}^{k} (-1)^{k-j} \binom{k}{j} f(x + jh). \tag{4.13}$$

The second says that for every vector $e$ and every $\tau > 0$, we have

$$\Delta_{\tau e}^k f(x) = \int_0^\tau \dots \int_0^\tau \frac{\partial^k}{\partial e^k} f(x + (t_1 + \ldots + t_k)e) dt_1 \ldots dt_k. \tag{4.14}$$

This is an easy consequence of the one-dimensional formula $\varphi(x + h) - \varphi(x) = \int_x^{x+h} f'(t)dt$.

The right-hand side of (4.13) involves the term $f(x)$ with coefficient $+1$ or $-1$, depending on the parity of $k$. Taking a unit vector $e$, we put $h = (t_1 + \ldots + t_k)e$ in (4.13), denote $t = t_1 + \ldots + t_k$, and rewrite (4.13) as follows:

$$f(x) - \sum_{j=0}^{k} c_{jk} f(x + jte) = \pm \Delta_{te}^k f(x). \tag{4.15}$$

Here, up to a sign depending on $k$, the coefficients $c_{jk}$ are the coefficients in the expansion (4.13).

Next, we integrate over $[0, 1]$ in each $t_j$, to obtain

$$f(x) - W^{(e)}f(x) = \pm \int_0^1 \dots \int_0^1 \Delta_{te}^k f(x) dt_1 \dots dt_k,$$

where the operator $W^{(e)}$ is defined by

$$W^{(e)}f(x) = \int_0^1 \dots \int_0^1 \sum c_{jk} f(x + jte) dt_1 \dots dt_k, \tag{4.16}$$

It follows that $|f(x) - W^{(e)}f(x)| \leq C$ (a constant independent of $f$) for $x \in K$.

On the other hand, differentiating (4.16) $k$ times with respect to $x$ in the direction $e$ and then taking (4.14) into account, we obtain[1]

$$\frac{\partial^k}{\partial e^k} W^{(e)} f(x) = \sum_{1 \leq j \leq k} c_{jk} \Delta_{je}^k f(x).$$

Since $j$ ranges only from 1 to $k$, it follows that $|\frac{\partial^k}{\partial e^k} W^{(e)} f(x)| \leq C$ for $x \in K$ (again, $C$ is independent of $f$).

Thus, we have obtained an approximant to $f$, within a required precision, such that the $k$th pure derivative of it in some direction is duly bounded. We recall that we need an approximant whose partial derivatives of order $k$ (pure or mixed) are *all* bounded on $K$. The following lemma helps to achieve this.

**Lemma.** *There exists a collection* $\{e_i\}_{1 \leq i \leq N}$ *of unit vectors in* $\mathbb{R}^n$ *such that for every multiindex* $m$ *with* $|m| = k$ *we have* $D^m = \sum_i b_{mi} (\partial/\partial e_i)^k$, *where* $b_{im}$ *are some numerical coefficients and* $N$ *depends only on* $k$ *and* $n$.

If the lemma is proved, we define the required approximant $u$ to $f$ as follows: $u = W^{(e_N)} \dots W^{(e_1)} f$. It should be noted that the operators $W^{(e_l)}$ commute with one another and with differentiations. Now, we have

$$f - u = (f - W^{(e_N)}f) + W^{(e_N)}(f - W^{(e_{N-1})}f) + \dots \\ + W^{(e_N)} \dots W^{(e_2)}(f - W^{(e_1)}f),$$

and from the above discussion it follows that $|f - u| \leq C$ on $K$, where $C$ is some constant independent of $f$. It remains to estimate all $k$th partial derivatives of $u$. By the lemma, for this it suffices to estimate the directional derivatives $(\partial/\partial e_i)^k u$. For every $i$, this derivative is equal to the result of application to $W^{(e_i)}(\partial/\partial e_i)^k f(\cdot)$

[1]The natural notation is misleading here: we *do not mean* differentiation of $W^{(e)}f(\cdot)$ in the variable $e$.

of the product of all $W^{(e_j)}$ with $j \neq i$, and again the above discussion provides a required upper bound (independent of $f$) for it on $K$.

It remains to prove the lemma. In the proof, we do not insist the vectors $e_j$ be normalized (we can normalize them at the end). The full truth is that any collection $\{e_i\}_{i \le N}$ in general position with $N$ sufficiently large fits, but, to lighten the presentation, we confine ourselves to the following.

1. If we know that for every two nonzero vectors $u$ and $v$ the differentiation operator $(\partial/\partial u)^{k-1}(\partial/\partial v)$ can be represented as a linear combination of the $k$th powers of a controlled set of directional derivatives, then the lemma easily follows by induction on the number of variables.

2. If $u$ and $v$ are two nonzero vectors, for any real $t$ we expand $((\partial/\partial u) + t(\partial/\partial v))^k$ by the binomial formula. Then we take $k+1$ different values of $t$, thus obtaining a system of linear relations from which the derivatives $(\partial/\partial u)^a(\partial/\partial v)^b$, $a+b=k$, can be expressed as linear combinations of directional derivatives of the form $(\partial/\partial(u+tv))^k$ (the Vandermonde determinant is involved). So, we are done. □

## 4.3 BMO and atomic $H^1$

Now we pass to certain supplementary facts about singular integral operators. The main results will be proved in a more general setting than was announced in Theorems 4.10 and 4.11. Particular assumptions will vary from one statement to another. We start with a result for which nothing beyond long-range $L^1$-regularity is required.

Till the end of this section, we use the following equivalent norm on BMO:

$$\|f\|_{\mathrm{BMO}} = \sup_Q \left( \frac{1}{|Q|} \int_Q |f - f_Q|^2 \right)^{1/2},$$

where the supremum is taken over all cubes $Q$.

In fact, BMO is a dual space. We are going to describe its predual explicitly.

**Definition 4.17.** A function $a \in L^2(\mathbb{R}^n)$ is called an atom (more precisely, a $(1,2)$-atom) if it is supported on some cube $Q$, has zero integral, and satisfies $\|a\|_{L^2} \le |Q|^{-\frac{1}{2}}$.

The Hölder inequality implies immediately the estimate $\|a\|_{L^1} \le 1$.

**Lemma 4.18.** *If $T$ is a long-range regular operator bounded on $L^2$ and $a$ is an atom, then $\|Ta\|_{L^1} \le C$, where the constant $C$ depends only on $T$, but not on $a$. In particular, this statement applies to Calderón-Zygmund singular integral operators.*

*Proof.* Let $Q$ denote the cube related to the atom $a$ as in the above definition. Then

$$\int |Ta| \le \int_{\mathbb{R}^n\setminus 2Q} |Ta| + \int_{2Q} |Ta| \le C\|a\|_{L^1} + |2Q|^{1/2}\left(\int_{\mathbb{R}^n} |Ta|^2\right)^{\frac{1}{2}}.$$

It remains to refer to the $L^2$-boundedness of $T$. □

This lemma suggests the definition of a new function space, which is denoted by $H^1_{at}$ and is called the atomic (Hardy class) $H^1$. We define

$$H^1_{at} = \{f:\ f(x) = \sum \lambda_i a_i(x),\ a_i \text{ are atoms},\ \sum |\lambda_i| < \infty\}.$$

The norm in this space is introduced by $\|f\| = \inf \sum |\lambda_i|$, where the infimum is taken over all representations of $f$ in the above form. Since the $L^1$-norm of any atom is at most 1, a natural norm 1 embedding $H^1_{at} \subset L^1$ arises. As we know, any $L^2$-bounded long-range regular operator $T$ is well defined on $L^1$ (with values in $L^{1,\infty}$). The embedding described above shows that $T$ is well defined also on $H^1_{at}$, and the lemma shows that $T$ takes $H^1_{at}$ to $L^1$ boundedly.

We show that BMO is the space dual to $H^1_{at}$. Let $Q$ be a cube, and let $L^2_0(Q)$ be the space of functions with zero average that belong to $L^2(Q)$. If $g \in L^2_0(Q)$, then the function $\|g\|^{-1}_{L^2(Q)}|Q|^{-1/2}g$ is an atom. So, we have a natural embedding $j_Q : L^2_0(Q) \to H^1_{at}$, and $\|j_Q\| \le |Q|^{1/2}$.

Suppose $\Phi \in (H^1_{at})^*$, then $\Phi \circ j_Q \in L^2_0(Q)^*$, so that there is a function $h_Q \in L^2(Q)$ such that

$$\Phi(g) = \int g h_Q \quad \text{for } g \in L^2_0(Q)$$

(we have written $\Phi(g)$ instead of the "correct" $\Phi(j_Q g)$ intentionally). The function $h_Q$ is not unique: it is determined up to a constant summand. Clearly,

$$|Q|^{\frac{1}{2}}\|\Phi\| \ge \|\Phi\|\,\|j_Q\| \ge \|\Phi j_Q\| = \inf_c \|h_Q - c\|_{L^2(Q)} = \inf_c \left(\int_Q |h_Q - c|^2\right)^{\frac{1}{2}} \tag{4.17}$$

(the infimum is taken over all constants $c$).

Suppose $K$ is another cube with $K \supset Q$. On $Q$, the function $h_K$ differs from $h_Q$ by a constant. Subtracting this constant from $h_K$, we can ensure that $h_K$ agree with $h_Q$ on $Q$.

Now, consider an expanding sequence $\{Q_j\}$ of cubes that covers $\mathbb{R}^n$. We define a function $h$ on $\mathbb{R}^n$ as follows. On $Q_1$, we put $h = h_{Q_1}$, then choose (a representative for) $h_{Q_2}$ that agrees with $h$ on $Q_1$ and put $h$ equal to this representative on $Q_2$, then choose (a representative for) $h_{Q_3}$ that agrees with $h$ on $Q_2$, and so on. Eventually, $h$ becomes extended to the entire space $\mathbb{R}^n$ in such a way that for every cube $Q$ and every $a \in L^2_0(Q)$ we have

$$\Phi(a) = \int_{\mathbb{R}^n} ah. \tag{4.18}$$

Clearly, this formula applies to all atoms $a$. Since the linear combinations of atoms are dense in $H^1_{at}$, we can say that $h$ *represents* the functional $\Phi$. Formula (4.17) shows that

$$\|\Phi\| \geq \sup_Q \inf_c \left( \frac{1}{|Q|} \int_Q |h-c|^2 \right)^{\frac{1}{2}}. \tag{4.19}$$

It is easily seen that we have equality in (4.19): it suffices to substitute in (4.18) an atom $a$ at which the norm of $\Phi$ is nearly attained. Finally, the infimum on the right in (4.19) is attained at $c = f_Q$ because, in $L^2(Q)$, the function $f - f_Q$ is orthogonal to constants.

We have proved that for every $\Phi \in (H^1_{at})^*$ there exists a function $h \in \mathrm{BMO}$ (unique up to a constant summand) that represents $\Phi$ in the sense of (4.18), with equality of norms. It is natural to ask whether every function $u \in \mathrm{BMO}$ represents some functional $\Psi$ on $H^1_{at}$. Surely, the integral as in (4.18) is finite for such $u$ in the role of $h$: $|\int au| \leq \|u\|_{\mathrm{BMO}}$ for every atom $a$. So, it is natural to put $\Psi(a) = \int au$ for atoms. However, it is not quite clear that this definition extends to the space $H^1_{at}$ (an atom $b$ may have a representation $b = \sum c_j a_j$, where the $a_j$ are atoms and $\sum |c_j|$ is small).

To cope with this obstruction, we observe that an arbitrary function $\varphi$ satisfying $|\varphi(x) - \varphi(y)| \leq |x - y|$, $x, y \in \mathbb{C}$, operates on BMO not increasing the norm. Indeed, for every constant $c$ we have

$$\left( \frac{1}{|Q|} \int_Q |\varphi(f(x)) - \varphi(c)|^2 dx \right)^{\frac{1}{2}} \leq \left( \frac{1}{|Q|} \int_Q |f(x) - c|^2 dx \right)^{\frac{1}{2}},$$

and everything becomes clear after passage to the infimum over $c$. Next, we introduce the function $\varphi_N(t)$ equal to $t$ for $|t| \leq N$ and to $Nt/|t|$ otherwise. It is easily seen that $|\varphi_N(x) - \varphi_N(y)| \leq |x - y|$, $x, y \in \mathbb{C}$. Put $u_N(\cdot) = \varphi_N(u(\cdot))$, then $\|u_N\|_{\mathrm{BMO}} \leq \|u\|_{\mathrm{BMO}}$. Since $u_N$ is bounded, it generates a bounded linear functional $\Psi_N$ on $L^1$, and, therefore, also on $H^1_{at}$, because this space embeds continuously in $L^1$. By the first part of the proof, $\|\Psi_N\|_{(H^1_{at})^*} = \|u_N\|_{\mathrm{BMO}} \leq \|u\|_{\mathrm{BMO}}$ for all $N$.

By the Lebesgue dominated convergence theorem, for every atom $a$ we have

$$\Psi_N(a) = \int a u_N \to \int au \text{ as } N \to \infty. \tag{4.20}$$

Thus, we have convergence also on the linear combinations of atoms. Since these linear combinations form a dense subset of $H^1_{at}$ and the norms of the functionals $\Psi_N$ are uniformly bounded, they converge pointwise to some functional $\Psi \in (H^1_{at})^*$. This functional is represented by $u$ in view of (4.20).

What has been proved constitutes the content of the following theorem.

**Theorem 4.19.** *We have* $(H^1_{at})^* = \mathrm{BMO}$ *with equality of norms.*

Now, let $T$ be a long-range regular operator bounded on $L^2$, so that it takes $H^1_{at}$ to $L^1$. Then, clearly, its adjoint $T^*$ maps $L^\infty$ to BMO continuously. This is the principal result of this section. Note that if $f \in L^\infty$ and $h = T^* f$, then, up to a constant summand, $h$ is completely determined by the relation

$$\int ha = \int fTa \quad \text{for all atoms } a. \tag{4.21}$$

Apparently, stronger results about the continuity of $T^*$ cannot be obtained under the mere assumption of long-range regularity for $T$. From now on we assume at least that $T$ is a singular integral operator in the sense of Definition 2.1 (more conditions will be imposed later). Let $K$ be the kernel of $T$. We start with the observation that then there is a formula for $h = T^* f$ more explicit than (4.21). Let $Q$ be a cube. We shall give a representation for the function $(h-c)\chi_Q$, where $c$ is a constant depending on $Q$. The properties of $K$ and the relation $\int a = 0$ allow us to rewrite (4.21) as follows:

$$\int ha = \int_{\mathbb{R}^n \setminus 2Q} f(x) \int_Q (K(x,y) - K(x,y_0))a(y)dydx + \int_{\mathbb{R}^n} (\chi_{2Q} f)Ta.$$

Here $y_0$ is the center of $Q$, and $a$ may run through arbitrary atoms supported on $Q$. In the first summand on the right, we can change the order of integration by the Fubini theorem (see condition (II) in Definition 2.1). In the second summand, both $\chi_{2Q} f$ and $a$ belong to $L^2$, and $T$ is bounded on $L^2$ by assumption. Thus, we arrive at

$$\int ha = \int_Q a(y) \left[ \int_{\mathbb{R}^n \setminus 2Q} (K(x,y) - K(x,y_0)) f(x)dx + T^*(\chi_{2Q} f) \right] dy$$

for all atoms $a$ supported on $Q$. It follows that there is a constant $c = c_Q$ with

$$h(y) - c = T^*(\chi_{2Q} f)(y) + \int_{\mathbb{R}^n \setminus 2Q} (K(x,y) - K(x,y_0)) f(x)dx, \quad x \in Q. \tag{4.22}$$

Here on the right $T^*$ means the adjoint of $T$ if the latter is viewed as an operator from $L^2(\mathbb{R}^n)$ to itself. The right-hand side of formula (4.22) can be (and usually is) taken for the *definition* of the function $T^* f$. It should only be kept in mind that this definition applies only to $y \in Q$. It is clear from the above that, if the interiors of two cubes $Q_1$ and $Q_2$ have nonempty intersection, then the corresponding formulas of type (4.22) agree on it up to an additive constant. However, this can be seen directly, without passage to duals: consider a big cube $Q$ including both $Q_1$ and $Q_2$ and do some elementary calculations (similar calculations will be presented explicitly in the next section). It is also instructive to look at the proof of the inclusion $h \in$ BMO on the basis of (4.22):

$$\left(\frac{1}{|Q|}\int_Q |h-c|^2\right)^{\frac{1}{2}} \leq \left(\frac{1}{|Q|}\int_{\mathbb{R}^n} |T^*(\chi_{2Q}f)|^2\right)^{\frac{1}{2}}$$
$$+\left(\frac{1}{|Q|}\int_Q\left(\int_{\mathbb{R}^n\setminus 2Q}|K(x,y)-K(x,y_0)|dx\right)^2 dy\right)^{\frac{1}{2}}\|f\|_{L^\infty}.$$

The second integral is controlled by $c\|f\|_\infty$ with the help of condition (II) in Definition 2.1. In the first, we use the fact that $T^*$ is bounded on $L^2$ to obtain

$$\int_{\mathbb{R}^n}|T^*(\chi_{2Q}f)|^2 \leq C\int_{\mathbb{R}^n}\chi_{2Q}|f|^2 \leq |2Q|\|f\|_\infty^2.$$

Taken together, these two estimates yield $\|h\|_{\mathrm{BMO}} \leq C_1\|f\|_\infty$, as required.

Returning to Theorem 3.21, now we can see that, as it has been promised, an operator $T$ does satisfy conditions (S1) and (S2) stated before that theorem if both $T$ and $T^*$ are singular integral operators with some additional properties (say, with those mentioned in Remark 2.3). Later in this chapter we shall discuss condition (S3).

## 4.4 Continuity of operators on BMO and Lipschitz spaces

Let again $T$ be a linear operator bounded on $L^2(\mathbb{R}^n)$ and long-range regular on $L^1(\mathbb{R}^n)$. We have seen that the adjoint $T^*$ takes $L^\infty$ to BMO. Is there a natural way to extend $T^*$ to an operator from BMO to BMO?

If such an extension exists, then, clearly, we must have

$$\|T^*f\|_{\mathrm{BMO}} \leq C\|f\|_{\mathrm{BMO}} \quad \text{for every} \quad f\in L^\infty(\mathbb{R}^n). \tag{4.23}$$

This implies, in particular, that $T^*1$ must be the zero element of BMO (i.e., the class of constant functions). Thus, the condition

$$T^*1 = 0 \quad \text{in} \quad \mathrm{BMO} \tag{4.24}$$

is necessary for (4.23). In fact, it is also sufficient if $T$ is a singular integral operator whose kernel $K$ satisfies a condition of the form (2.10):

$$|K(x,y)-K(x,y_0)| \leq c\frac{|y-y_0|^\beta}{|x-y_0|^{n+\beta}}, \quad |x-y_0|\geq 2|y-y_0|, \tag{4.25}$$

for some $\beta\in(0,1]$. The required extension will be based on formula (4.22).

However, it is convenient to define $T^*$ at once also on more general homogeneous spaces of smooth functions, such as $\dot{C}_p^{k,s}$. (Recall that here $k$ is a natural number.) Then the above condition on the kernel should be replaced with the following one: for every cube $Q$, there is a function $U_Q(x,y)$ such that, for every fixed $x$, it is a polynomial of degree strictly smaller than $k$ in $y$ and

$$|K(x,y) - U_Q(x,y)| \le C\frac{|y-y_0|^\beta}{|x-y_0|^{n+\beta}}, \quad x \notin 2Q, \quad y \in Q. \tag{4.26}$$

Here $y_0$ is the center of $Q$ and $\beta \in (k-1, k]$.

If $k = 1$, this is slightly weaker than (4.25) (but also guarantees that $T^*$ acts on BMO, see below). More generally, (4.26) is fulfilled with $\beta = k$ if $K$ is differentiable $k$ times outside the diagonal and

$$|\nabla_y^l K(x,y)| \le C|x-y|^{-n-l} \text{ for } l = 0, \dots, k. \tag{4.27}$$

This is a consequence of the Taylor formula. Indeed, denote by $T^k_{f,u}(\cdot)$ the Taylor polynomial of order $k-1$ for a function $f$ relative to the center $u$. We fix $x$ and write this polynomial for $f(\cdot) = K(x,\cdot)$, taking $u = y_0$:

$$U_Q(x,y) = T^k_{K(x,\cdot),y_0}(y). \tag{4.28}$$

Then $U_Q(x,y)$ is indeed a polynomial in $y$. Next, for $y \in Q$ and $x \notin 2Q$ we can write (the subscript 2 of the differentiation symbol $D$ in the next formula signalizes that we take derivatives of $K(x,y)$ is in the second variable $y$):

$$|K(x,y) - U_Q(x,y)| \le c \max_{|\alpha|=k} \sup_{\xi\in Q} |D_2^\alpha K(x,\xi)||y-y_0|^k \le c_1 \frac{|y-y_0|^k}{\sup_{\xi\in Q}|x-\xi|^{n+k}}$$

by (4.27). It remains to observe that $|x-\xi| \approx |x-y_0|$ uniformly in $\xi \in Q$.

Estimates (4.27) do occur, e.g., if $K(x,y) = a(x-y)$, where $a$ is homogeneous of degree $-n$ and is continuously differentiable $k$ times everywhere except for the point 0 (simply because any derivative of order $l$ will be homogeneous of degree $-n-l$). Another sufficient condition in terms of Fourier transforms was provided by Theorem 2.5.

Now, suppose that (4.26) is fulfilled. Let $f \in L^1_{\mathrm{loc}}$ be such that $|f|(1 + |x|^{n+\beta})^{-1}$ is integrable (temporarily, such functions will be called *suitable*; clearly, every polynomial in $\mathcal{P}_k$ is suitable). Let $Q$ be a cube. We put

$$\varphi^{Q,f}(y) = T^*(\chi_{2Q}f)(y) + \int_{\mathbb{R}^n\setminus 2Q} (K(x,y) - U_Q(x,y))f(x)\,dx, \quad y \in Q. \tag{4.29}$$

The (Lebesgue) integral on the right exists by (4.26) because $f$ is suitable. The next lemma shows that the functions $\varphi^{Q,f}$ for different $Q$ can be glued into one determined uniquely up to a summand in $\mathcal{P}_k$.

**Lemma 4.20.** *If $Q_1 \subset Q_2$ are two cubes and $f$ is suitable, then $\varphi^{Q_2,f} - \varphi^{Q_1,f}$ is a polynomial of degree strictly smaller than $k$ on $Q_1$.*

*Proof.* Indeed,

$$\varphi^{Q_2}(y) - \varphi^{Q_1}(y) = T^*(\chi_{2Q_2\setminus 2Q_1} f)(y) - \int\limits_{2Q_2\setminus 2Q_1} (K(x,y) - U_{Q_1}(x,y))f(x)\,dx$$
$$+ \int\limits_{\mathbb{R}^n\setminus 2Q_2} (U_{Q_1}(x,y) - U_{Q_2}(x,y))f(x)\,dx.$$

The third integral yields a polynomial belonging to $\mathcal{P}_k$. (This requires slight concentration: writing $U_{Q_1}(x,y) - U_{Q_2}(x,y) = \sum_{|l|<k} b_l(x)y^l$, it is easy to deduce from (4.26) that $b_l(x) = O((1+|x|)^{-n-\beta})$ for all $l$, and then the claim follows.) Next, since $y \in Q_1$, by the basic properties of the kernel (see Definition 2.1 and the discussion after it), the second term is equal to

$$-T^*(\chi_{2Q_2\setminus 2Q_1} f) + \int\limits_{2Q_2\setminus 2Q_1} U_{Q_1}(x,y)f(x)\,dx.$$

Again, the second summand here belongs to $\mathcal{P}_k$, and we are done. □

By the way, a similar but even simpler calculation shows that, modulo $\mathcal{P}_k$, the definition (4.29) does not depend on a particular choice of the functions $U_Q(x,y)$ satisfying (4.26), provided they are polynomials in $y$ of degree smaller than $k$.

Now it is clear how to choose a joint representative $\varphi$ of $\varphi^{Q,f}$ modulo $\mathcal{P}_k$ for all $Q$ at once. Consider an expanding sequence $\{Q_j\}_{j\in\mathbb{N}}$ of cubes. We start with putting $\varphi = \varphi^{Q_1,f}$ on $Q_1$. Then we put $\varphi = \varphi^{Q_2,f} + r$ on $Q_2$ where $r \in \mathcal{P}_k$ is chosen so that on $Q_1$ the definition agrees with that at the first step, then we extend $\varphi$ to $Q_3$ by the same procedure, and so on. Eventually, we extend $\varphi$ to $\mathbb{R}^n$, and, obviously, this function will do the job.

### 4.4.1 A pointwise estimate

We denote by $T^*f$ a joint representative (modulo $\mathcal{P}_k$) of all functions $\varphi^{Q,f}$. If we want to define $T^*$ on a space whose elements are determined modulo $\mathcal{P}_i$ so that it take values in a space whose elements are determined modulo $\mathcal{P}_k$, it is natural to impose the following condition in place of (4.24):

$$T^*u = 0 \pmod{\mathcal{P}_k} \quad \text{for every} \quad u \in \mathcal{P}_i. \tag{4.30}$$

Note that this condition is void if $i = 0$ (because the only function $u$ belonging to $\mathcal{P}_0$ is identically zero). For greater $i$, this condition may be violated (see, e.g., [M] for relevant examples). However, *it is always fulfilled for singular integral operators mentioned in the first section of this chapter* provided $k \geq i$. In that section we treated operators of two sorts: partial sums for wavelet expansions and multipliers with smooth symbols. In the former case, the explanation will be given in §5 below;

in the latter case, we present it now. The essence of the phenomenon is simple. Suppose an operator $S$ defined on the set of reasonable functions commutes with translations. If $g$ is a polynomial of degree strictly smaller than $i$, then $g(x+t)$ can be written as a linear combination of monomials $x^\alpha$ with coefficients $p_\alpha(t)$ that are polynomials in $t$. Now, if $h = Sg$, then $h(y+t) = [S(g(\cdot+t))] = \sum \varphi_\alpha(y)p_\alpha(t)$ and putting $y = 0$ we see that $h$ is also a polynomial of degree smaller than $i$.

Now, if $T$ is a Fourier multiplier, nearly this can be done for the operator defined by (4.29), which also commutes with translations if the term is understood properly. However, precise arguments are somewhat entangled, which is inevitable because so is formula (4.29) itself. We present them.

Let $T = T_m$ be a multiplier with an infinitely smooth symbol satisfying the assumptions of Theorem 2.4. By Theorem 2.5, the kernel of $T$ has the form $K(x,y) = a(x-y)$ and satisfies (4.27). We define the functions $U_Q(x,y)$ by (4.28) (recall that $k \geq i$).

Next, we refer to Proposition 2.7 to represent $a$ as the limit of some functions $a^{(j)}$ in the Schwartz class, as described in that statement. Let $T_j f = f * a^{(j)}$, then $K_j(x,y) = a^{(j)}(x-y)$ is the kernel of $T_j$. Again, we define $U^j_Q(x,y)$ by (4.28) with $K_j$ in place of $K$ and then apply the recipe (4.29) to $T_j$ and a *suitable* function $f$:

$$\varphi_j^{Q,f}(y) = T_j^*(\chi_{2Q}f)(y) + \int_{\mathbb{R}^n\setminus 2Q} (a^{(j)}(x-y) - U^j_Q(x,y))f(x)dx, \quad y \in Q.$$

Using Proposition 2.7, it is easy to realize that $\lim_{j\to\infty} \varphi_j^{Q,f}(y) = \varphi^{Q,f}(y)$ (say, in $L^2(Q)$).

Now, suppose $f$ is a polynomial of degree strictly smaller than $i$. We see that it suffices to prove the inclusion $\varphi_j^{Q,f} \in \mathcal{P}_k$ for every $j$. But both $a_j(x-y)$ and $U^j_Q(x,y)$ decay rapidly in $x$, so that we can integrate termwise in the second summand on the right in the last display, obtaining

$$\varphi_j^{Q,f}(y) = \int_{\mathbb{R}^n} a^{(j)}(x-y)f(x)dx - \int_{\mathbb{R}^n\setminus 2Q} U^j_Q(x,y)f(x)dx.$$

The second term on the right here is a polynomial of degree at most $k$ (because so is $U^j_Q(x,y)$ as a function of $y$). The first term is equal to $\int_{\mathbb{R}^n} a(\xi)f(\xi+y)d\xi$, which is a polynomial of degree at most $i$.

We remind the reader of the definition (4.4) of the maximal function $M_{k,p,s}$:

$$M_{k,p,s}f(x) = \sup_{Q\ni x} \frac{1}{|Q|^{\frac{s}{n}}} \inf_{g\in\mathcal{P}_k} \left(\frac{1}{|Q|}\int_Q |f-g|^p\right)^{1/p}, \tag{4.31}$$

where the supremum is taken over all cubes $Q$ containing $x$. Recall that here $k$ is a nonnegative integer, $s$ is a real number (which may be negative), and $1 \leq p < \infty$. Now, we can formulate the main estimate for $T^*$.

**Theorem 4.21.** *Let $s \in \mathbb{R}$, and suppose that condition* (4.30) *is fulfilled with some $i \geq s$, $i \in \mathbb{Z}_+$, whereas condition* (4.26) *is fulfilled with some $\beta > \max\{s, i-1\}$. If $T^*$ is bounded on $L^p(\mathbb{R}^n)$, then*

$$M_{k,p,s}(T^* f)(x) \leq C M_{i,p,s} f(x)$$

*for all $x$ for which the right-hand side is finite. Here $C$ is independent of $f$ and $x$, and $k$ is the natural number such that $\beta \in (k-1, k]$, see* (4.26).

*Remark* 4.22. The general theory in Chapter 2 shows that if $T^*$ is bounded on $L^p(\mathbb{R}^n)$, then it is bounded on $L^r(\mathbb{R}^n)$ for all $r \in [p, \infty)$. So, there is always some flexibility in the choice of $p$ in the above theorem. In many particular cases $p$ can be chosen arbitrarily close to 1 but, if we talk about a "genuine" singular integral, $p = 1$ is excluded. The condition $i \geq s$ is imposed to avoid degeneration (we refrain from more detailed explanations). Note that $k \geq i$ under the assumptions of the theorem. The calculations in the proof below will show (in passing) that every $f$ with $M_{i,p,s} f(x) < \infty$ for at least one $x$ is *suitable* in the sense discussed above.

Before we start to prove the theorem, we need some observations. First, if in (4.31) we restrict ourselves to cubes centered at $x$, we obtain an equivalent quantity (the controlling constant will depend only on the parameters involved and on the dimension). Precisely, Lemma 0.24 helps to see this. Second, instead of taking the infimum over $g$ in the same formula, we may content ourselves to choosing, "in a unified manner", some polynomial $u$ at which this infimum is attained roughly. We already discussed this at the beginning of Subsection 2.2 in Chapter 1. So, we put $f_Q = P_Q f$, where $P_Q$ is the projection defined there, and, instead of (4.31), we shall work with

$$\widetilde{M}_{k,p,s} f(x) = \sup_Q \frac{1}{|Q|^{s/n}} \left( \frac{1}{|Q|} \int |f - f_Q|^p \right)^{1/p}, \tag{4.32}$$

where the supremum is taken over all cubes centered at $x$.

*Proof of Theorem* 4.21. We take some point $y_0$ for which $\mu = \widetilde{M}_{i,p,s} f(y_0)$ is finite. If $Q$ is a cube centered at $y_0$, then

$$\frac{1}{|Q|^{\frac{s}{n}}} \left( \frac{1}{|Q|} \int_Q |f - f_Q|^p \right)^{1/p} \leq \mu. \tag{4.33}$$

Using formula (4.29), Lemma 4.20, and condition (4.30), we see that there exists a polynomial $v \in \mathcal{P}_k$ such that

$$\begin{aligned}(T^* f - v)(y) &= T^*(\chi_{2Q}(f - f_Q))(y) \\ &\quad + \int_{\mathbb{R}^n \setminus 2Q} (K(x,y) - U_Q(x,y))(f(x) - f_Q(x))\, dx, \quad y \in Q.\end{aligned}$$

Clearly, the theorem will be proved if we show that $|Q|^{-\frac{s}{n}}(|Q|^{-1}\int_Q |T^*f-v|^p)^{1/p} \leq C\mu$, where $C$ is independent of $f$ and $y_0$. To do this, we write

$$\frac{1}{|Q|^{s/n}}\left(\frac{1}{|Q|}\int_Q |T^*f-v|^p\right)^{1/p} \leq \frac{1}{|Q|^{s/n}}\left(\frac{1}{|Q|}\int_Q |T^*(\chi_{2Q}\cdot(f-f_Q))|^p\right)^{1/p}$$

$$+\frac{1}{|Q|^{s/n}}\left(\frac{1}{|Q|}\int_Q\left(\int_{\mathbb{R}^n\setminus 2Q} |K(x,y)-U_Q(x,y)|\,|f(x)-f_Q(x)|\,dx\right)^p dy\right)^{1/p}. \tag{4.34}$$

In the first summand, we extend integration to $\mathbb{R}^n$ and use the $L^p$-boundedness of $T^*$ to majorize this summand by

$$|Q|^{-\frac{s}{n}}\left(|Q|^{-1}\int_{2Q}|f-f_Q|^p\right)^{1/p} \leq C|2Q|^{-\frac{s}{n}}\left(|2Q|^{-1}\int_{2Q}|f-f_Q|^p\right)^{1/p}.$$

Now, by Lemma 0.24,

$$\|f_{2Q}(x)-f_Q(x)\|_{L^\infty(2Q)} \leq C\left(\frac{1}{|Q|}\int_Q |f_{2Q}-f_Q|^p\right)^{1/p}$$

$$= C\left(\frac{1}{|Q|}\int_Q |P_Q(f-f_{2Q})|^p\right)^{1/p} \leq C'\left(\frac{1}{|Q|}\int_Q |f-f_{2Q}|^p\right)^{1/p} \tag{4.35}$$

$$\leq C''\left(\frac{1}{|2Q|}\int_{2Q}|f-f_{2Q}|^p\right)^{1/p} \leq C''\mu|2Q|^{\frac{s}{n}}.$$

Combining this with the preceding estimate, we see that the first summand on the right in (4.34) is dominated by $C\mu$, and it remains to obtain a similar inequality for the second summand. But by (4.26), that summand does not exceed the quantity

$$C\frac{1}{|Q|^{\frac{s}{n}}}\left(\frac{1}{|Q|}\int_Q\left(\int_{\mathbb{R}^n\setminus Q}\frac{|y-y_0|^\beta}{|x-y_0|^{n+\beta}}|f(x)-f_Q(x)|\,dx\right)^p dy\right)^{1/p}$$

$$\leq C\frac{1}{|Q|^{\frac{s}{n}}}\left(|Q|^{\frac{\beta p}{n}}\left(\int_{\mathbb{R}^n\setminus 2Q}\frac{|f(x)-f_Q(x)|}{|x-y_0|^{n+\beta}}\,dx\right)^p\right)^{1/p}$$

(we have estimated $|y-y_0|$ by $|Q|^{\frac{1}{n}}$)

$$\leq C|Q|^{\frac{\beta-s}{n}} \sum_{j=1}^{\infty} |2^j Q|^{-\frac{n+\beta}{n}} \int\limits_{2^{j+1}Q\setminus 2^j Q} |f(x)-f_Q(x)|\,dx$$

$$\leq C'|Q|^{-1-\frac{s}{n}} \sum_{j=1}^{\infty} 2^{-j(n+\beta)} \int\limits_{2^{j+1}Q} |f-f_Q|. \tag{4.36}$$

The integral in the last expression is estimated as follows:

$$\int\limits_{2^{j+1}Q} |f-f_Q| \leq \int\limits_{2^{j+1}Q} |f-f_{2^{j+1}Q}| + \sum_{m=1}^{j} |2^{j+1}Q| \|f_{2^{m+1}Q}-f_{2^m Q}\|_{L^\infty(2^{j+1}Q)}.$$

Next, by Lemma 0.24 and formula (4.35), we have

$$\begin{aligned} \|f_{2^{m+1}Q}-f_{2^m Q}\|_{L^\infty(2^{j+1}Q)} &\leq C\left(\frac{|2^{j+1}Q|}{|2^m Q|}\right)^{i-1} \frac{1}{|2^m Q|} \int\limits_{2^m Q} |f_{2^{m+1}Q}-f_{2^m Q}| \\ &\leq C2^{(j-m)(i-1)} \left(\frac{1}{|2^m Q|} \int\limits_{2^m Q} |f_{2^{m+1}Q}-f_{2^m Q}|^p\right)^{1/p} \\ &\leq C|2^{m+1}Q|^{\frac{s}{n}} 2^{(j-m)(i-1)}\mu. \end{aligned}$$

This yields

$$\int\limits_{2^j Q} |f-f_Q| \leq C|2^{j+1}Q|\mu|Q|^{\frac{s}{n}} \sum_{m=1}^{j} 2^m(s-i+1)2^{j(i-1)},$$

and (4.36) can be continued as follows:

$$\cdots \leq C\mu \sum_{j=1}^{\infty} 2^{-j(n+\beta)} 2^{jn} 2^{j(i-1)} \sum_{m=1}^{j} 2^{m(s-i+1} = C\mu \sum_{j=1}^{\infty} 2^{-j(\beta-i+1)} \sum_{m=1}^{j} 2^{m(s-i+1)}.$$

The sum over $m$ is dominated by a constant if $s-i+1<0$, by $Cj$ if $s-i+1=0$, and by $C2^{j(s-i+1)}$ if $s-i+1>0$. We see that in any case the series in $j$ is convergent (recall that $\beta > \max\{s, i-1\}$ by assumption). It follows that the second summand on the right in (4.34) is also dominated by $C\mu$, and we are done. □

### 4.4.2 Norm estimates

Theorem 4.21 easily implies the boundedness of $T^*$ on various homogeneous spaces of smooth function. We start with the simplest case.

**Zero smoothness: the space BMO**

A function $f$ belongs to BMO if and only if $M_{1,p,0}f$ is a uniformly bounded function for some (equivalently, for all, see Chapter 1) $p \in [1, +\infty)$. Thus, *for* $T^*$ *to act boundedly from* BMO *to itself it suffices that* $T$ *be a singular integral operator satisfying* (4.24) *and* (4.25) *with some* $\beta \in (0, 1]$.

**Positive smoothness: Campanato spaces**

If $s > 0$, the space $\dot{C}_p^{s,l}$ is determined by the requirement $M_{l,p,s}f \in L^\infty$. Again, the space does not depend on $p \in [1, \infty)$, see Chapter 1. Theorem 4.21 shows that $T^*$ takes $\dot{C}_p^{s,i}$ to $\dot{C}_p^{s,k}$ boundedly provided $T$ satisfies (4.30), and (4.26) is fulfilled with $\beta > \max\{s, i-1\}$ ($k$ is the integer related to condition (4.26)). Of course, it would be good to take $k = i$, and this is possible indeed if $i > s$. In particular, we may do so whenever $s$ is not an integer. But if $s$ is an integer and $i = s$, the minimal value of $k$ allowed by Theorem (4.21) is $i+1$. For clarity, we say explicitly what this means for the classes $\mathrm{Lip}_s$, $0 < s \leq 1$. We know that $\mathrm{Lip}_s = \dot{C}_p^{s,1}$ (see Proposition 4.9). Next, by the same proposition, $\dot{C}_p^{1,2}$ is the Zygmund class $Z$ consisting of all locally integrable functions $f$ with $|\Delta_h^2 f(x)| \leq C|h|$ uniformly in $x$. The above discussion shows the following.

Suppose (4.30) is fulfilled with $i = 1$. If $s < 1$, then $T^*$ takes $\mathrm{Lip}_s$ to itself boundedly provided (4.26) is true with some $\beta \in (s, 1]$. If, moreover, (4.26) is true with some $\beta > 1$, then $T^*$ takes $\mathrm{Lip}_1$ to $Z$. If, in addition to this, (4.30) is fulfilled with $i = 2$, then $T^*$ takes $Z$ to $Z$.

**Negative smoothness: Morrey spaces**

For $s < 0$, the Morrey space $\dot{C}_p^{s,0}$ is determined by the requirement $M_{0,p,s}f \in L^\infty$. (We recall the agreement that by a polynomial of degree smaller than 0 we mean the zero function; $M_{0,p,s}$ is defined in accordance with that.) In this case, condition (4.30) should be imposed with $i = 0$, so it is void. In condition (4.26) we can now take any $\beta$ with $0 < \beta \leq 1$, and the outcome will be

$$\|M_{1,p,s}T^*f\|_{L^\infty} \leq C\|f\|_{\dot{C}_p^{s,0}}.$$

In this setting the function $T^*f$ is defined up to a constant summand. But Proposition 4.5 (see also Corollary 4.6) shows that *this constant summand can be chosen in such a way that the* $\dot{C}_p^{s,0}$*-norm of the corresponding representative of* $T^*f$ *is dominated by* $C'\|f\|_{\dot{C}_p^{s,0}}$; *moreover, this representative is unique.* This allows us to interpret $T^*$ as a bounded operator from $\dot{C}_p^{s,0}$ to itself (if, of course, $T^*$ is bounded on $L^p$ *a priori*).

## 4.5 Singular integrals related to wavelet expansions

In this section, for a function $\psi \in L^2(\mathbb{R})$, we write $\psi_{j,m}(t) = 2^{j/2}\psi(2^j t + l)$, $j, m \in \mathbb{Z}$. We say that $\{\psi_{j,m}\}$ is a *system of wavelets* with *generating wavelet* $\psi$ [2] if it is an orthogonal basis of $L^2(\mathbb{R})$. For instance, if we take $\psi = \chi_{[0,1/2]} - \chi_{(1/2,1]}$, then $\{\psi_{j,m}\}$ is really an orthonormal basis in $L^2(\mathbb{R})$, called the Haar system and known since very long ago. But at the end of the 1980s it was discovered that there exist wavelet systems for which both the generating wavelet and its Fourier transform are well localized and sufficiently smooth. (The term itself, originally *ondelettes* in French, arose at that time.) We refer the reader to [M1], [D], or [NPS] for detailed information on wavelets.

A feature of wavelet systems very important by itself but also of primary interest to us is that, usually, for every set $A$ of pairs $(j, m)$ of integers, the orthogonal projection $\pi_A$ onto the closed linear span of $\{\psi_{j,m}\}_{(j,m)\in A}$ is a singular integral operator. Returning to the Haar wavelets mentioned above, we can say that, strictly speaking, for them this is not true, but something quite similar occurs. Specifically, all operators $P_A$ are $L^2$-bounded with norm 1. For treating them as singular integrals, it would suffice to have long-range $L^1$-regularity: $\|P_A f\|_{L^1(\mathbb{R}\setminus 2I)} \le C\|f\|_{L^1}$ (preferably, with $C$ independent of $A$) whenever $f$ is supported on the interval $I$ and has zero average. In fact, this can be ensured only for *dyadic* intervals $I$. In principle, this suffices for applications. For instance, we recall that the weak type $(1, 1)$ estimate was deduced with the help of the Calderón–Zygmund procedure, which is maintained within the dyadic grid; etc.

Now, if $I$ is dyadic, in our case it is easy to realize that $\operatorname{supp} P_A(f) \subset I$ provided that $f$ is supported on $I$ and has zero average. So, there is no need whatsoever to estimate $P_A(f)$ outside $2I$!

It turns out that, for sufficiently smooth and well-localized wavelets, the operators $P_A$ are singular integrals indeed; moreover, they obey (4.26), and so (being selfadjoint) act also on many spaces of smooth functions. Not to plunge too deeply in the subject, we impose *excessive* assumptions on the wavelet system that will ensure this behavior.

**Theorem 4.23.** *Suppose we are given a wavelet system $\{\psi_{j,m}\}$ such that the Fourier transform of the generating wavelet $\psi$ belongs to $\mathcal{D}(\mathbb{R})$. Fix a numerical sequence $t_{i,m}$ uniformly bounded by 1 (say) in modulus, and define an operator $T$ on $L^2$ by the formula*

$$Tf = \sum_{j,m} t_{j,m}\langle f, \psi_{j,m}\rangle \psi_{j,m}.$$

*Then $T$ is a singular integral operator whose kernel*

$$K(x, y) = \sum_{j,m} t_{j,m}\psi_{j,m}(x)\psi_{j,m}(y)$$

[2]The usual term is "wavelet function", but it seems to be applied only when $\psi$ comes from a so-called "multiresolution analysis". Here we disregard the origin of wavelet systems.

*satisfies* (4.26) *with* $n = 1$ *and* $\beta = k$, *for every* $k \in \mathbb{N}$. *All estimational constants in the conditions in question depend only on the wavelet system, but not on the coefficients* $t_{j,m}$. *Next, all moments of* $\psi$ *(consequently, of all functions* $\psi_{j,m}$*) are zero.*

Before we pass to the proof, some discussion is in order. First, wavelet systems with the above properties do exist. For example, they can be found among the so-called Meyer wavelets, see, e.g., [NPS]. Second, $\psi$ decays faster than any power of $|x|$ at infinity and is infinitely differentiable. From the arguments below it will be clear that, to ensure (4.26) with a particular $k$, only the decay as *some* power and smoothness up to *some finite* order should be demanded for $\psi$. It is known that a generating wavelet with these two properties automatically has moments vanishing up to *some* order depending on the order of smoothness and rate of decay, see, e.g., Theorem 1.7.7 in [NPS]. So, under the assumptions of Theorem 4.23, all moments of $\psi$ do vanish, and we shall not dwell on that statement. The vanishing of moments will lead to condition (4.30) for $T^*$.

Since all moments of $\psi$ vanish, all derivatives of its Fourier transform $\hat{\psi}$ vanish at zero. (It is easy to realize that, in fact, these two properties are equivalent.) In other words, $\hat{\psi}$ remains infinitely differentiable after division by any natural power of the independent variable. This observation will be used in the study of the above operator $T$ on Sobolev spaces.

## 4.5.1 More general operators

It is convenient to prove a stronger version of Theorem 4.23. Some preparations are needed to formulate it. A bounded sequence $\{f_i\}$ of elements of a Hilbert space $H$ is said to be a *Hilbert system* if $(\sum |\langle x, f_i\rangle|^2)^{1/2} \le C\|x\|$ for every $x \in H$.[3] Here and below $\langle \cdot, \cdot \rangle$ denotes the scalar product in $H$.

**Lemma 4.24.** *If* $\{f_i\}$ *and* $\{g_i\}$ *are two Hilbert systems in* $H$ *and* $t_i$ *are uniformly bounded complex numbers, then the formula* $Sx = \sum_i t_i \langle x, f_i\rangle g_i$ *introduces a well-defined bounded operator on* $H$.

*Proof.* We estimate the corresponding bilinear form. Let $x, y \in H$, then

$$|\langle Sx, y\rangle| \le \sum |t_i||\langle x, f_i\rangle||\langle g_i, y\rangle| \le C\|x\|\,\|y\|$$

by the Cauchy inequality and the condition that both $\{f_i\}$ and $\{g_i\}$ are Hilbert systems. □

[3]Another term is "a weakly 2-summable sequence". Equivalently, a Hilbert system is the image of the unit vector basis of $l^2$ under a bounded linear operator from $l^2$ to $H$. Yet another trivial reformulation is that the norm of any linear combination of the $f_i$ is dominated by the $l^2$-norm of the sequence of coefficients. The opposite inequality between these two quantities distinguishes the class of so-called Bessel systems; we do not need them.

If $\psi \in L^2$, the system $\{\psi_{j,m}\}$ may happen to be Hilbert without being orthogonal (i.e., it need not be a wavelet system in order to be Hilbert). Specifically, the following is true.

**Theorem 4.25.** *If $\psi$ is an arbitrary function in $L^2$ with $|\hat{\psi}(\xi)| \le C|\xi|$ and $|\hat{\psi}(\xi)| \le C(1+|\xi|)^{-1/2-\varepsilon}$ for some $\varepsilon > 0$, then $\{\psi_{j,m}\}$ is a Hilbert system.*

We shall discuss the proof of a partial case of this statement later. See Lemma 2.7.3 in [NPS] for the general case.

Now, suppose we are given two functions $\alpha$ and $\beta$ whose Fourier transforms belong to $\mathcal{D}(\mathbb{R})$. Suppose also that all moments of $\alpha$ and $\beta$ vanish. By Theorem 4.25 and Lemma 4.24, for any bounded sequence $t_{j,m}$ of complex numbers, the operator $T$,

$$T(f) = \sum_{j,m} t_{j,m} \langle f, \alpha_{j,m} \rangle \beta_{j,m},$$

is bounded on $L^2$.

**Theorem 4.26.** *The operator $T$ possesses a kernel given by*

$$K(x,y) = \sum_{j,m} t_{j,m} \overline{\alpha_{j,m}(x)} \beta_{j,m}(y)$$

*and satisfying* (4.26) *(with $n = 1$) for every $k$. Moreover, the definition* (4.29) *yields zero modulo $\mathcal{P}_k$ for functions in $\mathcal{P}_k$.*

*Proof.* We verify inequality (4.27). For this, we show that

$$\sum_{j,m} \left| \alpha_{j,m}(x) \left( \frac{d}{dy} \right)^l \beta_{j,m}(y) \right| \le C_l |x-y|^{-l-1} \tag{4.37}$$

for every $l$. Surely, this will imply also the differentiability of $K$ with respect to $y$ (and with respect to $x$, by symmetry) outside the diagonal $x = y$. Next, if only finitely many of the numbers $t_{j,m}$ in the definition of $T$ are nonzero, then the function

$$g(y) = \sum_{j,m} t_{j,m} \int_{\mathbb{R}} f(x) \overline{\alpha_{j,m}(x)} dx \, \beta_{j,m}(y)$$

is well defined if $f$ grows at most polynomially, and $g$ differs from the function given by (4.29) by a polynomial of degree strictly smaller than $k$. Clearly, $g$ is zero if $f$ is a polynomial, so (4.30) is fulfilled for $T$ if the sum in the definition of $T$ is finite. In the case of general $T$, we simply pass to the limit (in the sense of convergence in measure on $Q$, say) in formula (4.29), approximating $T$ by finite sums; again, the limit passage is justified by (4.37) (also, (4.28) should be taken into account) and by the continuity of $T$ on $L^2$. Next, the same inequality (4.37) for $l = 0$, combined with the same limit passage from finite sums, makes it possible to prove that $K$ is a kernel of $T$ indeed, i.e., we have $Th(x) = \int K(x,y)h(y)dy$ for

$x$ at a positive distance from the support of $h$ whenever this support is bounded and $h$ is square integrable.

So, it remains to prove (4.37). The functions $\alpha(t)$ and $\beta(t)$ and all their derivatives decay at infinity faster than any power of $|t|$. Fixing $s > l+1$ and using the estimates $|\alpha(t)| \le C(1+|t|)^{-s}$ and $|\left(\frac{d}{dt}\right)^l \beta(t)| \le C(1+|t|)^{-s}$, we dominate the left-hand side of (4.37) by some constant times the quantity

$$\sum_j \sum_m 2^{j(l+1)}(1+|2^j x+m|)^{-s}(1+|2^j y+m|)^{-s}. \tag{4.38}$$

First, we estimate the sum of all terms with fixed $j$. For this, we introduce the function $\varphi(u,v) = \sum_m (1+|u+m|)^{-s}(1+|v+m|)^{-s}$. We find an integer $m_0$ such that $a = v+m_0$ does not exceed $1/2$ in the absolute value, put $b = u+m_0$, and observe that $\varphi(u,v) = \varphi(a,b)$. Next, we consider the cases where $|b| \le 2$ and $|b| > 2$ separately. In the first case, $\varphi(a,b)$ is dominated by a constant depending only on $s$, which, trivially, does not exceed $C(1+|a-b|/8)^{-s}$. But if $|b| > 2$, we write

$$\begin{aligned}\varphi(a,b) = &\sum_{|m|\le|b|/2} (1+|a+m|)^{-s}(1+|b+m|)^{-s}\\ &+ \sum_{|m|>|b|/2} (1+|a+m|)^{-s}(1+|b+m|)^{-s} = S_1 + S_2\end{aligned}$$

and estimate $S_1$ and $S_2$. We have

$$S_1 \le \left(1+\frac{|b|}{2}\right)^{-s} \sum_m (1+|a+m|)^{-s} \le C\left(1+\frac{|b|}{2}\right)^{-s} \le C\left(1+\frac{|a-b|}{4}\right)^{-s}$$

because $|a-b| \le |a|+|b| \le 2|b|$. To estimate $S_2$, we observe that $|a+m| \ge |m|-|a| \ge |b|/2-|a| \ge |b|/4$ in that case, whence

$$S_2 \le \sum_m (1+|b+m|)^{-s}\left(1+\frac{|b|}{4}\right)^{-s} \le C\left(1+\frac{|b|}{4}\right)^{-s} \le C\left(1+\frac{|a-b|}{8}\right)^{-s}.$$

Thus, in any case $\varphi(u,v) = \varphi(a,b) \le C_1\left(1+\frac{|a-b|}{8}\right)^{-s} = C_1\left(1+\frac{|u-v|}{8}\right)^{-s}$.

It follows that the quantity (4.38) is dominated by a constant times

$$\sum_j 2^{j(l+1)}(1+2^{j-3}|x-y|)^{-s} = \sum_{2^{j-3}\le|x-y|^{-1}} \cdots + \sum_{2^{j-3}>|x-y|^{-1}} \cdots .$$

In the first sum we replace the factor $(1+2^{j-3}|x-y|)^{-s}$ by 1, and in the second by $(2^{j-3}|x-y|)^{-s}$. We are left with two geometric series (recall that $s > l+1$) whose sums are $O(|x-y|^{-l-1})$. □

### 4.5.2 Consequences

So, Theorem 4.23 is also proved, and the results of the preceding section show that for the operators $T$ of that theorem both $T$ and $T^*$ can be extended naturally to Morrey–Campanato spaces. Surely, results proved even earlier in this book show that these operators are bounded on $L^p$, $1 < p < \infty$, and are of weak type $(1,1)$. (We emphasize that the properties of the kernel of $T$ are the same in both variables.) Let us show that $T$ (or, rather, $T^*$, which is equivalent by symmetry, but slightly more natural under our way of presentation) takes also homogeneous Sobolev spaces with $1 < p < \infty$ to themselves. We restrict ourselves to the case where all but a finite number of the coefficients $t_{j,m}$ ($|t_{j,m}| \leq 1$) in the definition of $T$ are zero and estimate the norm of $T^*$ by a quantity independent of the sequence $\{t_{j,m}\}$. Then the general case will follow by a simple limit passage. When estimating the norm of $T^*f$, we may assume that $f$ ranges over a dense subset of $\dot{W}_p^k(\mathbb{R})$, for which we take the set of functions whose $k$th derivative is in $\mathcal{D}(\mathbb{R})$. As we have seen, under all these assumptions we may use the "naive" definition of $T^*$ instead of (4.29), namely,

$$T^*f(y) = \sum_{j,m} t_{j,m} \int_{\mathbb{R}} f(x)\overline{\psi_{j,m}(x)}dx\psi_{j,m}(y).$$

Now, we observe that there exists a function $\alpha$ satisfying the same conditions as $\psi$ (i.e., having zero moments and with Fourier transform in $\mathcal{D}$) the $k$th derivative of which is equal to $\psi$. Indeed, the inverse Fourier transform of the function $\hat{\psi}(\xi)/(2\pi i\xi)^k$ has these properties (we recall that $\hat{\psi}$ has a "zero of infinite order" at 0). The function $\beta(y) = d^k\psi/dy^k$ also has the same properties as $\psi$ (again, obviously, its Fourier transform has zero of infinite order at 0, so that all moments of the function itself vanish). Differentiating $k$ times in the above displayed formula and applying integration by parts $k$ times to the integrals involved in it (observe that our assumptions ensure an at most polynomial growth of all derivatives of $f$), we arrive at

$$(T^*f)^{(k)} = \sum_{j,m}(-1)^k t_{j,m} \int_{\mathbb{R}} f^{(k)}(x)\overline{\alpha_{j,m}(x)}dx\, \beta_{j,m}.$$

Our task is to estimate the $L^p$-norm of the right-hand side in terms of the $L^p$-norm of $f^{(k)}$. This is possible indeed, because here we deal with an operator as in Theorem 4.26, and we have seen that it is bounded on $L^p$.

### 4.5.3 An omitted proof

For simplicity, we verify the claim of Theorem 4.25 under the assumptions $\hat{\psi} \in \mathcal{D}(\mathbb{R})$ and $|\hat{\psi}(\xi)| \leq C|\xi|$. Observe that we have used that theorem only in this partial case.

So, we must prove that $\{\psi\}_{j,m}$ is a Hilbert system, that is,

$$\sum_{j,m} |\langle f, \psi_{j,m}\rangle|^2 \le C\|f\|^2_{L^2}, \tag{4.39}$$

where we may assume that $f \in \mathcal{D}(\mathbb{R})$. First, we transform the left-hand side. Namely, the Plancherel theorem shows that

$$\langle f, \psi_{j,m}\rangle = 2^{-j/2}\int_{\mathbb{R}} \hat{f}(\xi)e^{-2\pi i\xi\frac{m}{2^j}}\overline{\hat{\psi}(\frac{\xi}{2^j})}\, d\xi = 2^{j/2}\int_{\mathbb{R}} \hat{f}(2^j\eta)e^{-2\pi i m\eta}\overline{\hat{\psi}(\eta)}\, d\xi.$$

We rewrite the last integral as $\sum_l \int\limits_l^{l+1} \dots$ and, since the exponential in it is 1-periodic in $\eta$, observe that the integral is none other than the $m$th Fourier coefficient of the 1-periodic function

$$h_j(\eta) = \sum_{l\in\mathbb{Z}} \hat{f}(2^j(\eta+l))\overline{\hat{\psi}(\eta+l)}.$$

Now, the Parceval identity shows that

$$\sum_{j,m} |\langle f, \psi_{j,m}\rangle|^2 = \sum_j 2^j \int\limits_0^1 |h_j(\eta)|^2 d\eta,$$

and we must estimate the right-hand side in terms of the $L^2$-norm of $f$. By the Cauchy inequality,

$$|h_j(\eta)|^2 \le \sum_{l\in\mathbb{Z}} |\hat{f}(2^j(\eta+l))|^2|\hat{\psi}(\eta+l)| \sum_{l\in\mathbb{Z}} |\hat{\psi}(\eta+l)|.$$

The second factor on the right is uniformly bounded because $\hat{\psi}$ is compactly supported, so we arrive at

$$\begin{aligned}\sum_{j,m} |\langle f, \psi_{j,m}\rangle|^2 &\le C\sum_j 2^j \sum_l \int\limits_0^1 |\hat{f}(2^j(\eta+l))|^2|\hat{\psi}(\eta+l)|\, d\eta \\ &= \sum_j 2^j \int_{\mathbb{R}} |\hat{f}(2^j\eta)|^2|\hat{\psi}(\eta)|\, d\eta = \int_{\mathbb{R}} |\hat{f}(\xi)|^2 \sum_j |\hat{\psi}(2^{-j}\xi)|\, d\xi.\end{aligned}$$

Now, if we show that the sum $\alpha(\xi) = \sum_j |\hat{\psi}(2^{-j}\xi)|$ is uniformly bounded on $\mathbb{R}$, the required estimate (4.39) will follow by the Plancherel theorem. Since $\alpha(\xi)=\alpha(2^n\xi)$ for every $n$, it suffices to prove that $\alpha$ is bounded for $\xi \in [-2,-1]\cup[1,2]$. But for such $\xi$ only finitely many summands with $j$ negative in the definition of $\alpha$ may be nonzero, and their number can be taken independent of $\xi$. For positive $j$, we simply observe that $|\hat{\psi}(2^{-j}\xi)| \le C2^{-j}|\xi|$ by assumption, etc.

## 4.6 Weak $L^1$-boundedness

We recall inequality (3.45), which expresses weak $L^1$-boundedness: there is a natural number $k$ such that

$$\left|\int T(f)(s)\varphi(s)ds\right| \le C\,\|f\|_{L^1}\,(\|\varphi\|_{L^\infty} + \sum_{j\le k}(\operatorname{diam} Q)^j\,\|\nabla^j\varphi\|_{L^\infty})$$

whenever $\varphi$ is a $C^\infty$-function supported on a cube $Q$, and $f \in L^1$ is such that $Tf$ is integrable over $Q$. Surely, this is interesting only if $T$ is not bounded on $L^1$. We split the verification of this inequality in two steps. The first step is subtle and leans upon Corollary 1.21. Specifically, we assume that $T$ is long-range regular on $L^1$ and bounded on $L^p$ with some $p > 1$. By that corollary, then $\int \varphi T f = \int T^*(\varphi) f$ provided $T^*\varphi$ is a bounded function. At the second step, putting $S = T^*$, we simply verify the inequality

$$\|S\varphi\|_{L^\infty} \le C(\|\varphi\|_{L^\infty} + \sum_{j\le k}(\operatorname{diam} Q)^j\,\|\nabla^j\varphi\|_{L^\infty}) \tag{4.40}$$

for some natural number $k$, whenever $\varphi$ is a $C^\infty$-function supported on the cube $Q$. For this, some additional conditions should be imposed on $S$.

Suppose $S$ is the adjoint to a singular integral operator. Then inequality (4.40) is fulfilled fairly often and for crude reasons. Usually, we may take $k = 1$. For instance, if $S$ is a convolution operator (in the principal value sense) with homogeneous kernel of degree $-n$ and the kernel is smooth everywhere except the origin and has zero integral over the unit sphere (see Chapter 2), then for every $C^\infty$-function $\varphi$ supported on $Q$ we have

$$(S\varphi)(x) = \int_{|x-y|\le 2^{n/2}|Q|^{1/n}} (\varphi(y) - \varphi(x))\frac{\Omega(x-y)}{|x-y|^n}dy.$$

(The integrand is summable. As in Chapter 2, here we denote by $|x - y|$ the Euclidean distance between $x$ and $y$.) It follows that

$$|(S\varphi)(x)| \le C\|\nabla\varphi\|_{L^\infty} \int_{|x-y|\le 2^{n/2}|Q|^{1/n}} |x-y|^{-n+1}dy,$$

which clearly implies (4.40) with $k = 1$.

More generally, (4.40) can be proved with $k = 1$ if $S$ takes $L^p$ to itself and $\mathrm{Lip}_\alpha$ to itself boundedly for some $p \in (1, \infty)$ and some $\alpha \in (0, 1)$. (We have discussed this type of behavior in this chapter.) We present the argument. Let $\tau$ be a homothety on $\mathbb{R}^n$, and let $\rho$ be the corresponding operator on functions: $\rho g(y) = g(\tau y)$. Clearly, the operator $h \mapsto \rho^{-1}S\rho$ has the same norms as $S$ both on $L^p$ and on $\mathrm{Lip}_\alpha$. Therefore, when proving (4.40), we may assume that the edge length of $Q$ is 1. Clearly, we may also assume that the $L^\infty$-norms of $\varphi$ and its first derivatives do not exceed 1. Then $|\varphi(s) - \varphi(t)| \leq \|s - t\| \leq \|s - t\|^\alpha$ for $s, t \in Q$. Since $\varphi$ is zero outside $Q$, we see that $\|\varphi\|_{\mathrm{Lip}_\alpha} \leq 1$. Also, $\|\varphi\|_{L^p} \leq 1$. It follows that $\|S\varphi\|_{\mathrm{Lip}_\alpha} \leq C$ and $\|S\varphi\|_{L^p} \leq C$. The first of these inequalities shows that if $|(S\varphi)(y)| > D$ for some $y$, then $|(S\varphi)(x)| > D/2$ whenever $\|x - y\| \leq \left(\frac{D}{2C}\right)^{1/\alpha}$. By the second inequality, this shows that $D$ must admit an upper estimate in terms of $C$, and we are done.

However, in principle there is no need to aim at small values of $k$ (the proofs of Theorem 3.21 and the stability theorems in Chapter 10 run equally well with any $k$). Estimate (4.40) becomes weaker as $k$ grows, and, indeed, its verification with large $k$ may be less involved. As an illustration, we prove the following statement.

*Let $m$ be a bounded measurable function on $\mathbb{R}^n$, and let $S$ be the corresponding Fourier multiplier: $\widehat{Sf} = m\hat{f}$ (say, for $f \in L^2$). Then $S$ satisfies* (4.40) *with $k = n + 1$.*

*Proof.* We have $\int |\varphi| \leq |Q|\, \|\varphi\|_{L^\infty}$ and $\int |(\nabla)^{n+1}\varphi| \leq |Q|\, \|(\nabla)^{n+1}\varphi\|_{L^\infty}$. Hence, $|\hat{\varphi}(\xi)|$ is dominated by $C|Q|\, \|\varphi\|_{L^\infty}$ and by $C|Q|\, \|(\nabla)^{n+1}\varphi\|_{L^\infty}|\xi|^{-n-1}$. This shows that

$$\int |\widehat{S\varphi}| \leq C(\|\varphi\|_{L^\infty} + (\operatorname{diam} Q)^{n+1}\|(\nabla)^{n+1}\varphi\|_{L^\infty})$$

(split the integral into two, over the domains $|\xi| < (\operatorname{diam} Q)^{-1}$ and $|\xi| \geq (\operatorname{diam} Q)^{-1}$, and estimate the integrand by one of the above expressions in each domain). After passage to the inverse Fourier transform, we prove the claim. □

# Notes and remarks

For us, homogeneous spaces of smooth functions present "raw material" for problems about near-minimizers, and, as has been claimed repeatedly, the boundedness of singular integrals on them is an important ingredient of stability theorems.

The minimal information we need about spaces of smooth functions can be classified as "well-known but difficult to find".[4] We were unable to cover the required facts with a *small* number of *sharply focused* references, and decided to include proofs. Calculations in the spirit of the proof of Proposition 4.5 can be found in [DeVSh] and in [ST].

Morrey spaces were introduced in [Mo] in connection with certain problems in elliptic PDE and calculus of variations. Subsequently, they found many important applications, see, e.g., [Ta], [Sh], [A]. Campanato spaces with positive smoothness date back to [C], and the space BMO to [JN].

The modern theory of Hardy spaces was started by the paper [FS1] by Fefferman and Stein. We say only that the atomic Hardy class introduced in §3 of Chapter 4 is closely related to the analytic Hardy spaces mentioned in Chapter 7. See, e.g., the book [GR].

[4]Especially, this is true for homogeneous spaces.

From the point of view developed in a relatively recent time, the boundedness on a Lipschitz space is a *more basic* property of a singular integral operator then the $L^p$-boundedness. See, e.g., [Me]. To the contrary, in our presentation in §4 of Chapter 4, $L^p$-boundedness for some $p$ is treated as an initial "axiom", from which continuity on Morrey-Campanato spaces is deduced. This way is quicker and still provides sufficiently many examples, so we have followed it. We do not know whether the pointwise inequality of Theorem 4.21 was mentioned in the literature earlier for the entire range of indices, but it has been well known at least for $s = 0$, $k = 1$, and $i = 0$ (a standard method of obtaining weighted $L^p$-inequalities for singular integrals is based on estimates of this type, see, e.g., [GR]). The case of general indices is verified with the help of the same ideas as in that well-known particular situation.

The results of §5 in Chapter 4 imply a case of the following general principle: a wavelet system is an unconditional basis in any "reasonable" function space possessing *some* unconditional basis. A minimal condition ensuring this type of behavior in $L^p$ was found by Wojtaszczyk. It requires only a moderate decay of the generating wavelet, but does not involve smoothness assumptions. See [Wo] and also the exposition in the monograph [NPS] (Theorem 11.1). Strictly speaking, under this condition partial sums may fail to be singular integral operators (at least in the sense we understand this term here). For assumptions close to minimal under which they still remain singular integrals, see [D].

# Chapter 5

# Some topics in interpolation

In this chapter we summarize some information from interpolation theory. In distinction to the preceding chapters, here we mainly avoid giving proofs of general results. Good references for that stuff are [BL], [BK], [BSh1], and [KPS]. To the contrary, we do give detailed proofs of many statements directly related to near-minimizers. It should be noted that, combined with explicit constructions of near-minimizers and with stability theorems presented in this book, these results give additional interesting information.

Inevitably, this chapter (like the preceding one) contains more technicalities and abstract considerations than the first three chapters.

## 5.1 Main notions

Let two (quasi-)Banach spaces $X_0$ and $X_1$ form a (quasi-)Banach couple.[1] This means that both are embedded linearly in some "big" linear space $\mathcal{X}$ in such a way that if a sequence $u_n \in X_0 \cap X_1$, $n = 1, 2, \ldots$, converges in $X_0$ and $X_1$, respectively, to elements $x_0$ and $x_1$, then $x_0 = x_1$.

It is well known that the initial objective of interpolation theory is to make it possible to conclude that an operator acts between some "complicated" spaces (so-called interpolation spaces) if its boundedness in some "simple" cases is given for granted. Real interpolation emerged once as a method to exhibit certain interpolation spaces between given $X_0$ and $X_1$. These interpolation spaces are defined in terms of careful measurement of the "quality" of splitting a vector $x \in X_0 + X_1$ in a sum of two parts (one in $X_0$ and the other in $X_1$). The most convenient tool for such measurement is the Peetre $K$-functional. We repeat its definition.

Suppose $x \in X_0 + X_1$ (the space $X_0 + X_1$ is well defined because $X_0$ and $X_1$ are embedded in $X$). For $t > 0$, the $K$-functional $K(t, x; X_0, X_1)$ is defined by the

[1] A quasi-Banach space is like $L^p$ with $p < 1$, i.e., this is a complete space with a homogeneous quasinorm.

formula

$$K(t, x; X_0, X_1) = \inf_{x_1 \in X_1} (\|x - x_1\|_{X_0} + t\, \|x_1\|_{X_1}). \tag{5.1}$$

It should be noted that, in the case of Banach spaces, for $t > 0$ fixed, the functional $K(t, \cdot; X_0, X_1)$ is a norm on $X_0 + X_1$, whereas for $x \in X_0 + X_1$ fixed, the function $K(\cdot, x; X_0, X_1)$ is nonnegative and concave on $\mathbb{R}_+ = (0, \infty)$.

The definition implies that the $K$-functional can be expressed in terms of the so-called $E$-functional, whose geometric meaning is slightly clearer. Specifically,

$$K(t, x; X_0, X_1) = \inf_{s>0} \, (E(s, x; X_0, X_1) + ts), \tag{5.2}$$

where

$$E(s, x; X_0, X_1) = \inf_{\|x_1\|_{X_1} \le s} \|x - x_1\|_{X_0} \,. \tag{5.3}$$

Since the $E$-functional is merely the distance in $X_0$ from $x$ to the $s$-ball $B_{X_1}(s)$ of $X_1$ centered at zero, we often use the more suggestive notation

$$E(s, x; X_0, X_1) = \mathrm{dist}_{X_0}(x, B_{X_1}(s)),$$

and the more suggestive term "the distance functional".

The $K$-functional possesses a useful and nontrivial property of $K$-divisibility to be described below. Suppose we are given a Banach couple $(X_0, X_1)$ and an element $x \in X_0 + X_1$. Suppose also we are given a family of nonnegative concave functions $\varphi_i$ on $\mathbb{R}_+$ such that

$$K(\cdot, x; X_0, X_1) \le \sum_{i=1}^{\infty} \varphi_i.$$

Then the following is true, see [BK].

**Theorem 5.1.** *If $\sum_{i=1}^{\infty} \varphi_i(1) < \infty$, then there exist elements $x_i \in X_0 + X_1$ such that $x = \sum_{i=1}^{\infty} x_i$ (convergence in $X_0 + X_1$) and*

$$K(\cdot, x_i; X_0, X_1) \le \gamma \varphi_i$$

*for all $i$, where $\gamma > 0$ is a universal constant.*

The sharp value of $\gamma$ is unknown. It is known only that $3 + 2\sqrt{2} + \varepsilon$ can be taken for $\gamma$, where $\varepsilon$ is an arbitrary positive constant.

We present the most well-known cases in which the $K$-functional is calculated explicitly.

(a) The couple $(L^1, L^\infty)$. Two formulas are known in this case. One of them is in terms of the nonincreasing rearrangement[2] of $f$:

$$K(t, f; L^1, L^\infty) = \int_0^t f^*(s)ds, \tag{5.4}$$

[2] This is a monotone decreasing function on $[0, \infty)$ equimeasurable with $f$; the standard notation is $f^*$.

and another one is approximate, in terms of the Hardy–Littlewood maximal function:

$$K(t, f; L^1, L^\infty) \approx t(Mf)^*(t). \tag{5.5}$$

(b) The couple $(L^1, \mathrm{BMO})$. Here the $K$-functional can be calculated approximately by a formula similar to (5.5), specifically

$$K(t, f; L^1, L^\infty) \approx t(f^\#)^*(t), \text{ where } f^\#(x) = \sup_{Q \ni x} \left( \inf_c \frac{1}{|Q|} \int_Q |f(x) - c|\, dx \right). \tag{5.6}$$

(Note that $f^\#$ coincides with $M_{1,1,0}f$, see (4.4).)

(c) The couple $(L^p, \dot{W}_p^k)$ , $1 < p < \infty$. In this case there is a formula in terms of the modulus of smoothness $\omega_k(t, f)_p$:

$$K(t, f; L^p, \dot{W}_p^k) \approx \omega_k(t^{\frac{1}{k}}, f)_p. \tag{5.7}$$

The modulus of smoothness (for $k = 1$ the usual term is "the modulus of continuity") can be most conveniently written in the case where $k = 1$:

$$\omega_1(t, f)_p = \sup_{|h| \le t} \left( \int_{\mathbb{R}^n} |f(x+h) - f(x)|^p \, dx \right)^{\frac{1}{p}}.$$

For general $k$, the $k$th difference is involved in the definition. The proofs of all these formulas (except (5.4)) are nontrivial. As to (5.4), this formula is deduced from (5.2) and the fact that the $E$-functional for this couple is obvious:

$$E(t, f; L^1, L^\infty) = \int_0^\infty \max(f^*(s) - t, 0) ds .$$

As has already been mentioned, the $K$-functional is employed to introduce the interpolation spaces $(X_0, X_1)_{\theta,q}$, $\theta \in (0,1), 1 \le q \le \infty$, of the real method. These spaces are determined by the following norms:

$$\|x\|_{\theta,q} = \left( \int_0^\infty (t^{-\theta} K(t, x; X_0, X_1))^q \frac{dt}{t} \right)^{\frac{1}{q}}, \tag{5.8}$$

with natural modifications for $q = \infty$.

If $T : X_0 + X_1 \to Y_0 + Y_1$ is a linear operator that takes boundedly $X_i$ to $Y_i$ $(i = 0, 1)$, then, clearly,

$$K(t, Tx; Y_0, Y_1) \le c(T) K(t, x; X_0, X_1)$$

and directly from the definitions we deduce that

$$\|Tx\|_{(Y_0,Y_1)_{\theta,q}} \le c(T) \, \|x\|_{(X_0,X_1)_{\theta,q}} .$$

Thus, $T$ maps $(X_0, X_1)_{\theta,q}$ to $(Y_0, Y_1)_{\theta,q}$ boundedly.

For many specific couples $X_0, X_1$, the spaces $(X_0, X_1)_{\theta,q}$ have been calculated. For instance, if $X_0 = L^p$ and $X_1 = \dot{W}_p^k$, then formula (5.7) leads to the Besov space $\dot{B}_p^{s,q}$:

$$(L^p, \dot{W}_p^k)_{\theta,q} = \dot{B}_p^{s,q} \text{ where } s = \theta k,\ 1 < p < \infty.$$

Next, formula (5.4) and the Hardy inequality show that $(L^1, L^\infty)_{\theta,q}$ is the Lorentz space $L^{p,q}$ if $\theta = 1 - \frac{1}{p}$. For $q = p$, the space $L^{p,q}$ coincides with $L^p$, therefore,

$$\|f\|_{L^p} \approx \left( \int_0^\infty (t^{-(1-\frac{1}{p})} K(t, f; L^1, L^\infty))^p \frac{dt}{t} \right)^{\frac{1}{p}}.$$

By the way, if we replace $K(t, f; L^1, L^\infty)$ on the right with the expression given by (5.5), we obtain the Hardy–Littlewood maximal theorem:

$$\|f\|_{L^p} \approx \left( \int_0^\infty (Mf)^*(t)^p dt \right)^{\frac{1}{p}} = \|Mf\|_{L^p}.$$

These examples show that the $K$-functional carries nontrivial information.

*Remark* 5.2. The spaces $X_{\theta,q} = (X_0, X_1)_{\theta,q}$ (and more general spaces $K_\Phi(X)$ of the real method, see [BK]) possess a series of nontrivial and useful properties. For example, they are stable under iteration of the construction (the so-called reiteration theorem):

$$(X_{\theta_0,q_0}, X_{\theta_1,q_1})_{\lambda,q} = X_{(1-\lambda)\theta_0+\lambda\theta_1,q}, \quad \theta_0 \neq \theta_1.$$

Also, the construction is compatible with duality (the duality theorem): if $X_0 \cap X_1$ is dense both in $X_0$ and in $X_1$, then

$$[(X_0, X_1)_{\theta,p}]' = (X_0', X_1')_{\theta,q} \quad \text{for } 1 \leq p < \infty \text{ and } \frac{1}{p} + \frac{1}{q} = 1.$$

If the restrictions of an operator $T : X_0 + X_1 \to Y_0 + Y_1$ to $X_i$ $(i = 0, 1)$ possess some additional properties, then usually $T$ has the same properties when viewed as an operator on $(X_0, X_1)_{\theta,q}$. For example, compactness is interpolated in this way, see [Cw] and [CFM].

In recent years, the $K$-functional arose in questions of applied origin, for example, in image processing. In this context, the second summand in the definition (5.1) is a "penalty term", and the problem is in finding a minimizer; this will be discussed in Chapter 6.

Now, we repeat a definition already given in this book at least twice.

**Definition 5.3.** A vector $x_t \in X_1$ is called a near-minimizer for the $K$-functional of an element $x \in X_0 + X_1$ if

$$\|x - x_t\|_{X_0} + t\,\|x_t\|_{X_1} \leq cK(t, x; X_0, X_1) \tag{5.9}$$

with a constant $c$ independent of $x$ and $t$. If $c = 1$, we talk about an exact minimizer.

## 5.2 Near-minimizers and interpolation

It turns out that near-minimizers for the couple $(X_0, (X_0, X_1)_{\theta,q})$ can easily be exhibited provided those for the couple $(X_0, X_1)$ are known. To illustrate the use of this observation, we recall that in this book we shall deal with near-minimizers in the case of the couple $(L^p, \dot{W}^k_q)$. If we know them, interpolation gives near-minimizers for the couple $(L^p, \dot{B}^s_r)$, where $\frac{1}{r} = \frac{1-\theta}{p} + \frac{\theta}{q}$, $s = \theta k$, and $\dot{B}^{\theta k}_r$ is the homogeneous Besov space of smoothness $\theta k$ built on the basis of $L^r$.

**Theorem 5.4.** *Suppose $x_t$ is a near-minimizer for the $K$-functional of an element $x$ relative to the couple $(X_0, X_1)$ at the point $t$, that is,*

$$\|x - x_t\|_{X_0} + t\,\|x_t\|_{X_1} \le cK(t, x; X_0, X_1).$$

*If $x \in X_0 + (X_0, X_1)_{\theta,q}$, then*

$$\|x - x_t\|_{X_0} + t^\theta\,\|x_t\|_{(X_0,X_1)_{\theta,q}} \le cK(t^\theta, x; X_0, (X_0, X_1)_{\theta,q}), \tag{5.10}$$

*that is, $x_t$ is a near-minimizer for the $K$-functional of $x$ relative to the couple $(X_0, (X_0, X_1)_{\theta,q})$ at the point $t^\theta$.*

For the reader's convenience, we include a complete proof of this statement, though, in fact, the arguments are much the same as in the proof of the Holmstedt formula (an important statement well-known in interpolation theory; see [BL]). We have already mentioned a remark in the same spirit in [JRW], where detailed calculations were omitted.

*Proof.* Suppose $x \in X_0 + (X_0, X_1)_{\theta,q}$ and consider a decomposition $x = x_0 + x_1$ at which $K(t^\theta, x; X_0, X_{\theta,q})$ is nearly attained. The inequality

$$K(s, x; X_0, X_1) \le cK(s, x_0; X_0, X_1) + cK(s, x_1; X_0, X_1)$$

($c$ can be greater than 1 in the case of quasi-Banach spaces) implies the following estimate:

$$\begin{aligned}
&t^\theta \left( \int_t^\infty \left(s^{-\theta} K(s, x; X_0, X_1)\right)^q \frac{ds}{s} \right)^{\frac{1}{q}} \\
&\le ct^\theta \left( \int_t^\infty \left(s^{-\theta} K(s, x_0; X_0, X_1)\right)^q \frac{ds}{s} \right)^{\frac{1}{q}} + ct^\theta \left( \int_t^\infty \left(s^{-\theta} K(s, x_1; X_0, X_1)\right)^q \frac{ds}{s} \right)^{\frac{1}{q}} \\
&\le c(\|x_0\|_{X_0} + t^\theta\,\|x_1\|_{X_{\theta,q}}) \le cK(t^\theta, x; X_0, X_{\theta,q})
\end{aligned}$$

(we have used the fact that $K(s, x_0; X_0, X_1) \le \|x_0\|_{X_0}$). So, it suffices to prove that

$$\|x - x_t\|_{X_0} + t^\theta\,\|x_t\|_{(X_0,X_1)_{\theta,q}} \le ct^\theta \left( \int_t^\infty \left(s^{-\theta} K(s, x; X_0, X_1)\right)^q \frac{ds}{s} \right)^{\frac{1}{q}}. \tag{5.11}$$

Since a $K$-functional is a monotone nondecreasing function of $t$, by (5.9) we see that

$$\|x - x_t\|_{X_0} \le cK(t, x; X_0, X_1) \le ct^\theta \left( \int_t^\infty \left(s^{-\theta} K(s, x; X_0, X_1\right)^q \frac{ds}{s} \right)^{\frac{1}{q}}. \quad (5.12)$$

Now, to prove (5.11), it remains to establish the inequality

$$t^\theta \|x_t\|_{(X_0,X_1)_{\theta,q}} \le ct^\theta \left( \int_t^\infty \left(s^{-\theta} K(s, x; X_0, X_1\right)^q \frac{ds}{s} \right)^{\frac{1}{q}}. \quad (5.13)$$

We observe that

$$t^\theta \|x_t\|_{(X_0,X_1)_{\theta,q}} = t^\theta \left( \int_0^\infty \left(s^{-\theta} K(s, x_t; X_0, X_1\right)^q \frac{ds}{s} \right)^{\frac{1}{q}}$$
$$\le ct^\theta \left( \int_0^t \left(s^{-\theta} K(s, x_t; X_0, X_1\right)^q \frac{ds}{s} \right)^{\frac{1}{q}} + ct^\theta \left( \int_t^\infty \left(s^{-\theta} K(s, x_t; X_0, X_1\right)^q \frac{ds}{s} \right)^{\frac{1}{q}}. \quad (5.14)$$

The first integral on the right is handled easily:

$$t^\theta \left( \int_0^t \left(s^{-\theta} K(s, x_t; X_0, X_1\right)^q \frac{ds}{s} \right)^{\frac{1}{q}} \le t^\theta \left( \int_0^t \left(s^{-\theta} s \|x_t\|_{X_1}\right)^q \frac{ds}{s} \right)^{\frac{1}{q}} \le ct \|x_t\|_{X_1}$$
$$\le cK(t, x; X_0, X_1) \le ct^\theta \left( \int_t^\infty \left(s^{-\theta} K(s, x; X_0, X_1\right)^q \frac{ds}{s} \right)^{\frac{1}{q}}.$$

To estimate the second integral, we employ (5.12):

$$\begin{aligned} & t^\theta \left( \int_t^\infty \left(s^{-\theta} K(s, x_t; X_0, X_1\right)^q \frac{ds}{s} \right)^{\frac{1}{q}} \\ & \le ct^\theta \left( \int_t^\infty \left(s^{-\theta} K(s, x; X_0, X_1\right)^q \frac{ds}{s} \right)^{\frac{1}{q}} \\ & \quad + ct^\theta \left( \int_t^\infty \left(s^{-\theta} K(s, x - x_t; X_0, X_1\right)^q \frac{ds}{s} \right)^{\frac{1}{q}} \\ & \le ct^\theta \left( \int_t^\infty \left(s^{-\theta} K(s, x; X_0, X_1\right)^q \frac{ds}{s} \right)^{\frac{1}{q}} + c\|x - x_t\|_{X_0} \\ & \le ct^\theta \left( \int_t^\infty \left(s^{-\theta} K(s, x; X_0, X_1\right)^q \frac{ds}{s} \right)^{\frac{1}{q}}. \end{aligned}$$

This proves (5.13) and, with it, the theorem. □

## 5.3 Near-minimizers for $L_{p,q}$- and $K$-functionals

As in [Pe], for fixed positive numbers $p$ and $q$, we define the $L_{p,q}$-functional by the formula

$$L_{p,q}(s,x;X_0,X_1) = \inf_{x_1\in X_1}(\|x-x_1\|_{X_0}^p + s\,\|x_1\|_{X_1}^q). \tag{5.15}$$

This is a natural generalization of the $K$-functional, coinciding with it if $p=q=1$.

**Definition 5.5.** A vector $x_{s,L}\in X_1$ is called a near-minimizer for the $L_{p,q}$-functional of an element $x\in X_0+X_1$ at the point $s>0$ if

$$\|x-x_{s,L}\|_{X_0}^p + s\,\|x_{s,L}\|_{X_1}^q \le cL_{p,q}(s,x;X_0,X_1). \tag{5.16}$$

Again, we talk about an exact minimizer if $c=1$.

Often, instead of the $K$-functional, it is convenient to calculate the $L_{p,q}$-functional and find near-minimizers for it. The question arises as to whether it is possible to find near-minimizers for the $K$-functional on this basis. It turns out that the answer is in the positive.

First, we establish an inequality relating $L_{a,b}$-functionals for various $a$ and $b$ to one another. To formulate it, we fix two couples of indices $(a,b)$ and $(p,q)$, and introduce the function

$$\varphi(t) = t^{\frac{q}{b}}L_{(a,b)}(t,x;X_0,X_1)^{\frac{p}{a}-\frac{q}{b}} = \left(\frac{t}{L_{(a,b)}(t,x;X_0,X_1)}\right)^{\frac{q}{b}} L_{(a,b)}(t,x;X_0,X_1)^{\frac{p}{a}}. \tag{5.17}$$

Basically, the following proposition is due to Peetre. See [Pe], where it was stated in a somewhat different form.

**Proposition 5.6.** *The function $\varphi(t)$ takes bijectively $\mathbb{R}_+$ onto itself, and*

$$2^{-\max(\frac{p}{a},\frac{q}{b})}L_{(a,b)}(t,x;X_0,X_1)^{\frac{p}{a}} \le L_{p,q}(\varphi(t),x;X_0,X_1) \le 2L_{(a,b)}(t,x;X_0,X_1)^{\frac{p}{a}}. \tag{5.18}$$

*Proof.* Since the functions

$$L_{(a,b)}(t,x;X_0,X_1) \quad\text{and}\quad \frac{t}{L_{(a,b)}(t,x;X_0,X_1)}$$

are continuous and monotone nondecreasing, $\varphi(t)$ is also monotone nondecreasing. If $t>s$ and $L_{(a,b)}(t,x;X_0,X_1) = L_{(a,b)}(s,x;X_0,X_1)$, then

$$\frac{t}{L_{(a,b)}(t,x;X_0,X_1)} > \frac{s}{L_{(a,b)}(s,x;X_0,X_1)};$$

consequently, $\varphi(t)$ is strictly monotone increasing. We show that

$$\lim_{t\to 0}\varphi(t)=0 \quad\text{and}\quad \lim_{t\to\infty}\varphi(t)=\infty.$$

Indeed, if
$$\lim_{t\to 0} L_{(a,b)}(t,x;X_0,X_1) > 0,$$
then
$$\lim_{t\to 0} \frac{t}{L_{(a,b)}(t,x;X_0,X_1)} = 0$$
and therefore, $\lim_{t\to 0}\varphi(t) = 0$. Similarly, if
$$\lim_{t\to\infty} L_{(a,b)}(t,x;X_0,X_1) < \infty,$$
then
$$\lim_{t\to\infty} \frac{t}{L_{(a,b)}(t,x;X_0,X_1)} = \infty$$
and therefore, $\lim_{t\to\infty}\varphi(t) = \infty$. Thus, $\varphi$ maps $\mathbb{R}_+$ onto itself bijectively (so, the inverse function $\varphi^{-1}$ also maps $\mathbb{R}_+$ onto itself bijectively). Now we prove (5.18). We write
$$L_{p,q}(\varphi(t),x;X_0,X_1) = \inf_{u\in X_1}\left(\|x-u\|_{X_0}^p + t^{q/b}L_{(a,b)}(t,x;X_0,X_1)^{p/a-q/b}\|u\|_{X_1}^q\right)$$
$$= L_{(a,b)}(t,x;X_0,X_1)^{p/a}\inf_{u\in X_1}\left(\frac{\|x-u\|_{X_0}^p}{L_{(a,b)}(t,x;X_0,X_1)^{p/a}} + \frac{t^{q/b}\|u\|_{X_1}^q}{L_{(a,b)}(t,x;X_0,X_1)^{q/b}}\right). \tag{5.19}$$

Choosing $u \in X_1$ to be an element at which $L_{(a,b)}(t,x;X_0,X_1)$ is almost attained, we arrive at
$$L_{p,q}(\varphi(t),x;X_0,X_1) \le 2L_{(a,b)}(t,x;X_0,X_1)^{p/a}.$$
Next, by the definition of the $L_{(a,b)}$-functional, for every $u \in X_1$ we have
$$\max\left(\frac{\|x-u\|_{X_0}^a}{L_{(a,b)}(t,x;X_0,X_1)}, \frac{t\,\|u\|_{X_1}^b}{L_{(a,b)}(t,x;X_0,X_1)}\right) \ge \frac{1}{2}.$$

So, (5.19) implies
$$L_{p,q}(\varphi(t),x;X_0,X_1) \ge 2^{-\max\left(\frac{p}{a},\frac{q}{b}\right)}L_{(a,b)}(t,x;X_0,X_1)^{\frac{p}{a}}.$$ □

Now, suppose we know the $L_{a,b}$-functional for some couple of spaces. Putting $p = q = 1$ in the above proposition and invoking the definition (5.8), we get a formula for interpolation spaces entirely in terms of $L_{a,b}$ (it should be emphasized that the function $\varphi$ is also defined in terms of $L_{a,b}$ in the setting chosen). See [BK], pp. 519-522, where a formula of this sort (due to Peetre) is written out explicitly. Though calculation of interpolation spaces by such formulas may be bulky, sometimes there are no better devices (this happens already for couples of weighted $L^p$-spaces). See, for instance, [BK], pp. 447-451.

Next, if we know near-minimizers for $L_{a,b}$, it is easy to realize that we know them for the $K$-functional. We give a precise statement.

**Proposition 5.7.** *Put $a = b = 1$ in Proposition 5.6 and define $\varphi$ accordingly, i.e.,*

$$\varphi(t) = t^q K(t, x; X_0, X_1)^{p-q},$$

*see (5.17). A vector $x_t$ is a near-minimizer for the $K$-functional of an element $x$ at the point $t$ if and only if it is a near-minimizer for the $L_{p,q}$-functional of $x$ at the point $\varphi(t)$.*

*Proof.* If $x_t$ is a near-minimizer for the $K$-functional of $x$ at $t$, then Proposition 5.6 shows that

$$\begin{aligned}
&\|x - x_t\|_{X_0}^p + \varphi(t)\,\|x_t\|_{X_1}^q \\
&= K(t, x; X_0, X_1)^p \left( \frac{\|x - x_t\|_{X_0}^p}{K(t, x; X_0, X_1)^p} + \frac{t^q\,\|x_t\|_{X_1}^q}{K(t, x; X_0, X_1)^q} \right) \\
&\le (c^p + c^q) K(t, x; X_0, X_1)^p \le 2^{\max(p,q)} (c^p + c^q) L_{p,q}(\varphi(t), x; X_0, X_1).
\end{aligned}$$

Thus, $x_t$ is a near-minimizer of the $L_{p,q}$-functional at $\varphi(t)$. Conversely, if $u$ is a near-minimizer for the $L_{p,q}$-functional at $\varphi(t)$, then, again by Proposition 5.6,

$$\|x - u\|_{X_0}^p + \varphi(t)\,\|u\|_{X_1}^q \le c L_{p,q}(\varphi(t), x; X_0, X_1) \le 2cK(t, x; X_0, X_1)^p.$$

Therefore,

$$\|x - u\|_{X_0}^p \le 2cK(t, x; X_0, X_1)^p$$

and

$$\varphi(t)\,\|u\|_{X_1}^q = t^q K(t, x; X_0, X_1)^{p-q}\,\|u\|_{X_1}^q \le 2cK(t, x; X_0, X_1)^p.$$

These two estimates imply the inequality

$$\|x - u\|_{X_0} + t\,\|u\|_{X_1} \le ((2c)^{\frac{1}{p}} + (2c)^{\frac{1}{q}}) K(t, x; X_0, X_1),$$

which means that $u$ is a near-minimizer for the $K$-functional of $x$ at $t$. □

The complicated and nonlinear nature of the transformation from "$K$" to "$L$" and back hints why an $L_{p,q}$-functional "adjusted" to a couple under study may happen to be handier than the $K$-functional.

## 5.4 Near-minimizers for $E$- and $K$-functionals

For certain couples, instead of the $K$-functional, it turns out to be easier to calculate the distance

$$\operatorname{dist}_{X_0}(x, B_{X_1}(t)) = \inf_{u \in B_{X_1}(t)} \|x - u\|_{X_0}. \tag{5.20}$$

Here $B_{X_1}(t)$ stands for the ball of radius $t > 0$ and centered at zero in $X_1$:

$$B_{X_1}(t) = \{x \in X_1 : \|x\|_{X_1} \le t\}.$$

This is the distance functional familiar to us from the Preface, but in interpolation theory it is usual to call it the $E$-functional and to denote it by $E(t, x; X_1, X_0)$ (by the way, we have already said all this before; also, usually the spaces are indicated in the reverse order in this notation, but we prefer the way it is written above). Clearly, this is a nonnegative and monotone nonincreasing function on $\mathbb{R}_+$, which can attain the value $+\infty$.

The distance functional and the $K$-functional are closely related. In particular, the following identity (already mentioned) is a direct consequence of the definitions:

$$K(t, x; X_0, X_1) = \inf_{s>0} (\operatorname{dist}_{X_0}(x, B_{X_1}(s)) + ts). \tag{5.21}$$

In the case of Banach spaces, there is also a reverse formula:

$$\operatorname{dist}_{X_0}(x, B_{X_1}(s)) = \sup_{t>0} (K(t, x; X_0, X_1) - ts).$$

Now, again, we recall a definition given before.

**Definition 5.8.** Suppose $t > 0$ is such that $\operatorname{dist}_{X_0}(x, B_{X_1}(t)) > 0$. An element $x_{t,E} \in X_1$ is called a near-minimizer for the distance functional of $x \in X_0 + X_1$ if

$$\|x_{t,E}\|_{X_1} \le ct \text{ and } \|x - x_{t,E}\|_{X_0} \le c \operatorname{dist}_{X_0}\left(x, B_{X_1}\left(\frac{t}{c}\right)\right). \tag{5.22}$$

To show the exact relationship between near-minimizers for the $E$- and $K$-functionals, we need the following statement.

**Proposition 5.9.** *Put*

$$s = \frac{K(t, x; X_0, X_1)}{t}.$$

*Then*

$$\operatorname{dist}_{X_0}(x, B_{X_1}(2s)) \le K(t, x; X_0, X_1) \le 2 \operatorname{dist}_{X_0}\left(x, B_{X_1}\left(\frac{s}{2}\right)\right). \tag{5.23}$$

*Proof.* Let $u_\varepsilon \in X_1$ be an element at which $K(t, x; X_0, X_1)$ is attained roughly, i.e.,

$$\|x - u_\varepsilon\|_{X_0} + t\,\|u_\varepsilon\|_{X_1} \le K(t, x; X_0, X_1) + \varepsilon.$$

So, for a sufficiently small $\varepsilon > 0$ we have

$$\|u_\varepsilon\|_{X_1} \le 2\frac{K(t, x; X_0, X_1)}{t}.$$

Therefore,

$$\operatorname{dist}_{X_0}(x, B_{X_1}(2s)) \le \|x - u_\varepsilon\|_{X_0} \le K(t, x; X_0, X_1) + \varepsilon.$$

Letting $\varepsilon$ tend to zero, we obtain the left inequality in (5.23). In order to prove the right inequality, we observe that, for arbitrary $u$ satisfying

$$\|u\|_{X_1} \le \frac{s}{2} = \frac{K(t,x;X_0,X_1)}{2t},$$

the estimate

$$K(t,x;X_0,X_1) \le \|x-u\|_{X_0} + t\,\|u\|_{X_1}$$

implies

$$\|x-u\|_{X_0} \ge K(t,x;X_0,X_1) - t\,\|u\|_{X_1} \ge \frac{1}{2}K(t,x;X_0,X_1).$$

Since $u$ is an arbitrary element of $B_{X_1}(\frac{s}{2})$, we arrive at

$$\operatorname{dist}_{X_0}\left(x, B_{X_1}\left(\frac{s}{2}\right)\right) \ge \frac{1}{2}K(t,x;X_0,X_1),$$

which is the right inequality in (5.23). □

It should be noted that the transformation that takes near-minimizers of the functional $\operatorname{dist}_{X_0}(x, B_{X_1}(t))$ to near-minimizers for the $K$-functional of $x$ depends on constants involved in the definition of near-minimizers. We state this in precise terms.

**Proposition 5.10.** (a) *Suppose we are given a family $u_t$ of near-minimizers for the distance functional* $\operatorname{dist}_{X_0}(x, B_{X_1}(t))$ *with a constant $\gamma > 1$, that is*

$$\|u_t\|_{X_1} \le \gamma t \text{ and } \|x-u_t\|_{X_0} \le \gamma \operatorname{dist}_{X_0}\left(x, B_{X_1}\left(\frac{t}{\gamma}\right)\right).$$

*Consider the transformation*

$$\varphi(t) = \frac{2\gamma K(t,x;X_0,X_1)}{t}.$$

*Then the element $u_{\varphi(t)}$ is a near-minimizer for the $K$-functional of $x$ at $t$.*

(b) *Conversely, if $x_t$ a near-minimizer for the $K$-functional of $x$ at $t$ with constant $c_*$, i.e.,*

$$\|x-x_t\|_{X_0} + t\,\|x_t\|_{X_1} \le c_* K(t,x;X_0,X_1), \tag{5.24}$$

*then $x_t$ is a near-minimizer for the functional* $\operatorname{dist}_{X_0}(x, B_{X_1}(\varphi(t)))$ *with $\gamma = 2c_*$, where*

$$\varphi(t) = \frac{c_* K(t,x;X_0,X_1)}{t}.$$

*Proof.* (a) By Proposition 5.9, we have

$$\begin{aligned}
&\left\|x - u_{\varphi(t)}\right\|_{X_0} + t\left\|u_{\varphi(t)}\right\|_{X_1} \leq \gamma\operatorname{dist}_{X_0}(x, B_{X_1}(\varphi(t)/\gamma)) + t\gamma\varphi(t))\\
&\leq \gamma\operatorname{dist}_{X_0}\left(x, B_{X_1}\left(\frac{2K(t,x;X_0,X_1)}{t}\right)\right) + 2\gamma^2 K(t,x;X_0,X_1)\\
&\leq (\gamma+2\gamma^2)K(t,x;X_0,X_1).
\end{aligned}$$

(b) It is clear that

$$\|x_t\|_{X_1} \leq c_*\frac{K(t,x;X_0,X_1)}{t} = \varphi(t).$$

So, by (5.24) and Proposition 5.9,

$$\begin{aligned}
&\|x - x_t\|_{X_0} \leq c_* K(t,x;X_0,X_1)\\
&\leq 2c_*\operatorname{dist}_{X_0}\left(x, B_{X_1}\left(\frac{K(t,x;X_0,X_1)}{2t}\right)\right) \leq 2c_*\operatorname{dist}_{X_0}\left(x, B_{X_1}\left(\frac{\varphi(t)}{2c_*}\right)\right).
\end{aligned}$$

Thus, $x_t$ is a near-minimizer for the functional $\operatorname{dist}_{X_0}(x, B_{X_1}(s))$ at $s = \varphi(t)$ with constant $2c_*$. □

*Remark* 5.11. We supplement the converse statement in the above proposition with the following claim. Let $x_s$ be a near-minimizer for the $K$-functional of $x$ at $s$. If $\operatorname{dist}_{X_0}(x, B_{X_1}(t)) > 0$ and

$$\tau(t) = \frac{1}{t}\operatorname{dist}_{X_0}(x, B_{X_1}(t)), \tag{5.25}$$

then $x_{\tau(t)}$ is a near-minimizer for the functional $\operatorname{dist}_{X_0}(x, B_{X_1}(t))$.

*Proof.* We have

$$K(\tau,x;X_0,X_1) = \inf_{\rho>0}(\operatorname{dist}_{X_0}(x, B_{X_1}(\rho)) + \tau\rho).$$

Putting $\tau = \tau(t)$ and $\rho = t$ in the expression under the infimum sign, we obtain

$$K(\tau(t),x;X_0,X_1) \leq 2\operatorname{dist}_{X_0}(x, B_{X_1}(t)).$$

We also have

$$\left\|x - x_{\tau(t)}\right\|_{X_0} + \tau(t)\left\|x_t\right\|_{X_1} \leq cK(\tau(t),x;X_0,X_1).$$

Combining this with the preceding inequality, we prove the claim. □

## 5.5 The elementary stability theorem revisited

Here we present the proof of Theorem 1.17, which was omitted in Chapter 1.[3]

The statement is reproduced below; we recall that it was used in one of two proofs of Theorem 1.18.

**Theorem 5.12.** *Suppose that* $\operatorname{dist}_{L^1}(f, B_{L^p}(s)) > 0$. *Consider the function defined by* (1.49), *that is,*

$$u^{(s)} = f_{\varphi(\tau(s)^{1/\theta})},$$

*where* $\theta = 1 - 1/p$, $\tau(s) = s^{-1}\operatorname{dist}_{L^1}(f, B_{L^p}(s))$, *and* $\varphi$ *is defined by* (1.47):

$$\varphi(t) = \frac{2^{n+3}K(t, f; L^1, L^\infty)}{t}.$$

*Then*

$$\|u^{(s)}\|_{L^p} \le Cs \text{ and } \|f - u^{(s)}\|_{L^1} \le C\operatorname{dist}_{L^1}(f, B_{L^p}(s)),$$

*where* $C > 0$ *is independent of* $f$ *and* $s$. *Moreover, the Calderón–Zygmund cubes involved in the definition of* $u^{(s)}$ *satisfy*

$$\sum |Q_i| \le C\left(\frac{1}{s}\operatorname{dist}(f, B_{L^p}(s))\right)^{\frac{1}{\theta}}. \tag{5.26}$$

*Proof.* First, we outline the arguments. We know that the Calderón–Zygmund procedure yields a near-minimizer for the distance functional for the couple $(L^1, L^\infty)$ with a certain estimate of the cubes' volume. From it we pass to a near-minimiser for the $K$-functional for the same couple at a different value of the parameter "$t$" (see Proposition 5.10, (a)). Another recalculation (see Theorem 5.4) gives a near-minimizer for the $K$-functional for $(L^1, L^p)$ (again at a different value of the parameter), from which we return to the distance functional of the same couple (Proposition 5.10, (b)). In doing this, we must carefully trace all parameter changes. Ultimately, the required estimate for the total volume of the Calderón–Zygmund cubes is also deduced from the known estimate in the case of the distance functional for $(L^1, L^\infty)$.

So, we preceed to formal details. By Remark 5.11, a near-minimizer for the functional $\operatorname{dist}_{L^1}(f, B_{L^p}(s))$ can be found in the form $f^{(\tau(s))}$, where $\tau(s) = \frac{1}{s}\operatorname{dist}_{L^1}(f, B_{L^p}(s))$ (see (5.25)), and the family $f^{(t)}$ satisfies

$$\|f - f^{(t)}\|_{L^1} + t\|f^{(t)}\|_{L^p} \le CK(t, f; L^1, L^p).$$

[3]The argument was promised to require no advanced knowledge, but to be bulky. It is complicated indeed. The reason is that some $K$-functional enters in-between though ultimately we estimate one distance functional in terms of another distance functional. We have seen that these two types of functionals are related in a somewhat tricky nonlinear way. The fact that Theorem 5.4 with its lengthy calculations is also involved then aggravates things only slightly.

Next, we invoke Theorem 5.4 to conclude that, with $\theta = 1 - 1/p$, it suffices to produce a family $f^{(t)}$ satisfying

$$\|f - f^{(t)}\|_{L^1} + t^{\frac{1}{\theta}} \|f^{(t)}\|_{L^\infty} \leq CK(t^{\frac{1}{\theta}}, f; L^1, L^\infty).$$

(We remind the reader that, up to renorming, $L^p = (L^1, L^\infty)_{\theta,p}$.) By Proposition 5.10 (see also item A in Section 3 of Chapter 1), this is achieved if we put $f^{(t)} = f_{\varphi(t^{\frac{1}{\theta}})}$, where $\varphi$ was defined in the theorem we are proving. Recall that a *lower* index of $f$ indicates Calderón–Zygmund decomposition.

All these recipes in the reverse order constitute the definition (1.49) of $u^{(s)}$:

$$u^{(s)} = f_{\varphi(\tau(s)^{\frac{1}{\theta}})}.$$

To finish the proof, it remains to estimate the total volume of the Calderón–Zygmund cubes $Q_i$ involved in the construction of $f_{\varphi(\tau(s)^{\frac{1}{\theta}})}$. By item B) in Section 3, Chapter 1, we have:

$$\sum |Q_i| \leq C\varphi(\tau(s)^{\frac{1}{\theta}})^{-1} \operatorname{dist}(f, B_{L^\infty}(\frac{1}{2}\varphi(\tau(s)^{\frac{1}{\theta}}))). \tag{5.27}$$

We estimate the distance functional on the right-hand side in terms of a $K$-functional as follows. We have

$$K(\tau(s)^{\frac{1}{\theta}}, f; L^1, L^\infty) = \inf_{\rho>0}(\operatorname{dist}(f, B_{L^\infty}(\rho)) + \tau(s)^{\frac{1}{\theta}}\rho).$$

If $\rho > \frac{1}{2}\varphi(\tau(s)^{\frac{1}{\theta}})$, then the second term under the infimum sign dominates the $K$-functional on the left, so we can write

$$K(\tau(s)^{\frac{1}{\theta}}, f; L^1, L^\infty) \geq C_1 \inf_{0<\rho\leq\frac{1}{2}\varphi(\tau(s)^{\frac{1}{\theta}})} \operatorname{dist}(f, B_{L^\infty}(\rho)).$$

Clearly, the last infimum is attained at $\rho = \frac{1}{2}\varphi(\tau(s)^{\frac{1}{\theta}})$. Collecting the estimates, we see that (5.27) implies

$$\sum |Q_i| \leq C\varphi(\tau(s)^{\frac{1}{\theta}})^{-1}K(\tau(s)^{\frac{1}{\theta}}, f; L^1, L^\infty) = C_2\tau(s)^{\frac{1}{\theta}} = \left(\frac{1}{s}\operatorname{dist}_{L^1}(f, B_{L^p}(s))\right)^{\frac{1}{\theta}} \tag{5.28}$$

by the definition of the functions $\varphi$ and $\tau$. □

## 5.6 $K$-closed subcouples and stability

Consider an operator $T$ bounded on $L^p(\mathbb{R}^n)$ for some $p > 1$ and long-range regular on $L^1(\mathbb{R}^n)$. With this operator, we associate the spaces $\mathcal{X}^r = \{f \in L^r : Tf \in L^r\}$, $1 \leq r \leq p$. We know that an operator with these properties is bounded on $L^r$ for

$1 < r \le p$, so that $\mathcal{X}^r = L^r$ for such $r$. But in interesting cases $\mathcal{X}^1 \neq L^1$. The question arises as to how we can interpolate between $\mathcal{X}^1$ and $\mathcal{X}^p = L^p$ by the real method.

It turns out that $(\mathcal{X}^1, L^p)_{\theta,r} = (L^1, L^r)_{\theta,r} = L^r$ provided $r^{-1} = (1-\theta)+\theta p^{-1}$, that is, interpolation is not sensitive to the replacement of $L^1$ with $\mathcal{X}^1$. We shall explain in this section that this is a consequence of the stability theorem (Theorem 1.18). In fact, that theorem implies a subtler property we are going to discuss in an abstract setting. However, before this we make an observation concerning the above specific situation.

**Lemma 5.13.** *The intersection $\mathcal{X}^1 \cap L^p$ is dense both in $\mathcal{X}^1$ and in $\mathcal{X}^p$.*

*Proof.* Density in $\mathcal{X}^1$ follows directly from the (nontrivial) Corollary 1.20. To prove density in $\mathcal{X}^p = L^p$, we shall again use Calderón–Zygmund decomposition.

Specifically, given $f \in L^p$, we can write $f = u + v$ where $v \in L^1 \cap L^p$ and $\|v\|_{L^p}$ is as small as we wish. Next, we apply the Calderón–Zygmund procedure with parameter $t > 0$ to $v$, taking $t$ *small.* This will result in the representation $v = v_t + w$, where $v_t$ is the "good" part of $v$ (see §1 in Chapter 1). Since $|v_t| \le Ct$, we have

$$\|v_t\|_{L^p}^p \le C_1 \int |v_t| t^{p-1} = C_1 t^{p-1} \|v_1\|_{L^1},$$

implying that $w$ approximates $f$ in $L^p$ within a precision we like.

It remains to show that $w \in \mathcal{X}^1 \cap L^p$. Clearly, $w \in L^1 \cap L^p$, and we must only prove that $Tw \in L^1$. But $Tw \in L^p$, so it is integrable over any set of finite measure. In particular, it is integrable over $\Omega = \cup_j 2Q_j$, where the $Q_j$ are the Calderón–Zygmund cubes, cf. §1 in Chapter 1. But $Tw$ is also integrable over the complement of $\Omega$ by long-range regularity: recall that $w = \sum b_j$, each $b_j$ being supported on $Q_j$ and having zero average, whence

$$\int_{\mathbb{R}^n \setminus \Omega} |Tw| \le \sum_j \int_{\mathbb{R}^n \setminus 2Q_j} |Tb_j| \le C \sum \int_{Q_j} |b_j| = \|w\|_{L^1}. \qquad \square$$

It should be noted that the lemma is useful for understanding the situation, but, in fact, is not of critical need. Indeed, so far as we are dealing with interpolation, we could have replaced $\mathcal{X}^1$ and $\mathcal{X}^p = L^p$ by the closures $\mathcal{X}_0^1$ and $\mathcal{X}_0^p$ of $\mathcal{X}^1 \cap L^p$ in $L^1$ and $L^p$ (respectively), even if we would have ignored what these closures are. In general, such a replacement does not influence interpolation spaces of the real method.

Now, we pass to the abstract setting. Suppose we have a couple $(E, F)$ of quasi-Banach spaces embedded continuously in some linear topological space $\mathcal{G}$. By an *operator on this couple* we mean a continuous linear operator from $E+F$ to $\mathcal{G}$.[4] If $T$ is an operator on $(E, F)$, we can introduce a new couple $(X, Y)$ of spaces by putting $X = \{x \in E : Tx \in E\}$, $\|x\|_{E(T)} = \|x\|_E + \|Tx\|_E$, and similarly for $Y$

[4] As usual, the sum $E + F$ is endowed with the quasinorm $K(1, \cdot; E, F)$.

(substitute $F$ for $E$ throughout). Next, we denote by $E(T)$ and $F(T)$ the closures of $X \cap Y$ in $X$ and $Y$, respectively.

Now, as before, we put the following question. Suppose we know the interpolation spaces for the couple $(E, F)$, how to calculate them for the couple $(E(T), F(T))$?

Surely, the answer is trivial if $T$ takes boundedly both $E$ and $F$ to themselves. In what follows we assume only that $T$ takes $F$ to itself. In place of the boundedness on $E$, we assume only that $T$ has a stability property in the spirit of §4 in Chapter 1 (similar patterns will be discussed in Chapter 10).

**Definition 5.14.** Suppose $t > 0$ and $x \in E$ is such that $x \notin B_F(t)$ and $Tx \in E$. Let $x_t$ be a near-minimizer for the distance functional of $x$ at $t$, i.e.,

$$\|x_t\|_E \leq ct \text{ and } \|x - x_t\|_F \leq c\,\mathrm{dist}_E\left(x, B_F\left(\frac{t}{c}\right)\right).$$

If, moreover,

$$\|Tx - Tx_t\|_E \leq c(\mathrm{dist}_F(Tx, B_E(t/c)) + \mathrm{dist}_{X_0}(x, B_F(t/c))) \tag{5.29}$$

(here $c$ must not depend on the vectors involved), then the near-minimizer $x_t$ is said to be *$T$-stable.*

If $T$-stable near-minimizers exist for every $x$ and $t$, we say that *the distance functional is $T$-stable.* Note that, since $T$ takes $F$ to itself boundedly, the element $Tx_t$ lies (along with $x_t$) in the ball of radius roughly $t$ in $F$, and the second formula says that $Tx_t$ approximates $Tx$ nearly within the same precision as $x_t$ approximates $x$, up to the inevitable penalty term $\mathrm{dist}_E(Tx, B_F(t/c))$.

In a similar way, we can introduce the notion of $T$-stability for the $K$- and more general $L_{p,q}$-functionals. All these properties lead to a fundamental notion called $K$-closedness. Proposition 5.15 below shows that, in the context of $K$-closedness, replacing one functional by another is fairly easy and, unlike the previous material of this chapter, does not involve complicated nonlinear transformations. Now we explain what $K$-closedness is.

Let $(X_0, X_1)$ be a (quasi-)Banach couple, and this time let $Y_0$ and $Y_1$ be *closed* linear subspaces in $X_0$ and $X_1$. To avoid degeneration, we assume that

$$Y_i = X_i \cap (Y_1 + Y_2), \quad i = 0, 1. \tag{5.30}$$

Again, we are interested in finding the interpolation spaces $(Y_0, Y_1)_{\theta,q}$ if the corresponding spaces $(X_0, X_1)_{\theta,q}$ are known. More specifically, it is natural to ask whether the following is true:

$$(Y_0, Y_1)_{\theta,q} = (Y_0 + Y_1) \cap (X_0, X_1)_{\theta,q}. \tag{5.31}$$

In fact, this formula may perfectly be false. However, if we look at the definition (5.8), we easily conclude that (5.31) is definitely true provided the $K$-functionals of the couples $(Y_0, Y_1)$ and $(X_0, X_1)$ are equivalent on $Y_0 + Y_1$ uniformly in the parameter $t$.

If such equivalence does occur, the couple $(Y_0, Y_1)$ is said to be *K-closed* in $(X_0, X_1)$. This notion has been investigated quite thoroughly through recent years and has turned out to be quite convenient. Moreover, it does occur for many classical couples and some important subcouples of them. See, e.g., [P], [KX] and [Kis] for the relevant material.

The property of being $K$-closed admits several reformulations collected in the following statement. The proof will support the above claim about convenience: the interchange of various functionals of interpolation theory is done practically for free here.

**Proposition 5.15.** *Let $(Y_0, Y_1)$ be a subcouple of a quasi-Banach couple $(X_0, X_1)$ formed by closed subspaces. Under condition (5.30), the following statements are equivalent.*

1) *The couple $(Y_0, Y_1)$ is $K$-closed in $(X_0, X_1)$.*
2) *There exists a constant $c > 0$ such that*
$$K(t, y; Y_0, Y_1) \le cK(t, y; X_0, X_1)$$
*for every $t > 0$ and $y \in Y_0 + Y_1$.*
3) *For some (equivalently, for all) $a, b > 0$ there is a constant $c$ such that*
$$L_{a,b}(t, y; Y_0, Y_1) \le cL_{a,b}(t, y; X_0, X_1)$$
4) *There exists a constant $c > 0$ such that*
$$\operatorname{dist}_{Y_0}(y, B_{Y_1}(cs)) \le c \operatorname{dist}_{X_0}(y, B_{X_1}(s))$$
*for every $s > 0$ and $y \in Y_0 + Y_1$.*
5) *There is a constant $c > 0$ such that for every $y \in Y_0 + Y_1$ and every representation $y = x_0 + x_1$ with $x_i \in X_i$, $i = 0, 1$, there is a representation $y = y_0 + y_1$ with $y_i \in Y_i$ and $\|y_i\| \le c\|x_i\|$, $i = 0, 1$.*
6) *The same as 5), but only for $y$ in some dense linear subspace of $Y_0 + Y_1$.*

*Proof.* The equivalence of 5) and 6) is an easy and standard exercise, and is left to the reader. Clearly, 2) is equivalent to 1), because the reverse inequality $K(t, y; Y_0, Y_1) \ge K(t, y; X_0, X_1)$ for $y \in Y_0 + Y_1$ and $t > 0$ is obvious. It is also clear that 4) and 5) tell the same, and that 5) implies 2) and also 3) with arbitrary $a$ and $b$. It remains to show that 3) with fixed $a$ and $b$ implies 5) (note that the last statement includes also the fact that 1) implies 5)).

Suppose that $y$, $x_0$, and $x_1$ are as in 5). If $x_0 = 0$ or $x_1 = 0$, there is nothing to prove by (5.30). Otherwise we put $t = \|x_0\|^a / \|x_1\|^b$ and observe that

$L_{a,b}(t,y;X_0,X_1) \le \|x_0\|^a_{X_0} + t\|x_1\|^b_{X_1} = 2\|x_0\|^a_{X_0}$. Then 3) says that $y$ admits a representation $y = y_0 + y_1$ with $y_i \in Y_i$ and $\|y_0\|^a_{Y_0} + t\|y_1\|^b_{Y_1} \le 4\|x_0\|^a_{X_0}$. Each summand on the left is also dominated by the right-hand side, and it remains to recall what $t$ is. □

It should also be noted that condition 5) in the above proposition *implies* (5.30).

Now, we return to an operator $T$ on a couple $(E,F)$ and suppose, as before, that $T$ is bounded on $F$ and the distance functional is $T$-stable. We show that then a $K$-closed couple is related in a natural way to $(E(T),F(T))$. Specifically, put $X_0 = E \times E$ and $X_1 = F \times F$ (the norm on the direct product is introduced as the sum of the norms on the components). Then $E(T)$ and $F(T)$ are identified isometrically with certain subspaces $Y_0$ and $Y_1$ of $X_0$ and $X_1$. These subspaces are formed by the pairs $(x,Tx)$, where $x$ runs through $E(T)$ or $F(T)$.

**Theorem 5.16.** *Under the above assumptions about $T$, the couple $(Y_0,Y_1)$ is $K$-closed in $(X_0,X_1)$.*

*Proof.* Observe that, by the definition of $E(T)$ and $F(T)$, $Y_0$ is dense in $Y_0+Y_1$, therefore, by 6) in Proposition 5.15, it suffices to verify condition 5) in the same proposition for $y \in Y_0$. So, let $y = (w,Tw) = U+V$, where $w \in E(T)$ and $U \in X_0$, $V \in X_1$. This means that *both $w$ and $Tw$* are within the $E$-distance $\|U\|_{X_0}$ from the $\|V\|_{X_1}$-ball of $F$, and the claim is a direct consequence of (5.29): the two summands on the right in that formula are dominated by $\|U\|_{X_0}$, etc. □

In particular, formula (5.31) holds true, and it is easy to realize that, indeed, basically it expresses $(E(T),F(T))_{\theta,r}$ in terms of $(E,F)_{\theta,r}$. Surely, all this applies to the case of a long-range regular operator $T$ on the couple $(L^1,L^p)$, which opened this section. More examples are provided by Chapter 10. Among other things, it will be proved there that the distance functional for the couple $(L^1,\mathrm{Lip}_\alpha)$ is stable under an arbitrary singular integral operator with certain additional properties. This also leads to $K$-closedness and to some interpolatory consequences.

We finish the section with a description of another (related and, in fact, more general) setting. Suppose $P$ is an operator on $(E,F)$ bounded on $F$, and suppose that the distance functional is $P$-stable. Next, assume that $P$ is a projection, that is, $PPx = x$ whenever $x$ and $Px$ are in $E+F$. Put $W = \{x \in E \cap F : Px = x\}$, and let $E_0$ and $F_0$ be the closures of $W$ in $E$ and $F$, respectively. *Then the couple $(E_0,F_0)$ is $K$-closed in $(E,F)$.* Again, this is a consequence of (5.29) for $P$ in place of $T$: the property is used in the case where $Px = x$, so that the two terms on the right in (5.29) are the same in essence.

In the previous situation, a projection was also involved, namely, the operator $(a,b) \mapsto (a,Ta)$. That is why it can be included in what is described in the last paragraph.

## 5.7 Linearization

Sometimes, the $K$-functional and near-minimizers can be calculated with the help of linear operators. Here various patterns are possible. Historically, the first one was suggested by Peetre in 1963, but it is too restrictive for our purposes. The following framework is more flexible. Note that the first examples of linearization in the sense described below already occurred in §3, C of Chapter 1 and in §2 of Chapter 3 (look at the operators denoted by $P_{f,t}$ in both places and compare with what follows). Both cases are related to Calderón–Zygmund decomposition. Also note that the abstract considerations presented below will not be used until Chapter 10.

**Definition 5.17.** We say that a family $V_{x,t} : X_0 + X_1 \to X_1$ of linear operators *linearizes* the $K$-functional if

$$\|x - V_{x,t}x\|_{X_0} + t\,\|V_{x,t}x\|_{X_1} \le cK(t, x; X_0, X_1) \tag{5.32}$$

and

$$\|V_{x,t}\|_{X_0 \to X_0} \le c,\ \|I - V_{x,t}\|_{X_1 \to X_0} \le ct \tag{5.33}$$

with a constant $c \ge 1$ independent of $x$ and $t$.

We formulate a condition sufficient for linearizability.

**Proposition 5.18.** *The inequalities*

$$\|I - V_{x,t}\|_{X_0 \to X_0} \le c,\ \|I - V_{x,t}\|_{X_1 \to X_0} \le ct,$$
$$\|V_{x,t}\|_{X_1 \to X_1} \le c,\ \|V_{x,t}\|_{X_0 \to X_1} \le c\frac{1}{t},$$

*with a constant $c > 0$ independent of $t > 0$ and $x$ imply that the family $V_{x,t}$ is a linearization for the $K$-functional.*

*Proof.* Choose a decomposition $x = x_0 + x_1$ in such a way that

$$\|x_0\|_{X_0} + t\,\|x_1\|_{X_1} \le 2K(t, x; X_0, X_1).$$

Then the first summand in (5.32) is estimated as follows:

$$\begin{aligned}\|(I - V_{x,t})x\|_{X_0} &\le \|(I - V_{x,t})x_0\|_{X_0} + \|(I - V_{x,t})x_1\|_{X_0}\\ &\le c\left(\|x_0\|_{X_0} + t\,\|x_1\|_{X_1}\right) \le 2cK(t, x; X_0, X_1).\end{aligned}$$

The second summand is treated similarly. □

Now we show how linearizability behaves under interpolation. Consult, e.g., [BL] for the Hölder-type inequality for operator norms to be used in the proof.

**Proposition 5.19.** *If a family $V_{x,t}$ linearizes the $K$-functional for the couple $(X_0, X_1)$, then so does the family $V_{x,t^{\frac{1}{\theta}}}$ for the couple $(X_0, (X_0, X_1)_{\theta,q})$.*

*Proof.* Condition (5.33) is obvious. Theorem 5.4 shows that (5.32) is fulfilled for the operators $V_{x,t^{\frac{1}{\theta}}}$ and the couple $(X_0, (X_0, X_1)_{\theta,q})$. Therefore, it suffices to prove that

$$\left\|I - V_{x,t^{\frac{1}{\theta}}}\right\|_{(X_0,X_1)_{\theta,q}\to X_0} \le ct.$$

By (5.33), we have

$$\begin{aligned}&\left\|I - V_{x,t^{\frac{1}{\theta}}}\right\|_{(X_0,X_1)_{\theta,q}\to X_0}\\ &\le \left\|I - V_{x,t^{\frac{1}{\theta}}}\right\|_{X_0\to X_0}^{1-\theta} \left\|I - V_{x,t^{\frac{1}{\theta}}}\right\|_{X_1\to X_0}^{\theta} \le c\left(t^{\frac{1}{\theta}}\right)^{\theta} \le ct.\end{aligned}$$ □

For some couples, linearization for the distance functional is available. Then the $K$-functional also becomes linearized, after a change of variables. Here is the precise statement.

**Proposition 5.20.** *Suppose that for every $x \in X_0 + X_1$ there exists a family of linear operators $P_{x,t} : X_0 + X_1 \to X_1$ such that*

$$\|P_{x,t}x\|_{X_1} \le \gamma t \text{ and } \|x - P_{x,t}x\|_{X_0} \le \gamma \operatorname{dist}_{X_0}\left(x, B_{X_1}\left(\frac{t}{\gamma}\right)\right)$$

*and*

$$\|P_{x,t}\|_{X_0\to X_0} \le \gamma \text{ and } \|I - P_{x,t}\|_{X_1\to X_0} \le \frac{\gamma}{t}\operatorname{dist}_{X_0}\left(x, B_{X_1}\left(\frac{t}{\gamma}\right)\right). \tag{5.34}$$

*Put*

$$\varphi(t) = \frac{2\gamma K(t, x; X_0, X_1)}{t}.$$

*Then the operators*

$$V_{x,t} = P_{x,\varphi(t)}$$

*linearize the $K$-functional for the couple $(X_0, X_1)$.*

*Proof.* By (5.34) and Proposition 5.10, the element

$$x_t = P_{x,\varphi(t)}x$$

is a near-minimizer for the $K$-functional of the couple $(X_0, X_1)$, that is

$$\|x - x_t\|_{X_0} + t\,\|x_t\|_{X_1} \le cK(t, x; X_0, X_1).$$

Moreover,

$$\|V_{x,t}\|_{X_0\to X_0} = \left\|P_{x,\varphi(t)}\right\|_{X_0\to X_0} \le \gamma,$$

and (5.23) shows that

$$\begin{aligned}\|I - V_{x,t}\|_{X_1\to X_0} &\le \frac{\gamma}{\varphi(t)}\operatorname{dist}_{X_0}\left(x, B_{X_1}\left(\frac{\varphi(t)}{\gamma}\right)\right)\\ &\le t\frac{\operatorname{dist}_{X_0}\left(f, B_{X_1}\left(\frac{2K(t,x;X_0,X_1)}{t}\right)\right)}{2K(t, x; X_0, X_1)} \le \frac{t}{2}.\end{aligned}$$ □

There is a similar result for the $L_{p,q}$-functional.

**Proposition 5.21.** *Suppose that $p \neq q$ and for any $x \in X_0 + X_1$ there exists a family of linear operators $P_{x,t} : X_0 + X_1 \to X_1$ such that*

$$\|x - P_{x,t}x\|_{X_0}^p + t^{p-q}\|P_{x,t}x\|_{X_1}^q \leq cL_{p,q}(t^{p-q}, x; X_0, X_1)) \tag{5.35}$$

*and*

$$\|P_{x,t}\|_{X_0 \to X_0} \leq c \text{ and } \|I - P_{x,t}\|_{X_1 \to X_0} \leq c\left(\frac{L(t^{p-q}, x; X_0, X_1)}{t^p}\right)^{\frac{1}{p}-\frac{1}{q}}.$$

*Put*

$$\varphi(t) = t^q K(t, x; X_0, X_1)^{p-q}.$$

*Then the operators*

$$V_{x,t} = P_{x,\varphi(t)^{\frac{1}{p-q}}}$$

*linearize the $K$-functional for the couple $(X_0, X_1)$.*

*Proof.* By (5.35) and Proposition 5.7 we see that the element

$$x_t = P_{x,\varphi(t)^{\frac{1}{p-q}}}x$$

is a near-minimizer for the $K$-functional of the couple $(X_0, X_1)$, that is

$$\|x - x_t\|_{X_0} + t\|x_t\|_{X_1} \leq cK(t, x; X_0, X_1).$$

Moreover,

$$\|V_{x,t}\|_{X_0 \to X_0} = \left\|P_{x,\varphi(t)^{\frac{1}{p-q}}}\right\|_{X_0 \to X_0} \leq c.$$

Now, (5.18) with $a = b = 1$ shows that

$$\begin{aligned}\|I - V_{x,t}\|_{X_1 \to X_0} &\leq c\left(\frac{L(\varphi(t), x; X_0, X_1)}{\varphi(t)^{\frac{p}{p-q}}}\right)^{\frac{1}{p}-\frac{1}{q}} \\ &\leq c\left(\frac{2K(t, x; X_0, X_1)^p}{(t^q K(t, x; X_0, X_1)^{p-q})^{\frac{p}{p-q}}}\right)^{\frac{1}{p}-\frac{1}{q}} = c2^{\frac{1}{p}-\frac{1}{q}}t. \qquad \square\end{aligned}$$

## Notes and remarks

The information collected in §1 of the chapter is important but not critical for understanding the main body of the book. The $K$-divisibility Theorem 5.1 can be used in various ways. First, sometimes a $K$-functional happens to coincide with some significant quantity of independent origin, as in (5.5) and especially in (5.7). In all such cases $K$-divisibility says something nontrivial and interesting about such quantities. Second (and main), there is a theory of so-called "general $K$-spaces", that is, spaces defined by analogy with (5.8). Specifically, let $\Phi$ be a Banach function lattice on $(0, \infty)$, which means that it is a Banach space of measurable functions with the following properties: 1) if $f \in \Phi$, $g$ is measurable, and $|g| \leq |f|$, then $g \in \Phi$ and $\|g\|_\Phi \leq \|f\|_\Phi$; 2) $\min(1, t) \in \Phi$. For a Banach couple $(X_0, X_1)$, define the space $K_\Phi(X_0, X_1)$ to be the set of all $x \in X_0 + X_1$ such that

$$\|x\|_{K_\Phi(X_0,X_1)} = \|K(\cdot, x; X_0, X_1)\|_\Phi < \infty.$$

It should be noted that, for many important Banach couples, *all* interpolation spaces are $K$-spaces, see, e.g., [CNS] and the references therein.

Nowadays, the theory of general $K$-spaces is as complete as that of the spaces defined by (5.8), see [BK]; see also the more recent paper [K2] for a short survey. This development would be impossible without the $K$-divisibility theorem and the methods of its proof, which are used critically to deduce, say, the reiteration and duality theorems and other indispensable statements about $K$-spaces.

The abstract "recalculation" results of §§2, 3, and 4, when combined with explicit constructions of near-minimizers (see especially Chapter 9), mean each time something specific. The same can be said about the results of §7 (we emphasize that a linearization procedure occurs in *all* our explicit constructions of near-minimizers). The proof of Theorem 1.17 presented in §5 is due to Kruglyak and was published before only in the preprint [KK0], in a somewhat different form.

A statement in the spirit of Lemma 5.13 can be found in [Kis] (see Lemma 4.2 there).

The property of being $K$-closed (see §6) passes from a cubcouple to the subcouple of annihilators in the dual couple, and *vice versa*. This simplifies greatly the treatment of duality in interpolation questions, provided $K$-closedness occurs. See, e.g., the survey [Kis]. See also the paper [Kis1] for this duality phenomenon in a somewhat unusual setting.

# Chapter 6

# Regularization for Banach spaces

Here we want to discuss a field irrelevant to interpolation, but in which a near-minimizer for an $L_{p,q}$-functional also arises in a natural way. In the theory of ill-posed problems (see, e.g., [Ki]), important part is played by the Tikhonov regularization method described below. Suppose $A : X \to Y$ is an injective and bounded but, in general, not boundedly invertible linear operator from a Hilbert space $X$ to a Hilbert space $Y$. Suppose also that we observe some quantity

$$y = Ax_* + \eta, \tag{6.1}$$

where $\eta \in Y$ is the "noise" of small $Y$-norm:

$$\|\eta\|_Y \leq \varepsilon. \tag{6.2}$$

The problem is to find the "exact" solution $x_* \in X$ approximately. Surely, an attempt to define $x_*$ as $A^{-1}y$ is improper, because $y$ may fail to belong to the image of $A$, and even if it belongs, the noise may influence the accuracy enormously in case $A$ has small singular numbers.

The Tikhonov regularization method suggests the following: as an approximate solution to (6.1), take an element $\tilde{x}_* \in X$ that minimizes the Tikhonov functional

$$T(t, y) = \inf_{x \in X} (\|y - Ax\|_Y^2 + t \|x\|_X^2). \tag{6.3}$$

The value of $t > 0$ is chosen on the basis of some *a priori* information about the noise $\eta$ and the exact solution $x_*$.

In this prescription, we easily recognize the $L_{2,2}$-functional for the Hilbert couple $(Y, A(X))$, where the image $A(X)$ of $X$ under the action of $A$ is endowed with the norm

$$\|u\|_{A(X)} = \left\|A^{-1}u\right\|_X.$$

Indeed, the obvious relations

$$
\begin{aligned}
L_{2,2}(t, y; Y, A(X)) &= \inf_{u \in A(X)} \left( \|y-u\|_Y^2 + t\, \|u\|_{A(X)}^2 \right) \\
&= \inf_{x \in X} \left( \|y-Ax\|_Y^2 + t\, \|x\|_X^2 \right) \\
&= \|y - A\tilde{x}_*\|_Y^2 + t\, \|\tilde{x}_*\|_X^2 = \|y - A\tilde{x}_*\|_Y^2 + t\, \|A\tilde{x}_*\|_{A(X)}^2
\end{aligned}
$$

show that an element $\tilde{x}_* \in X$ minimizes the Tikhonov functional if and only if the value of the $L_{2,2}$-functional of $y$ for the above Hilbert couple is attained at $A\tilde{x}_*$.

In connection with the Tikhonov functional (6.3), it should be noted that the Hilbert space framework leads to a very nice formula convenient for calculation of an exact minimizer:

$$\tilde{x}_* = (A^*A + tI)^{-1} A^* y. \tag{6.4}$$

The most fascinating feature here is that, sometimes, it turns out to be possible to estimate the error $\|x_* - \tilde{x}_*\|_X$. Specifically, this can be done under the assumption that the solution $x_*$ belongs to a certain Hilbert space $X_0 \subset X$ (for instance, $A^*(Y)$ may be taken as $X_0 \subset X$).

In recent years, similar regularization procedures emerged repeatedly in connection with problems of image processing (see, e.g., [SGGHL]), [CDPH]), and even in the case where the spaces $X$ and $Y$ are not necessarily Hilbertian. We mention the quite popular Rudin–Osher–Fatemi denoising model (see [CS]). In order to free an image $y \in L^2$ from the noise, this model suggests employing a minimizer for the $L_{2,1}$-functional of the couple $(L^2, \mathrm{BV})$ (BV is the space of functions of bounded variation, see Chapter 11 for the definition; note that in the same chapter explicit near-minimizers for this functional are constructed). That is, for an approximate solution of (6.1) we take a function $x$ that minimizes the expression under the infimum sign below:

$$L_{2,1}(t, y; L^2, \mathrm{BV}) = \inf_{x \in \mathrm{BV}} (\|y - x\|_{L^2}^2 + t\, \|x\|_{\mathrm{BV}}).$$

Construction of (near-)minimizers for $L_{2,1}(t, y; L^2, \mathrm{BV})$ has been discussed quite thoroughly in the literature (see, e.g., the recent paper [BDKPW] and the book [CS]).

We want to give a more formal treatment to the extension of Tikhonov's method to the case of general Banach spaces. The guideline for us will be the fact (see Lemma 6.3 below) that the *exact* minimizer $\tilde{x}_*$ for the Tikhonov functional (6.1) is a *near*-minimizer for the functional

$$\tilde{T}(t, y) = \inf_{x \in A^*(Y)} (\|y - Ax\|_Y^2 + t^2\, \|x\|_{A^*(Y)}^2). \tag{6.5}$$

Note that here we use $t^2$ in place of $t$ in (6.1). We shall need several supplementary conditions, whose meaning for the Hilbert spaces will be explained later. First, we suppose that the precise solution $x_*$ belongs to a certain Banach space $X_0$, which is embedded in $X$ linearly and continuously (in symbols: $X_0 \subset X$). Also, since $A : X \to Y$ is injective, we can consider the Banach spaces $A(X_0) \subset A(X) \subset Y$ endowed with the norms

$$\|u\|_{A(X_0)} = \left\|A^{-1}u\right\|_{X_0}, \quad \|u\|_{A(X)} = \left\|A^{-1}u\right\|_X .$$

We impose the following two conditions.

**Condition A.** *For some $\theta \in (0,1)$ we have*

$$A(X) \supset (Y, A(X_0))_{\theta,1}.$$

To state the second condition, consider the couple $(Y, A(X_0))$ and define the $L_{p,q}$-functional of an element $y \in Y$ by the formula

$$L_{p,q}(t, y; Y, A(X_0)) = \inf_{u \in A(X_0)} (\|y - u\|_Y^p + t\,\|u\|_{A(X_0)}^q).$$

**Condition B.** *For arbitrary $y \in Y$ and the $L_{p,q}$-functional of the couple $(Y, A(X_0))$ with some $p$ and $q$, we know a procedure to construct near-minimizers, that is, a family $y_t \in A(X_0)$ with*

$$\|y - y_t\|_Y^p + t\,\|y_t\|_{A(X_0)}^q \le c L_{p,q}(t, y; Y, A(X_0)),$$

*where the constant $c > 0$ does not depend on $y \in Y$ and $t > 0$ .*

*Remark* 6.1. (a) It is known that the embedding $(Y, A(X_0))_{\theta,1} \subset A(X)$ is equivalent to the multiplicative inequality

$$\|u\|_{A(X)} \le c\,\|u\|_Y^{1-\theta}\,\|u\|_{A(X_0)}^{\theta}$$

for all $u \in A(X_0)$ with a constant $c > 0$ independent of $u$; see [BL].

(b) As has already been explained, the $K$-functional can be expressed in terms of the $L_{p,q}$-functional, but a nonlinear dependence on $y$ arises, see Propositions 5.6 and 5.7. Therefore, sometimes it is more convenient to find near-minimizers for the $L_{p,q}$-functional rather than for the $K$-functional.

The following simple theorem shows why near-minimizers are so important for regularization.

**Theorem 6.2.** *Under conditions A and B, suppose also that we know an estimate for the precise solution, specifically,*

$$\|x_*\|_{X_0} \le E.$$

*Then, taking the element $A^{-1}y_t$ with $t = \frac{\varepsilon^p}{E^q}$ for the role of an approximate solution, we arrive at the following error estimate:*

$$\left\|x_* - A^{-1}y_t\right\|_X \le c\varepsilon^{1-\theta}E^{\theta} \tag{6.6}$$

*with a constant $c > 0$ independent of $y$, $x_*$, $A$, $E$ and $\varepsilon$.*

*Proof.* Since $Ax_*$ and $y_t$ are elements of $A(X_0)$, Condition A and Remark 6.1 imply

$$\left\|x_* - A^{-1}y_t\right\|_X = \|Ax_* - y_t\|_{A(X)} \le c\,\|Ax_* - y_t\|_Y^{1-\theta}\,\|Ax_* - y_t\|_{A(X_0)}^{\theta},$$

and we need to estimate the factors on the right. Since

$$\begin{aligned}\|y - y_t\|_Y^p + t\,\|y_t\|_{A(X_0)}^q &\le cL_{p,q}(t, y; Y, A(X_0)) \le c\left(\|y - Ax_*\|_Y^p + t\,\|Ax_*\|_{A(X_0)}^q\right)\\ &\le c(\varepsilon^p + tE^q) \le c\varepsilon^p,\end{aligned}$$

by the choice of $t$ we have

$$\|y - y_t\|_Y \le c\varepsilon \text{ and } \|y_t\|_{A(X_0)} \le cE.$$

Consequently,

$$\|Ax_* - y_t\|_Y \le \|Ax_* - y\|_Y + \|y - y_t\|_Y \le \varepsilon + c\varepsilon \le c\varepsilon$$

and

$$\|Ax_* - y_t\|_{A(X_0)} \le \|Ax_*\|_{A(X_0)} + \|y_t\|_{A(X_0)} \le E + cE \le cE.$$

Therefore,

$$\left\|x_* - A^{-1}y_t\right\|_X \le c\varepsilon^{1-\theta}E^{\theta}.$$ □

In reality, quite often the value of the parameter $E$ is unknown. In this case, $t$ is chosen in such a way that the quantity $\|y_t\|_{A(X_0)}$ be minimal and the following inequality be fulfilled: $\|y - y_t\|_Y \le \varepsilon$. This is the so-called "discrepancy principle of Morozov". We refrain from giving an error estimate for this method. Instead, we show that the theorem proved above implies a well-known result of the classical regularization theory. Usually, this result is proved under the assumption that $A$ is compact (see [Ki]). We shall see that this assumption is redundant. For this, a representation theorem for selfadjoint operators on Hilbert space (see [DS]) will be employed.

So, let $A : X \to Y$ be a bounded linear operator from a Hilbert space $X$ to a Hilbert space $Y$. We assume that $A$ is injective and its range is dense in $Y$. Thus, $\operatorname{Ker} A = \{0\}$ and $\operatorname{Ker} A^* = \{0\}$. For the role of the space $X_0 \subset X$ containing the exact solution $x_*$, we take the space $A^*(Y)$ endowed with the norm

$$\|x\|_{X_0} = \left\|(A^*)^{-1}x\right\|_Y .$$

We show that then Condition A is fulfilled with $\theta = \frac{1}{2}$. Indeed, let $u \in A(X_0)$. By Remark 6.1, it suffices to verify the inequality

$$\|u\|_{A(X)} \le \|u\|_Y^{\frac{1}{2}}\,\|u\|_{A(X_0)}^{\frac{1}{2}}. \tag{6.7}$$

Since $u \in A(X_0)$, the definition of $X_0$ implies the existence of an element $z \in Y$ with

$$u = AA^*z \quad\text{and}\quad \|u\|_{A(X_0)} = \|z\|_Y .$$

So, inequality (6.7) is equivalent to

$$(A^*z, A^*z)_X \leq \|AA^*z\|_Y \cdot \|z\|_Y \ .$$

But the last inequality is obvious:

$$(A^*z, A^*z)_X = (z, AA^*z)_Y \leq \|AA^*z\|_Y \cdot \|z\|_Y \ .$$

Now, we verify Condition B. Consider the $L_{2,2}$-functional for the Hilbert couple $(Y, A(X_0))$, i.e.,

$$L_{2,2}(t, y; Y, A(X_0)) = \inf_{u \in A(X_0)} (\|y-u\|_Y^2 + t\,\|u\|_{A(X_0)}^2).$$

Remarkably, the Hilbert space framework allows us to write a simple formula for a near-minimizer of this functional. Put

$$y_t = A(A^*A + \sqrt{t}I)^{-1}A^*y.$$

**Lemma 6.3.** *The element $y_t \in A(X_0)$ is a near-minimizer with constant 2 for $L_{2,2}$:*

$$\|y-y_t\|_Y^2 + t\,\|y_t\|_{A(X_0)}^2 \leq 2L_{2,2}(t, y; Y, A(X_0)).$$

*Proof.* By the spectral theorem for selfadjoint operators, there is no loss of generality in assuming that $Y$ is $L^2(\mu)$ for some measure $\mu$ and $AA^*$ is multiplication by a bounded positive function $\varphi$ (because $AA^*$ is bounded and has trivial kernel). Since

$$A(X_0) = AA^*(Y),$$

we see that

$$\|u\|_{A(X_0)}^2 = \int \frac{u(s)^2}{\varphi(s)^2}\,d\mu$$

and

$$\begin{aligned} L_{2,2}(t, y; Y, A(X_0)) &= \inf_{u \in A(X_0)} \left( \int (y(s)-u(s))^2\,d\mu + t\int \frac{u(s)^2}{\varphi(s)^2}\,d\mu \right) \\ &= \inf_{u \in A(X_0)} \int \left( (y(s)-u(s))^2 + t\frac{u(s)^2}{\varphi(s)^2} \right) d\mu. \end{aligned}$$

Minimizing the integrand for every fixed $s$, we obtain

$$u(s) = \frac{\varphi(s)^2 y(s)}{\varphi(s)^2 + t} \in A(X_0),$$

whence

$$L_{2,2}(t, y; Y, A(X_0)) = \int \frac{ty^2(s)}{\varphi(s)^2 + t}\,d\mu.$$

Since $A(A^*A+\sqrt{t}I)^{-1}=(AA^*+\sqrt{t}I)^{-1}A$, we have $y_t=(AA^*+\sqrt{t}I)^{-1}AA^*y$, and therefore

$$y_t(s)=\frac{\varphi(s)y(s)}{\varphi(s)+\sqrt{t}}.$$

Consequently,

$$\begin{aligned}
&\|y-y_t\|_Y^2+t\,\|y_t\|_{A(X_0)}^2\\
&=\int (y(s)-\frac{\varphi(s)y(s)}{\varphi(s)+\sqrt{t}})^2\,d\mu+t\int (\frac{\varphi(s)y(s)}{\varphi(s)+\sqrt{t}})^2\cdot\frac{1}{\varphi^2(s)}\,d\mu\\
&=2\int\frac{ty^2(s)}{(\varphi(s)+\sqrt{t})^2}\,d\mu\le 2\int\frac{ty^2(s)}{\varphi(s)^2+t}\,d\mu=2L_{2,2}(t,y;Y,A(X_0)).
\end{aligned}$$ □

**Theorem 6.4.** *Let $X$ and $Y$ be Hilbert spaces, and let $X_0=A^*(Y)$. Suppose we know the following estimate for the precise solution:*

$$\|x_*\|_{X_0}\le E.$$

*As an approximation to the precise solution, we take the element*

$$\tilde{x}_*=(A^*A+\frac{\varepsilon}{E}I)^{-1}A^*y.$$

*Then*

$$\|x_*-\tilde{x}_*\|_X\le 3\varepsilon^{\frac12}E^{\frac12}. \tag{6.8}$$

*Proof.* Since Conditions A and B are fulfilled, the above Theorem 6.2 applies. It remains to observe that, for $t=\frac{\varepsilon^2}{E^2}$, the element $A^{-1}y_t$ is equal to $\tilde{x}_*$. The constant 3 in (6.8) (in place of $c$, see (6.6)) arises from the proof of Theorem 6.2, because Condition A is fulfilled with constant 1 and Condition B is fulfilled with constant 2. □

# Notes and remarks

The exposition and results of this short chapter are due to Asekritova and Kruglyak and were not published before. The chapter was written merely to attract the reader's attention to the relationship between near-minimizers for $L$-functionals and the theory of regularization of ill-posed problems. We repeat that problems of this sort emerge, in particular, under image processing, and near-minimizers for the couple ($L^2(\Omega)$, the space of functions of bounded variation on $\Omega$) for a planar domain $\Omega$ are important in this context. In this connection, see Chapter 11.

It should be noted that an approximation error for the exact solution can be estimated at the expense of a "smootheness loss" (the error was estimated in the space $X$, whereas the exact solution belongs to $X_0$) and a slower convergence ($\varepsilon^{1-\theta}$ arises in the norm estimate in place of $\varepsilon$).

In the case of Hilbert scales, this "smoothness loss effect" is not quite transparent because, by Lemma 6.3, the role of a near-minimizer for the Tikhonov functional for the couple $(X_0, Y)$ (see (6.5)) can be played by the exact minimizer for the couple $(X, Y)$ (see (6.4)).

# Chapter 7

# Stability for analytic Hardy spaces

As we have seen (and much more evidence to this will be given in Part 2 of this book), Calderón–Zygmund-type decompositions are a universal tool for constructing near-minimizers and proving their stability. However, there is a price to be paid for generality: some natural questions remain beyond the capacity of this method. We have already encountered a question of this sort, namely, the case of an infinite exponent in the scale of Lebesgue spaces, see §3 in Chapter 3.

We want to say that in specific situations some other approaches may be applicable, which may lead to deeper results. These methods cannot be expected to turn out as general, though. We illustrate this by the scale of analytic Hardy spaces. In this case, there is a theory related to the preceding material and giving answers to many difficult questions (especially, pertaining to $K$-closed couples) that cannot, as far as we know, be treated with the help of Calderón–Zygmund decompositions. The main results are here due to S. Kislyakov and Q. Xu. See the survey [Kis] and the references therein, and also the more recent papers [Kis1], [Kis2], and [KX1]. Here we present a principal ingredient of that study in the simplest case.

On $L^2(\mathbb{R})$, consider the operator $\mathbb{P}$, $\mathbb{P}f = \frac{1}{2}(f + iHf)$, where $H$ is the Hilbert transformation. Passing to Fourier transforms, we see that $\widehat{\mathbb{P}f} = \chi_{(0,+\infty)} \cdot \hat{f}$ a.e., so that $\mathbb{P}$ is a projection; usually it is called the *Riesz projection*. By the properties of $H$, $\mathbb{P}$ is bounded on $L^p(\mathbb{R})$ for $1 < p < \infty$ and is long-range regular on $L^1(\mathbb{R})$. So, the stability result proved in §1.4 shows that if a function $f$ lies in $L^1(\mathbb{R})$ and satisfies $f = \mathbb{P}f$, then for every $s > 0$ its $L^1$-distance to the ball $B_{L^p}(s)$ is roughly the same as its $L^1$-distance to $\mathbb{P}B_{L^p}(s)$ $(1 < p < \infty)$.

We introduce the *Hardy classes* $H^p$ by $H^p = \{f \in L^p : \mathbb{P}f = f\}$, $1 \le p < \infty$. The discussion in §5.6 shows that, for every $p \in (1, \infty)$, the above result means that the couple $(H^1, H^p)$ is $K$-closed in $(L^1, L^p)$. Revision of the proof shows that the crucial point is Calderón–Zygmund decomposition, which provides $\mathbb{P}$-stable near-minimizers for the distance functional of the couple $(L^1, L^p)$.

We want to explain that in this situation there is another approach to results of this type, which gives even more and is based on a version of formula (0.1). Unfortunately, this approach is quite specific and is hardly applicable beyond the present framework. It is based on the fact that $H^p$ is a boundary class for some space of functions analytic on the upper half-plane $\mathbb{C}_+$, and therefore multiplication by (the boundary values of) analytic "cut-off" functions is admissible.

So, let $\mathbb{C}_+$ be the upper half-plane $\{z = x + iy \in \mathbb{C} : y > 0\}$. The class $H^p(\mathbb{C}_+)$, $0 < p \leq \infty$, is defined to be the collection of all functions $f$ analytic on $\mathbb{C}_+$ and satisfying

$$\|f\|_{H^p(\mathbb{C}_+)} = \sup_{y>0} \|f(\cdot + iy)\|_{L^p} < +\infty$$

(thus, $H^\infty$ is the space of bounded analytic functions on $\mathbb{C}_+$). We refer the reader to [Koo] for the theory of the classes $H^p(\mathbb{C}_+)$. Here we only recall some facts about them necessary for what follows.

First, for every $f \in H^p(\mathbb{C}_+)$, the boundary values $\lim_{y\to 0_+} f(x + iy)$ exist for a.e. $x \in \mathbb{R}$, and the boundary function (usually denoted by the same latter $f$) belongs to $L^p(\mathbb{R})$. Moreover, the $L^p$-norm of this boundary function coincides with $\|f\|_{H^p(\mathbb{C}_+)}$. Next, for $1 \leq p < \infty$ the set of boundary functions for $H^p(\mathbb{C}_+)$ coincides with the class $H^p$ introduced above. We extend this definition to $p = \infty$ and $p < 1$, putting $H^p = H^p(\mathbb{C}_+)|_{\mathbb{R}}$ for these $p$.

We shall employ the term "analytic" for the functions on $\mathbb{R}$ belonging to the linear hull of $\bigcup_{0<p\leq\infty} H^p$. There is an important statement called "the boundary maximum principle" saying, in particular, that any analytic function in $L^p(\mathbb{R})$ belongs to $H^p$. See [Koo].

Now, we state the result.

**Theorem 7.1.** *The couple $(H^1, H^\infty)$ is $K$-closed in $(L^1(\mathbb{R}), L^\infty(\mathbb{R}))$.*

We choose $H^\infty$ as the second space in the couple for several reasons. First, stability in the case of an infinite exponent presents fundamental difficulties for the method based on Calderón–Zygmund decomposition. Second, in fact, the result for the couple $(H^1, H^p)$ with $p < \infty$ can be deduced from Theorem 7.1 by some "abstract" interpolation tricks. Finally, a modification of the arguments below gives the result for the couple $(H^1, H^p)$ directly.

As we have already seen, Theorem 7.1 can be interpreted as a stability result: the distance in $L^1$ from an analytic function to a ball of $L^\infty$ is roughly attained at a function in $H^\infty$, i.e., at a function that is also analytic.

Before passing to the proof, we give some hints. Recall formula (0.1): $g = f\cdot\Phi$, where $\Phi = \beta^{-1}$, $\beta = \max(1, t^{-1}|f|)$. We shall proceed similarly, but we want $g$ to be analytic if $f$ is. Surely, for that we need to modify $\Phi$. It turns out that taking $\Phi = \alpha^{-1}$ with $\alpha = \beta + iH\beta = 1 + [(\beta - 1) + iH(\beta - 1)]$ is a happy idea, but some refinement of this is required. See below.

*Proof.* Suppose $f \in H^1 + H^\infty$ is decomposed as follows: $f = g + h$, where $g \in L^1$, $\|g\|_{L^1} \le a$, $h \in L^\infty$, $\|h\|_{L^\infty} \le b$, $b > 0$. We need to replace $g$ and $h$ by functions in $H^1$ and $H^\infty$, respectively, that obey roughly the same estimates for the norms. We put $\beta = \max(1, (2b)^{-1}|f|)^{1/2}$, then $\beta - 1 \in L^2$, and $H(\beta - 1)$ is also in $L^2$, see §1 in Chapter 2. So, $\varphi = 1 + (\beta - 1) + iH(\beta - 1)$ is an analytic function, and its real part is $\beta$, which is greater than or equal to 1 a.e. It follows that $\Psi = \varphi^{-1}$ is also an analytic function and, since $|\Psi| \le \beta^{-1} \le 1$, we see that $\Psi \in H^\infty$. We put $\Phi = 1 - (1 - \Psi^2)^2$ and claim that $f = (1 - \Phi)f + \Phi f$ is a required decomposition. Note that the summands are analytic, and we need only estimate their norms.

First, $|\Phi| = |\Psi|^2|2 - \Psi^2| \le 3|\Psi|^2 \le \frac{3}{\beta^2}$, whence $|\Phi f| \le 6b$. Next,

$$|1 - \Phi| = |1 - \Psi^2|^2 \le 2|1 - \Psi| = \left|1 - \frac{1}{\beta + iH(\beta - 1)}\right|^2$$

$$= \left|\frac{\beta - 1 + iH(\beta - 1)}{\beta + iH(\beta - 1)}\right|^2 \le (|\beta - 1| + |H(\beta - 1)|)^2.$$

We put $e = \{x \in \mathbb{R} : |f(x)| \ge 2b\}$, then $|g| \ge b$ on $e$ so that $|e| \le ab^{-1}$. Observing also that $|\Phi| \le 3$, we can write

$$\int_{\mathbb{R}} |(1 - \Phi)f| \le 4 \int_e (|g| + b) + 2b \int_{\mathbb{R}\setminus e} (|\beta - 1| + |H(\beta - 1)|)^2$$

$$\le 4 \int_{\mathbb{R}} |g| + 4b|e| + 2b(\|\beta - 1\|_{L^2} + \|H(\beta - 1)\|_{L^2})^2$$

$$\le 8a + 8b \int_{\mathbb{R}} |\beta - 1|^2$$

because the norm of $H$ on $L^2$ is 1. Now,

$$\int_{\mathbb{R}} |\beta - 1|^2 \le \int_e \frac{|f|}{2b} \le \int_e \frac{|g| + b}{2b} \le \frac{1}{2b} a + \frac{1}{2}|e| \le \frac{a}{b}.$$

Collecting the estimates, we see that $\int |(1 - \Phi)f| \le 16a$, as required. □

# Notes and remarks

Certain ramifications and extensions of the method described above work very well also for weighted Hardy classes of analytic functions (more generally, for the subspaces of analytic functions in various quasi-Banach lattices of measurable functions on $\mathbb{R}$). See the papers mentioned at the beginning of this chapter. We give some more details about this stuff.

Let $w$ be a positive measurable function on $\mathbb{R}$. It is convenient to define the weighted space $L^p(w)$ to be the set of all measurable $f$ on $\mathbb{R}$ such that $f/w \in L^p(\mathbb{R})$. The function $w$ is called a *weight* if $(\log w(x))(1+x^2)^{-1}$ is integrable.[1] If $w$ is a weight, there is a standard procedure to introduce the weighted Hardy space $H^p(w)$ as a certain closed subspace of $L^p(w)$, which is also the boundary class for a certain space of analytic functions on the upper half-plane (to be specific, $H^p(w) = WH^p$, where $W$ is the outer function whose modulus is $w$; see, e.g., [Kis], [Kis1] for some more details).

**Theorem.** *Let $w_0$ and $w_1$ be weights, and let $0 < p_0, p_1 \leq \infty$. Then the couple $(H^{p_0}(w_0), H^{p_1}(w_1))$ is $K$-closed in the couple $(L^{p_0}(w_0), L^{p_1}(w_1))$ if and only if* $\log(w_0/w_1) \in \mathrm{BMO}$.

The "if" part can be proved by a variation of the above argument. The "only if" part is established by using the so-called Garcia norm on BMO; the method goes back to [CMW], where a particular case was analyzed. See [KX2] for a complete proof.

The following theorem (see [KX1]) also yields the "if" part in the above statement, but this time the proof is more sophisticated. We state this theorem here because, ideologically, it is in some relationship with linearisation (§7 in Chapter 5).

**Theorem.** *Suppose that* $\log(w_0/w_1) \in \mathrm{BMO}$. *If $f \in H^{p_0}(w_0) + H^{p_1}(w_1)$, then there is a linear operator $T$ that fixes $f$ and takes boundedly $L^{p_j}(w_j)$ to $H^{p_j}(w_j)$, $j = 1, 2$, with norms independent of $f$.*

This is interesting even if the two weights are identically equal to 1.

Now, we want to say some more words about comparison of the method based on Calderón–Zygmund decompositions and that based on multiplication by analytic cut-off functions in the context of analytic Hardy spaces. In this book, the emphasis is on the *universality* and *explicit form* of Calderón–Zygmund minimizers. The claim at the beginning of this chapter that the Calderón–Zygmund theory is not quite adequate if we want to prove the $K$-closedness of the couple $(H^1, H^\infty)$ in $(L^1, L^\infty)$ should be expanded as follows. This theory does fail totally in the case of weighted spaces with arbitrary weights. However, if the weights are identically equal to 1, it can still be used via duality and some interpolation tricks, so that only the explicit form of the relevant stable near-minimizers is lost. (At the same time, analytic cut-off functions do give explicit stable near-minimisers.) The approach via duality, interpolation, and the Calderón–Zygmund theory was developed in full detail in [KX] and [Kis], but the pioneering paper is [Bo].

Unfortunately, analytic cut-off functions are applicable only in the framework of one complex variable. We mention, however, a result for two variables, proved in [KX]. It concerns the Hardy spaces on the two-dimensional torus $\mathbb{T}^2$ (which, of

[1] In the papers cited above, the exposition is for the functions on the unit circle rather then on $\mathbb{R}$. A weight on the circle is a function with integrable logarithm; the factor of $(1+x^2)^{-1}$ arises as a result of the standard fractional-linear mapping of the unit disk onto the upper half-plane.

course, are identified with certain classes of functions analytic on the bidisk). See [Ru] for the definitions and all particulars about these classes.

**Theorem.** *For every $p \in (0, \infty)$, the couple $(H^p(\mathbb{T}^2), H^\infty(\mathbb{T}^2))$ is $K$-closed in $(L^p(\mathbb{T}^2), L^\infty(\mathbb{T}^2))$.*

In the proof, both Calderón–Zygmund decompositions and analytic cut-off functions are used critically (as well as duality, interpolation, and real-variable theory of Hardy spaces). It is not known whether the same is true for tori of dimension greater than 2. If the two integrability indices are finite, the corresponding $K$-closedness result is known in arbitrary dimension, see [Xu].

Returning to dimension 1, we remark that it is possible to do something beyond the framework of weighted Hardy spaces. For every reasonable Banach lattice (ideal space) $X$ of measurable functions on the real line, there is a way to define its *analytic subspace* $X_A$. (Surely, when applied to $L^p(w)$, the procedure yields $H^p(w)$.) Next, the lattice $X$ is said to be BMO-regular if for every nonzero function $f \in X$ there exists a function $g \in X$ with $|f| \leq g$, $\|g\|_X \leq C\|f\|_X$, and $\|\log g\|_{\mathrm{BMO}} \leq C$ with a constant $C$ independent of $f$.

It turns out (see [Kis], [Kis1], [Kis2]) that BMO-regularity occurs quite often. In particular, $L^p(w)$ is BMO-regular if and only if $\log w \in \mathrm{BMO}$. On the other hand, the following is true.

**Theorem.** *If $X$ and $Y$ are* BMO*-regular lattices of measurable functions on the real line, then the couple $(X_A, Y_A)$ is $K$-closed in $(X, Y)$.*

The main open problem in this range of questions is whether some version of BMO-regularity is necessary for $K$-closedness (only some partial statements in support of a positive answer are available as yet). On the other hand, much is known about the property of BMO-regularity itself. For example, there is a difficult theorem that this property passes from a lattice to its order dual. This was first proved in [Kis1] with the use of complex variable methods. A purely real proof can be found in [Rut]. That paper also contains solutions of many other problems of a general nature concerning various forms of BMO-regularity in many dimensions.

# Part II

# Advanced theory

The core of this part is formed by the first three chapters, where we prove some fine covering theorems, employ them for construction of Calderón–Zygmund type decompositions in new cases, specifically, for couples of the form

$$(L^p, \text{ some space of smooth functions}),$$

construct near-minimizers for such couples, and finally, prove that these near-minimizers are stable under the action of singular integral operators.

The remaining content (Chapter 11 and the Appendix) can be viewed as a supplement to the mainstream. There we consider some subtler settings in which the methods under study are still applicable. Chapter 11 is devoted to a limiting case in the theory of near-minimizers in Chapter 9. Technically, this topic turns out to be quite involved; moreover, an answer is obtained only in dimension 2, and the case of bigger dimensions remains open, though there are some natural conjectures. In the Appendix, we show how the methods can be generalized beyond the framework of Chapter 9, where one of the spaces in the couple is obligatory $L^p$.

# Chapter 8

# Controlled coverings

Central in this chapter are Theorems 8.27 and 8.39 on controlled extension and controlled contraction. Given a family of cubes, they make it possible to construct a new one with fine properties, which is controlled in a sense by the initial family. In the next chapter, these results will be used for construction of near-minimizers for various couples involving a space of smooth functions.

Before passing to new covering theorems, for the reader's convenience we summarize some covering results and their consequences proved in Part 1. In what follows, $Q_x$ stands for a closed cube centered at $x$ and with faces parallel to coordinate hyperplanes. We remind the reader that $r(Q)$ is the radius of a cube $Q \subset \mathbb{R}^n$, i.e., half its edge length. Next, $Q(x, r)$ will denote the cube centered at $x$ and of radius $r$. Recall also that the distance between two points $x, y \in \mathbb{R}^n$ is defined by

$$\|y - x\| = \max_{i=1,\ldots,n} |y_i - x_i|, \quad x = (x_1, \ldots, x_n), \quad y = (y_1, \ldots, y_n).$$

The balls in this metric are cubes:

$$y \in Q_x \Leftrightarrow \|y - x\| \leq r(Q_x).$$

Now, we start the summary promised above.

**Lemma 8.1** (Whitney). *Let $\Omega \subset \mathbb{R}^n$ be an open set with nonempty complement $F = \mathbb{R}^n \backslash \Omega$. Then there exists a family $\{Q_i\}_{i \in I}$ of dyadic cubes with nonintersecting interiors such that*

$$\bigcup_{i \in I} Q_i = \bigcup_{i \in I} 2Q_i = \Omega \tag{8.1}$$

*and*

$$5Q_i \cap F \neq \varnothing, \ i \in I. \tag{8.2}$$

We reserve the term *Whitney cubes* for the cubes of this family $\{Q_i\}_{i\in I}$. It should be recalled that (8.1) and (8.2) imply that the distance from a Whitney cube $Q_i$ to the boundary of $\Omega$ in the metric $\|y-x\| = \max_{i=1,\dots,n}|y_i - x_i|$ is roughly proportional to $r(Q_i)$.

In many important applications of the Whitney lemma, instead of the Whitney cubes $\{Q_i\}_{i\in I}$, the modified cubes $\{K_i = (1+\frac{1}{6})Q_i\}_{i\in I}$ are employed. We saw that these cubes have some remarkable properties. They are summarized in the next definition-reminder.

**Definition 8.2.** A family $\{K_i\}_{i\in I}$ of cubes is called a WB-covering of a set $\Omega$ (in case $\Omega$ is not indicated, it is assumed to be merely the union of the $K_i$) if it has the following properties:

- Whitney property

$$\bigcup_{i\in I}\Big(\frac{1}{1+\delta}K_i\Big) = \bigcup_{i\in I} K_i = \Omega; \tag{8.3}$$

- finite multiplicity property

$$\sum_{i\in I}\chi_{K_i} \le M; \tag{8.4}$$

- strong engagement property

$$\text{if } K_i \cap K_j \neq \varnothing, \text{ then } |K_i\cap K_j| \ge \varepsilon \max(|K_i|, |K_j|), \qquad i,j\in I. \tag{8.5}$$

Here the positive constants $\delta$, $M$, and $\varepsilon$ are independent of $i, j \in I$. They are called the Whitney, multiplicity, and engagement constants, respectively.

In fact (see Lemma 3.13), (8.4) is always fulfilled in a stronger form of the

- finite overlap property

$$\operatorname{card}\{j : Q_j \cap Q_i \neq \varnothing\} \le \nu \tag{8.6}$$

with a quantity $\nu$ independent of $i$ and called the overlap number.

We recall a simple example of a WB-covering of $\mathbb{R}^n$ (see Chapter 3).

*Example* 8.3. Split $\mathbb{R}^n$ into cubes $\{Q_i\}_{i\in I}$ of equal volume and put $K_i = (1+\frac{1}{2})Q_i$. Then the family $\{K_i\}_{i\in I}$ is a WB-covering of $\mathbb{R}^n$ with $\delta = \frac{1}{2}$, $M = 2^n$, $\varepsilon = \frac{1}{3^n}$, and $\nu = 3^n$.

We also recall the theorem about the modified Whitney cubes proved in Chapter 3.

**Theorem 8.4.** *Let $\{Q_i\}_{i\in I}$ be the family of Whitney cubes for an open set $\Omega$ with nonempty complement $\mathbb{R}^n\setminus\Omega$. Then the cubes $K_i = (1+\frac{1}{6})Q_i$, $i\in I$, constitute a* WB*-covering of $\Omega$. For this covering, the overlap constant $\nu$, the multiplicity constant $M$, and the engagement constant $\varepsilon$ depend only on the dimension $n$, and $\delta = \frac{1}{6}$ in the Whitney property* (8.3).

It is important that a WB-covering $\{K_i\}_{i\in I}$ makes it possible to construct a $C^\infty$-partition of unity adjusted to it. See Capter 3. We reproduce the statement.

**Theorem 8.5.** *Suppose that a family $\{K_i\}_{i\in I}$ of cubes is a WB-covering with constants $\delta$, $M$, and $\varepsilon$. Then there exists a family $\psi_i \in C^\infty$, $i \in I$, of nonnegative functions that form a partition of unity for the set $\cup K_i$, i.e.,*

$$\sum_{i\in I} \psi_i = \chi_{\cup K_i}, \tag{8.7}$$

*with the following properties:*

$$\operatorname{supp} \psi_i \subset \frac{1+\frac{\delta}{2}}{1+\delta} K_i, \quad i \in I, \tag{8.8}$$

$$\psi_i(x) \geq \frac{1}{M} \text{ for } x \in \frac{1}{1+\delta} K_i, \quad i \in I, \tag{8.9}$$

*and*

$$\left|D^k \psi_i\right| \leq c_k \frac{1}{|K_i|^{\frac{|k|}{n}}}, \qquad |k| = \sum_{j=1}^n k_j, \qquad D^k = \frac{\partial^k}{\partial^{k_1} \partial^{k_2} \ldots \partial^{k_n}}, \tag{8.10}$$

*for all nonnegative multiindices $k = (k_1, \ldots, k_n) \in Z^n$, with a constant $c_k > 0$ depending only on $k$, $\delta$, $M$, and $\varepsilon$, but independent of the particular cubes in the family $\{K_i\}_{i\in I}$.*

Finally, we recall the Besicovitch covering theorem (see Theorem 3.4).

**Theorem 8.6.** *Let $\Omega \subset \mathbb{R}^n$. Consider a family $\{Q_x\}_{x\in\Omega}$ of cubes satisfying*

$$\sup_{x\in\Omega} r(Q_x) < \infty.$$

*Then this family has an at most countable subfamily $\{Q_{x_i}\}_{i\in I}$ such that $\Omega \subset \bigcup_{i\in I} Q$ and $\sum_{i\in I} \chi_{Q_{x_i}} \leq M$. The multiplicity constant $M$ depends only on the dimension $n$.*

## 8.1 Whitney lemma and a theorem about Lipschitz families

It is natural to ask whether WB-coverings can be constructed under the assumptions of the Besicovitch theorem. Clearly, even the initial settings of the Whitney lemma and the Besicovitch theorem are quite different. In the Whitney lemma, we start with an open set with nonempty complement, whereas in the Besicovitch theorem an almost arbitrary family of cubes is treated. In this section we shall show that the existence of a WB-covering under the assumptions of the Whitney lemma is a particular case of a general result for Lipschitz families of cubes. This bridges the gap between the patterns (a family of cubes is considered instead of an open set) and indicates a way for constructing a WB-covering in the framework of the Besicovitch theorem.

**Definition 8.7.** A family of cubes $\{Q_x\}_{x\in\Omega}$ is said to be *Lipschitz with constant q* if for every $x, y \in \Omega$ we have

$$|r(Q_x) - r(Q_y)| \leq q\,\|x-y\|\,. \tag{8.11}$$

The next definition describes an important subclass of Lipschitz families.

**Definition 8.8.** A family of cubes $\{Q_x\}_{x\in\Omega}$ is said to be *strongly Lipschitz with constant q* if the function

$$F(x) = \begin{cases} r(Q_x), & x \in \Omega, \\ 0, & x \in \mathbb{R}^n \setminus \Omega \end{cases}$$

satisfies the Lipschitz condition with constant $q$:

$$|F(x) - F(y)| \leq q\,\|x-y\|\,, \quad x, y \in \mathbb{R}^n. \tag{8.12}$$

There are two trivialities to be said explicitly. First, a family $\{Q_x\}_{x\in\Omega}$ of cubes is strongly Lipschitz if the function $F(x) = r(Q_x)$ becomes Lipschitz (with constant $q$) when extended to $\mathbb{R}^n$ by 0 outside $\Omega$. Second, a nonnegative Lipschitz function $F$ on $\mathbb{R}^n$ *generates* the strongly Lipschitz family of cubes $\{Q_x\}_{x\in\Omega}$ with $r(Q_x) = F(x)$ and $\Omega = \{x : F(x) > 0\}$.

It turns out that, expanding slightly the cubes of some subfamily $\{Q_{x_i}\}_{i\in I}$ of the initial family $\{Q_x\}_{x\in\Omega}$, we can construct a WB-covering of the set $\Omega$ of centers. The precise statement is given below. We postpone the proof slightly to give room for applications and discussion.

**Theorem 8.9** (On strongly Lipschitz families)**.** *Suppose we are given a strongly Lipschitz family $\{Q_x\}_{x\in\Omega}$ of cubes with constant $q < 1$. Then there exist points $x_i \in \Omega$ and numbers $1 < \lambda_i < \frac{1}{q}$, $i \in I$, such that the family $\{K_{x_i} = \lambda_i Q_{x_i}\}_{i\in I}$ is a* WB*-covering of $\Omega$.*

*Remark* 8.10. The Whitney constant $\delta$ of the WB-covering mentioned in the preceding theorem depends only on $q$ whereas the overlap number $\nu$ and the engagement constant $\varepsilon$ depend on $q$ and the dimension $n$.

We explain how the theorem on Lipschitz families implies the existence of a WB-covering under the assumptions of the Whitney lemma. First, under these assumptions, a strongly Lipschitz family of cubes arises naturally.

**Proposition 8.11.** *Suppose $\Omega \subset \mathbb{R}^n$ is an open set and $\mathbb{R}^n \setminus \Omega \neq \varnothing$. Fix a positive number $q$, and for every $x \in \Omega$ consider the cube $Q_x$ centered at $x$ and of radius $r(Q_x) = q \cdot \rho_x$, where*

$$\rho_x = \inf_{u\in\mathbb{R}^n\setminus\Omega} \|x-u\|$$

*is the distance from $x$ to $\mathbb{R}^n \setminus \Omega$. Then the family $\{Q_x\}_{x\in\Omega}$ is strongly Lipschitz with constant $q$.*

*Proof.* Indeed, if $x, y \in \Omega$, then $|\rho_x - \rho_y| \leq \|x - y\|$ (because, clearly, $\rho_x \leq \|x - y\| + \rho_y$ ). Therefore, the family $\{Q_x\}_{x\in\Omega}$ is Lipschitz with constant $q$. The Lipschitz property of the function

$$F(x) = \begin{cases} q \cdot \rho_x, & x \in \Omega, \\ 0, & x \notin \Omega \end{cases}$$

follows from the observation that, if $x \in \Omega$ and $y \notin \Omega$, then

$$|r(Q_x) - 0| = q \cdot \rho_x \leq q \|x - y\| . \qquad \square$$

Now the existence of a WB-covering under the conditions of the Whitney lemma follows easily.

**Theorem 8.12.** *Suppose $\Omega \subset \mathbb{R}^n$ is open and $\mathbb{R}^n \setminus \Omega \neq \varnothing$. Then there exists a family $\{K_{x_i}\}_{i\in I}$ of cubes that is a* WB*-covering of $\Omega$ and satisfies $5K_{x_i} \cap (\mathbb{R}^n \setminus \Omega) \neq \varnothing$.*

*Proof.* Take the family $\{Q_x\}_{x\in\Omega}$ constructed in Proposition 8.11 with $q = \frac{1}{5}$, and apply the theorem on Lipschitz families (Theorem 8.9) to it. This results in a WB-covering $\{K_{x_i}\}_{i\in I}$ of $\Omega$. The property $5K_{x_i} \cap (\mathbb{R}^n \setminus \Omega) \neq \varnothing$ follows from the inclusion $Q_{x_i} \subset \lambda_i Q_{x_i} = K_{x_i}$ and the fact that, by construction, we have $5Q_x \cap (\mathbb{R}^n \setminus \Omega) \neq \varnothing$. $\square$

In the case of a Lipschitz family of cubes satisfying also the Besicovitch condition $\sup_{x\in\Omega} r(Q_x) < \infty$, Theorem 8.9 can be refined. The statement is presented below, the proof will also be postponed.

**Theorem 8.13.** *Suppose we are given a strongly Lipschitz family of cubes $\{Q_x\}_{x\in\Omega}$ with constant $q < \sqrt{2} - 1$ and such that*

$$\sup_{x\in\Omega} r(Q_x) < \infty.$$

*Then there exists a subfamily $\{K_{x_i} = \lambda_i Q_{x_i}\}_{i\in I}$ with $1 < \lambda_i < \frac{1}{q}$ that is a* WB*-covering of the set $\Omega$ and, moreover, has the property that for every cube $Q_x$ there is an index $i = i(x)$ with $Q_x \subset K_{x_i}$.*

*Remark* 8.14. The Whitney constant $\delta$ of the WB-covering constructed in this theorem is $1 + q$, and the multiplicity and engagement constants $M$ and $\varepsilon$, as well as the overlap number $\nu$ depend only on $q$ and the dimension $n$.

Concluding these introductory remarks, we discuss the necessity of the condition $\sup_{x\in\Omega} r(Q_x) < \infty$ in the Besicovitch theorem. In general, it cannot be lifted, as is shown by the following example.

*Example* 8.15. For the set $\Omega$ of centers of our cubes, we take the set $\mathbb{R}_+$ of positive numbers. The cubes (intervals) are determined by

$$r(Q_x) = x$$

for $x \in \mathbb{R}_+$. Since the origin belongs to all $Q_x$, any finite multiplicity subfamily of this family is finite and cannot cover $\Omega = \mathbb{R}_+$.

The family of cubes in this example is Lipschitz with constant $q = 1$. However, it turns out that for Lipschitz families with $q < 1$, the Besicovitch theorem remains true without the restriction $\sup_{x\in\Omega} r(Q_x) < \infty$; moreover, the finite overlap property can be ensured for the resulting cubes in place of the mere finite multiplicity.

**Theorem 8.16** (Besicovitch theorem for Lipschitz families). *Suppose $\Omega \subset \mathbb{R}^n$ and a family $\{Q_x\}_{x\in\Omega}$ of cubes is Lipschitz with constant $q < 1$. Then this family has an at most countable subfamily $\{Q_{x_i}\}_{i\in I}$ that covers $\Omega$:*

$$\Omega \subset \bigcup_{i\in I} Q_{x_i}, \tag{8.13}$$

*and is of finite overlap:*

$$\operatorname{card}\{i \in I : Q_{x_i} \cap Q_{x_j} \neq \varnothing\} \leq \nu \text{ for every } j \in I \tag{8.14}$$

*with $\nu$ depending only on $q$ and the dimension $n$.*

Again, we postpone the proof.

### 8.1.1 Auxiliary lemmas

The proof of the theorems on Lipschitz families stated above leans upon two lemmas (about finite overlap and the engagement property) stated and proved below.

First, we observe that a strongly Lipschitz family of cubes always possesses the Whitney property (see (8.3)).

**Lemma 8.17** (On the Whitney property). *Suppose a family $\{Q_x\}_{x\in\Omega}$ is strongly Lipschitz with constant $q < 1$. Then for every $\varepsilon_0 > 0$ such that $1 + \varepsilon_0 < \frac{1}{q}$ we have*

$$\bigcup_{x\in\Omega} Q_x = \bigcup_{x\in\Omega} (1+\varepsilon_0) Q_x = \Omega. \tag{8.15}$$

*Proof.* By definition, the nonnegative function $F$ on $\mathbb{R}^n$ determined by the conditions $\Omega = \{x \in \mathbb{R}^n : F(x) > 0\}$ and $r(Q_x) = F(x)$ for all $x \in \Omega$ is Lipschitz with constant $q$:

$$|F(x) - F(y)| \leq q \|x - y\|, \quad x, y \in \mathbb{R}^n.$$

We prove (8.15). Since $\Omega \subset \bigcup_{x\in\Omega} Q_x$, it suffices to show that $(1+\varepsilon_0)Q_x \subset \Omega$ for all $x \in \Omega$. Suppose the contrary, i.e., there exists $y \in (1+\varepsilon_0)Q_x$ with $y \notin \Omega$. Then $F(y) = 0$ and

$$\|y - x\| \le (1+\varepsilon_0)r(Q_x) = (1+\varepsilon_0)F(x).$$

Therefore,

$$F(x) = |F(x) - F(y)| \le q\,\|x - y\| \le q(1+\varepsilon_0)r(Q_x) = q(1+\varepsilon_0)F(x) < F(x),$$

because $1+\varepsilon_0 < \frac{1}{q}$. This is impossible. □

**Definition 8.18.** A family $\{Q_i\}_{i\in I}$ of cubes is said to be almost disjoint if for every $i, j \in I$, $i \ne j$, we have

$$\min\left(r(Q_i), r(Q_j)\right) < \|x_i - x_j\|\,, \tag{8.16}$$

where $x_i$ and $x_j$ are the centers of $Q_i$ and $Q_j$.

The above condition means that the center of at least one of the cubes $Q_i$, $Q_j$ does not belong to the other cube. The term "almost disjoint" is motivated by the following statement.

**Proposition 8.19.** *Suppose a family $\{Q_i\}_{i\in I}$ of cubes is Lipschitz with constant $q$ and almost disjoint. Then for every $i, j \in I$, $i \ne j$, we have*

$$\frac{1}{2+q}Q_i \cap \frac{1}{2+q}Q_j = \varnothing.$$

*Proof.* Let $x_i$ and $x_j$ denote the centers of $Q_i$ and $Q_j$. Since these cubes are almost disjoint, we may assume that $x_j \notin Q_i$. If there exists $z \in \frac{1}{2+q}Q_i \cap \frac{1}{2+q}Q_j$, the inequality $r(Q_j) - r(Q_i) \le q\,\|x_j - x_i\|$ implies

$$\begin{aligned}\|x_j - x_i\| \le \|x_j - z\| + \|z - x_i\| &\le \frac{1}{2+q}r(Q_j) + \frac{1}{2+q}r(Q_i)\\ &\le \frac{1}{2+q}(r(Q_i) + q\,\|x_j - x_i\|) + \frac{1}{2+q}r(Q_i).\end{aligned}$$

Therefore,

$$(1 - \frac{q}{2+q})\,\|x_j - x_i\| \le \frac{2}{2+q}r(Q_i),$$

and, consequently, $\|x_j - x_i\| \le r(Q_i)$, which is impossible because $x_j \notin Q_i$. □

## 8.1.2 Finite overlap

We shall need the following statement.

**Proposition 8.20.** *Suppose the cubes $\{Q_i\}_{i\in I}$ are centered at $x_i$, $i \in I$, and form a Lipschitz family with constant $q < 1$. If $\varepsilon_0 > 0$ satisfies $1 + \varepsilon_0 < \frac{1}{q}$, then the relation $(1+\varepsilon_0)Q_i \cap (1+\varepsilon_0)Q_j \neq \varnothing$ implies*

$$\max(r(Q_i), r(Q_j)) \leq c \min(r(Q_i), r(Q_j)), \tag{8.17}$$

*where*

$$c = \frac{1 + (1+\varepsilon_0)q}{1 - (1+\varepsilon_0)q}.$$

*Proof.* Let $r(Q_i) \geq r(Q_j)$ for definiteness, and let $z \in (1+\varepsilon_0)Q_i \cap (1+\varepsilon_0)Q_j$. By the Lipschitz condition, we obtain

$$\begin{aligned} r(Q_i) - r(Q_j) &\leq q\,\|x_i - x_j\| \leq q(\|x_i - z\| + \|z - x_j\|) \\ &\leq (1+\varepsilon_0)q(r(Q_i) + r(Q_j)). \end{aligned}$$

Therefore,

$$(1 - (1+\varepsilon_0)q)\, r(Q_i) \leq (1 + (1+\varepsilon_0)q)\, r(Q_j),$$

and (8.17) follows. □

The next important lemma shows that, under certain conditions, the finite overlap property is preserved after dilation.

**Lemma 8.21.** (On finite overlap) *Let a family $\{Q_i\}_{i\in I}$ of cubes be Lipschitz with constant $q < 1$ and almost disjoint. Then for every $\varepsilon_0 > 0$ with $1 + \varepsilon_0 < \frac{1}{q}$, the family $\{(1+\varepsilon_0)Q_i\}_{i\in I}$ is of finite overlap. Moreover, for fixed $\varepsilon_0$ and $q$, its overlap constant $\nu$ depends only on the dimension $n$.*

*Proof.* Let $x_i$, $i \in I$, denote the centers of the cubes $Q_i$. We fix an index $i_0$ and estimate the number of indices $i$ such that $(1+\varepsilon_0)Q_i$ intersects $(1+\varepsilon_0)Q_{i_0}$. By (8.17), we see that

$$r(Q_i) \leq \frac{1 + (1+\varepsilon_0)q}{1 - (1+\varepsilon_0)q} r(Q_{i_0}).$$

Therefore,

$$\|x_i - x_{i_0}\| \leq (1+\varepsilon_0)\left(\frac{1 + (1+\varepsilon_0)q}{1 - (1+\varepsilon_0)q} r(Q_{i_0}) + r(Q_{i_0})\right) = (1+\varepsilon_0)\frac{2r(Q_{i_0})}{1 - (1+\varepsilon_0)q}.$$

Thus, the centers $x_i$ of the cubes $Q_i$ lie in the cube $K$ centered at $x_{i_0}$ and of radius

$$r(K) = (1+\varepsilon_0) \cdot \frac{2r(Q_{i_0})}{1 - (1+\varepsilon_0)q}.$$

Next, by (8.17) we have

$$r(Q_i) \geq \frac{1 - (1+\varepsilon_0)q}{1 + (1+\varepsilon_0)q} r(Q_{i_0}).$$

Consequently, each cube $\frac{1}{2+q}Q_i$ cuts a volume of at least

$$V = \frac{1}{(2+q)^n}\left(\frac{1-(1+\varepsilon_0)q}{1+(1+\varepsilon_0)q}r(Q_{i_0})\right)^n$$

from $K$. By Proposition 8.19, the cubes $\frac{1}{2+q}Q_i$ do not intersect, so their number does not exceed the quantity

$$\nu = \frac{|K|}{V} = \frac{2^n\left((1+\varepsilon_0)\cdot\frac{2}{1-(1+\varepsilon_0)q}\right)^n}{\frac{1}{(2+q)^n}\left(\frac{1-(1+\varepsilon_0)q}{1+(1+\varepsilon_0)q}\right)^n} = 4^n\,(2+q)^n\,(1+\varepsilon_0)^n\frac{(1+(1+\varepsilon_0)q)^n}{(1-(1+\varepsilon_0)q)^{2n}}.$$

This is the desired estimate for the overlap number $\nu$. □

### 8.1.3 Meshing algorithm and the strong engagement lemma

We need the following statement.

**Proposition 8.22.** *Consider two cubes $K_u$ and $K_v$ (centered at $u$ and $v$) such that $\gamma K_u \cap K_v \neq \varnothing$ for some $\gamma < 1$. Then*

$$|K_u \cap K_v| \geq \left(\frac{1-\gamma}{2}\right)^n \min(|K_u|, |K_v|). \tag{8.18}$$

*Proof.* First, we treat the case where $v \notin K_u$. We take a point $z \in \gamma K_u \cap K_v$ and consider the cube $Q_z$ centered at $z$ and of radius $(1-\gamma)r(K_u)$. Clearly, $Q_z \subset K_u$. Consider also the "octant" relative to the origin $z$ that contains $v$. Now, $v$ is not in $K_u$ and, thus, not in $Q_z$. But $z \in K_v$, so the intersection of the above "octant" with $Q_z$ is included in $K_v$ and, therefore, in $K_u \cap K_v$. It follows that

$$|K_u \cap K_v| \geq (1-\gamma)^n r(K_u)^n = \left(\frac{1-\gamma}{2}\right)^n |K_u|,$$

which implies (8.18).

Now, we treat the case where $v \in K_u$. If $r(K_u) \leq r(K_v)$, then we take the "octant" with origin at $u$ that contains $v$. Clearly, the intersection of this "octant" with $K_u$ is included in $K_u \cap K_v$ and, consequently, $|K_u \cap K_v| \geq 2^{-n}\min(|K_u|, |K_v|)$. But if $r(K_u) > r(K_v)$, we consider the "octant" with $v$ as the origin that contains $u$. The intersection of this "octant" with $K_v$ is included in $K_u \cap K_v$ and, consequently, $|K_u \cap K_v| \geq 2^{-n}\min(|K_u|, |K_v|)$. Thus, we have proved (8.18) also in the case where $v \in K_u$. □

The proof of the next lemma involves an important algorithm.

**Lemma 8.23** (Meshing algorithm). *Suppose that $\{U_i\}_{i\in I}$ is a family of cubes with overlap number of at most $\nu$. Let $q_0 < 1$. Then we can find numbers $\mu_i \in \left(q_0^{\nu+1}, 1\right]$, $i \in I$, such that the relation $\mu_i U_i \cap \mu_j U_j \neq \varnothing$ implies*

$$|\mu_i U_i \cap \mu_j U_j| \geq \left(\frac{1-q_0}{2}\right)^n \min(|\mu_i U_i|, |\mu_j U_j|). \tag{8.19}$$

*Proof.* The family $\{U_i\}_{i\in I}$ is at most countable (because it splits in finitely many families of disjoint cubes, see Proposition 3.5), therefore we may assume that it is indexed by the natural numbers. Suppose we have already defined $\mu_i \in \left(q_0^{\nu+1}, 1\right]$ for $i = 1, \dots, N$. We describe the rule of finding $\mu_{N+1}$. Split the interval $\left(q_0^{\nu+1}, 1\right]$ into $\nu+1$ intervals $\left(q_0^{k+1}, q_0^k\right]$ with $k = 0, 1, \dots, \nu$ and for each $i = 1, \dots, N$ define $\gamma_i$ to be the smallest $\gamma \geq 0$ for which

$$\gamma U_{N+1} \cap \mu_i U_i \neq \varnothing. \tag{8.20}$$

Since the overlap number of the family $\{U_i\}_{i\in I}$ is at most $\nu$, among the $\gamma_i$ there may not be more than $\nu$ numbers smaller than or equal to 1. Therefore, there is at least one interval $\left(q_0^{k+1}, q_0^k\right]$ $(0 \leq k \leq \nu)$ free of the $\gamma_i$. Take such an interval and the corresponding $k$ and put

$$\mu_{N+1} = q_0^k.$$

Since none of the $\gamma_i$ lies in $\left(q_0^{k+1}, q_0^k\right]$, this choice of $\mu_{N+1}$ ensures the following: if $i = 1, \dots, N$ and $\mu_{N+1} U_{N+1} \cap \mu_i U_i \neq \varnothing$, then for all $\gamma \in (q_0, 1]$ we have

$$\gamma \mu_{N+1} U_{N+1} \cap \mu_i U_i \neq \varnothing.$$

Therefore, Proposition 8.22 implies that if $\mu_{N+1} U_{N+1} \cap \mu_i U_i \neq \varnothing$, then we have (8.19). □

The next lemma shows that application of the meshing algorithm to a Lipschitz and almost disjoint (see (8.16)) family yields strongly engaged cubes.

**Lemma 8.24** (On strong engagement). *Suppose the family $\{Q_i\}_{i\in I}$ of cubes is Lipschitz with constant $q < 1$ and almost disjoint. Take $\varepsilon_0 > \varepsilon_1 > 0$ with $1 + \varepsilon_0 < \frac{1}{q}$. Then there exist numbers $\lambda_i \in [1 + \varepsilon_1, 1 + \varepsilon_0]$, $i \in I$, such that the family $\{K_i = \lambda_i Q_i\}_{i\in I}$ has the strong engagement property, i.e., the relation $K_i \cap K_j \neq \varnothing$ implies*

$$|K_i \cap K_j| \geq \varepsilon \max(|K_i|, |K_j|). \tag{8.21}$$

*If $\varepsilon_0$, $\varepsilon_1$ and $q$ are fixed, the engagement constant $\varepsilon > 0$ depends only on $n$.*

*Proof.* By Lemma 8.21, the family $\{(1 + \varepsilon_0) Q_i\}_{i\in I}$ has finite overlap. We denote the overlap number by $\nu$ and apply Lemma 8.23, in which we put

$$U_i = (1 + \varepsilon_0) Q_i,\ i \in I, \text{ and } q_0 = \left(\frac{1 + \varepsilon_1}{1 + \varepsilon_0}\right)^{\frac{1}{\nu+1}}. \tag{8.22}$$

This lemma yields certain numbers $\mu_i$. We define $\lambda_i = \mu_i(1 + \varepsilon_0)$. Since $\mu_i \in \left(q_0^{\nu+1}, 1\right]$, $i \in I$, we see that $\lambda_i \in [1 + \varepsilon_1, 1 + \varepsilon_0]$, $i \in I$, and also the following is true: if $\lambda_j Q_j \cap \lambda_i Q_i \neq \varnothing$, then

$$\begin{aligned} |\lambda_j Q_j \cap \lambda_i Q_i| &\geq \left(\frac{1 - q_0}{2}\right)^n \min(|\lambda_j Q_j|, |\lambda_i Q_i|) \\ &\geq \left(\frac{1 - q_0}{2}\right)^n (1 + \varepsilon_1)^n \min(|Q_j|, |Q_i|). \end{aligned}$$

It remains to observe that, if $\lambda_j Q_j \cap \lambda_i Q_i \neq \varnothing$, then $(1+\varepsilon_0)Q_j \cap (1+\varepsilon_0)Q_i \neq \varnothing$. Therefore, by Proposition 8.20 we obtain

$$\begin{aligned}\min(|Q_j|, |Q_i|) &\geq \left(\frac{1-(1+\varepsilon_0)q}{1+(1+\varepsilon_0)q}\right)^n \max(|Q_j|, |Q_i|) \\ &\geq \left(\frac{1-(1+\varepsilon_0)q}{1+(1+\varepsilon_0)q}\right)^n \frac{1}{(1+\varepsilon_0)^n} \max(|\lambda_j Q_j|, |\lambda_i Q_i|).\end{aligned}$$

Thus,

$$|\lambda_j Q_j \cap \lambda_i Q_i| \geq \left(\frac{1-q_0}{2}\right)^n \left(\frac{1-(1+\varepsilon_0)q}{1+(1+\varepsilon_0)q}\right)^n \left(\frac{1+\varepsilon_1}{1+\varepsilon_0}\right)^n \max(|\lambda_j Q_j|, |\lambda_i Q_i|),$$

which is (8.21). □

### 8.1.4 Modified Besicovitch $q$-process

In what follows, we always deal with a family of cubes $\{Q_x\}_{x\in\Omega}$ that satisfies the Lipschitz condition with constant $q<1$.

In two statements to be proved first, it is not assumed that $\sigma = \sup_{x\in\Omega} r(Q_x) < \infty$. So, we modify the Besicovitch process described in Chapter 3 (see Algorithm 3.1).

Since now $r(Q_x)$ may be an arbitrary positive number, we split $\mathbb{R}_+$ into half-open intervals $\left(\frac{1}{(1+q)^k}, \frac{1}{(1+q)^{k-1}}\right]$, $k \in \mathbb{Z}$. We emphasize that here $k$ is an arbitrary integer (rather than a natural number, as it was in Chapter 3). We repeat the construction of Algorithm 3.1, "interlacing" positive and negative $k$, that is, taking consecutively $k = 0, 1, -1, 2, -2, \ldots$. Specifically, we do the following.

Take the first interval $(1, 1+q]$, corresponding to $k=0$, and consider all cubes $Q_x \in \{Q_x\}_{x\in\Omega}$ with $r(Q_x) \in (1, 1+q]$. The set of such cubes is denoted by $\Omega_1$. In $\Omega_1$, we choose a cube $Q_{x_1}$ arbitrarily and put the origin at its center $x_1$, i.e., we assume without loss of generality that $x_1 = 0$. Then we choose $Q_{x_2} \in \Omega_1$ with center $x_2$ *outside* $Q_{x_1}$ and "almost" the closest to zero, i.e.,

$$\|x_2\| \leq 2 \inf\left\{\|x\| : Q_x \in \Omega_1; x \notin Q_{x_1}\right\}.$$

Then we choose $Q_{x_3} \in \Omega_1$ with center $x_3$ outside the cubes already chosen (i.e., $Q_{x_1} \cup Q_{x_2}$) and such that $x_3$ is "almost" the closest to zero. Continuing in the same manner, we obtain a certain family $\{Q_{x_i}\}_{i\in I_1}$ in $\Omega_1$. At the second step, we consider all cubes $Q_x$ such that $r(Q_x)$ belongs to the interval $\left(\frac{1}{1+q}, 1\right]$, corresponding to $k=1$. The set of such cubes is denoted by $\Omega_2$. To $\Omega_2$, we apply the same procedure as above, with the difference that the center of every cube chosen newly must lie outside *all* cubes chosen before (including those chosen at the first step). This will lead to a family $\{Q_{x_i}\}_{i\in I_2}$ of cubes. Repeating the procedures, we shall arrive finally at a family $\{Q_{x_i}\}_{i\in I}$ of cubes (the index set $I$ is the union of the sets $I_k$ obtained at all steps).

The family $\{Q_{x_i}\}_{i\in I}$ is Lipschitz with constant $q < 1$ (because the initial family $\{Q_x\}_{x\in\Omega}$ possesses the same property) and almost disjoint (because the center of each cube is outside all cubes chosen earlier). Moreover,

$$\Omega \subset \bigcup_{i\in I} Q_{x_i}. \tag{8.23}$$

To prove this, we first observe that, since the family $\{Q_{x_i}\}_{i\in I}$ is Lipschitz with constant $q$ and almost disjoint, Proposition 8.19 implies

$$\frac{1}{2+q}Q_{x_i} \cap \frac{1}{2+q}Q_{x_j} = \varnothing, i \neq j. \tag{8.24}$$

Consequently, if the family of cubes chosen at the $k$th step is infinite, then their centers tend to infinity:

$$\lim_{i\in I_k} \|x_i\| = \infty. \tag{8.25}$$

Indeed, the radii of all cubes of the family $\{Q_{x_i}\}_{i\in I_k}$ are at least $\frac{1}{(1+q)^{k-1}}$. Therefore, (8.24) and volume considerations show that any cube centered at zero may contain at most finitely many cubes among $\{Q_{x_i}\}_{i\in I_k}$. So, (8.25) follows.

Now, we prove (8.23). Take an arbitrary element $x \in \Omega$ and take $k$ with $r(Q_x)$ belonging to the interval $\left(\frac{1}{(1+q)^k}, \frac{1}{(1+q)^{k-1}}\right]$. Suppose $Q_x$ was not chosen when we treated all cubes with radii in the same interval. Since throughout we choose cubes "almost" closest to zero, by (8.25) it follows that there exists $Q_{x_i}$ with $x \in Q_{x_i}$. Thus, (8.23) is fulfilled.

In the sequel, the process described above will be called the modified Besicovitch $q$-process .

### 8.1.5 Proof of Theorem 8.16

We recall the statement.

**Theorem.** *Suppose $\Omega \subset \mathbb{R}^n$ and a family $\{Q_x\}_{x\in\Omega}$ of cubes is Lipschitz with constant $q < 1$. Then this family has an at most countable subfamily $\{Q_{x_i}\}_{i\in I}$ that covers $\Omega$:*

$$\Omega \subset \bigcup_{i\in I} Q_{x_i},$$

*and is of finite overlap:*

$$\operatorname{card}\{i \in I : Q_{x_i} \cap Q_{x_j} \neq \varnothing\} \leq \nu \text{ for every } j \in I$$

*with $\nu$ depending only on $q$ and the dimension $n$.*

*Proof.* By assumption, the family $\{Q_x\}_{x\in\Omega}$ is Lipschitz with constant $q < 1$. Applying the modified Besicovitch $q$-process, we obtain a family $\{Q_{x_i}\}_{i\in I}$ of cubes that covers $\Omega$ (see (8.23)). Since this family is almost disjoint and Lipschitz with constant $q < 1$, Lemma 8.21 implies the finite overlap property. □

### 8.1.6 Proof of Theorem 8.9

Again, we recall the statement.

**Theorem.** *Suppose we are given a strongly Lipschitz family $\{Q_x\}_{x\in\Omega}$ of cubes with constant $q < 1$. Then there exist points $x_i \in \Omega$ and numbers $1 < \lambda_i < \frac{1}{q}$, $i \in I$, such that the family $\{K_{x_i} = \lambda_i Q_{x_i}\}_{i\in I}$ is a* WB*-covering of $\Omega$.*

*Proof.* By assumption, the family $\{Q_x\}_{x\in\Omega}$ has the property that the function $F(x) = r(Q_x)$ becomes Lipschitz with constant $q < 1$ after extension by zero from $\Omega$ to $\mathbb{R}^n$:

$$|F(x) - F(y)| \le q\,\|x - y\|\,,\ x, y \in \mathbb{R}^n.$$

We must construct a family $\{K_{x_i} = \lambda_i Q_{x_i}\}_{i\in I}$ with $1 < \lambda_i < \frac{1}{q}$ that is a WB-covering of $\Omega$. Since we do not assume that $\sigma = \sup_{x\in\Omega} r(Q_x) < \infty$, we use the modified Besicovitch $q$-process, which results in a family $\{Q_{x_i}\}_{i\in I}$ of cubes that is Lipschitz with constant $q < 1$ and almost disjoint, and satisfies

$$\Omega \subset \bigcup_{i\in I} Q_{x_i}.$$

The required family $\{K_{x_i}\}_{i\in I}$ is constructed in two steps. First, we fix $\varepsilon_0 > \varepsilon_1 > 0$ such that

$$1 + \varepsilon_0 < \frac{1}{q} \tag{8.26}$$

and consider the family $\{U_{x_i} = (1+\varepsilon_0)Q_{x_i}\}_{i\in I}$. Since the family $\{Q_{x_i}\}_{i\in I}$ is Lipschitz with constant $q < 1$ and almost disjoint, the finite overlap Lemma 8.21 shows that the family $\{(1+\varepsilon_0)Q_{x_i}\}_{i\in I}$ is of finite overlap with a constant $\nu$ depending only on $q$, $\varepsilon_0$, and the dimension $n$. Next, to the family $\{U_{x_i}\}_{i\in I}$ we apply the meshing algorithm (Lemma 8.23) with

$$q_0 = \left(\frac{1+\varepsilon_1}{1+\varepsilon_0}\right)^{\frac{1}{\nu+1}}$$

(here $\nu$ is the overlap number mentioned above). As a result (see Lemma 8.24), we obtain a family $\{K_{x_i} = \lambda_i Q_{x_i}\}_{i\in I}$ of cubes with $\lambda_i \in [1+\varepsilon_1, 1+\varepsilon_0] \subset \left(1, \frac{1}{q}\right)$.

We show that the family $\{K_{x_i}\}_{i\in I}$ is a WB-covering of $\Omega$. The strong engagement property was proved in Lemma 8.24. Finite multiplicity (in fact, even finite overlap) follows because this occurs already for the family $\{(1+\varepsilon_0)Q_{x_i}\}_{i\in I}$ (recall that $K_{x_i} \subset (1+\varepsilon_0)Q_{x_i}$). So, only the Whitney property needs a proof.

First, we show that

$$\bigcup_{i\in I} Q_{x_i} = \bigcup_{i\in I} (1+\varepsilon_0)Q_{x_i} = \Omega. \tag{8.27}$$

Since the family $\{Q_x\}_{x\in\Omega}$ is strongly Lipschitz, Lemma 8.17 implies the inclusion

$$(1+\varepsilon_0)Q_{x_i} \subset \Omega.$$

Therefore, (8.27) is a consequence of (8.23).

Since $K_{x_i} = \lambda_i Q_{x_i}$ and $\lambda_i \in [1+\varepsilon_1, 1+\varepsilon_0]$, from (8.27) we deduce that

$$\bigcup_{i\in I} \frac{1}{1+\varepsilon_1} K_i = \bigcup_{i\in I} K_i = \Omega.$$

So, the Whitney property is fulfilled with the constant $\delta = \varepsilon_1$. The statement concerning the other constants follows from the proof. □

### 8.1.7 Proof of Theorem 8.13

The statement is repeated below.

**Theorem.** *Suppose we are given a strongly Lipschitz family of cubes $\{Q_x\}_{x\in\Omega}$ with constant $q < \sqrt{2}-1$ and such that*

$$\sup_{x\in\Omega} r(Q_x) < \infty.$$

*Then there exists a subfamily $\{K_{x_i} = \lambda_i Q_{x_i}\}_{i\in I}$ with $1 < \lambda_i < \frac{1}{q}$ that is a* WB-*covering of the set $\Omega$ and has the property that for every cube $Q_x$ there is an index $i = i(x)$ with $Q_x \subset K_{x_i}$.*

*Proof.* Now the radii are uniformly bounded, and we may apply the usual Besicovitch $q$-process to our family. This results in a family $Q_i = Q_{x_i}$, $i \in I$, of cubes, to which we apply the constructions from the preceding proof. Then property (3.3) of Besicovitch cubes shows that for every $x \in \Omega$ there exists $i \in I$ such that

$$Q_x \subset (2+q)Q_i = \frac{2+q}{\lambda_i} K_i \subset \frac{2+q}{1+\varepsilon_1} K_i.$$

Thus, if we take $\varepsilon_1 = 1+q$, then $Q_x \subset K_i$. We observe that the relation $\varepsilon_1 = 1+q$ imposes a restriction on $q$, namely,

$$2+q = 1+\varepsilon_1 < \frac{1}{q},$$

which is equivalent to $q < \sqrt{2}-1$. Then $\varepsilon_0$ must obey the condition

$$1+q < \varepsilon_0 < \frac{1}{q} - 1.$$

□

## 8.2 Controlled extension and preservation of the $\alpha$-capacity

In order to state the result, we need several definitions.

Below we denote by $\Omega$ and $U$ some subsets of $\mathbb{R}^n$. In distinction to the Whitney lemma, they are not necessarily open and may coincide with $\mathbb{R}^n$. Next, $Q_x$ and $K_u$ stand for cubes centered at $x$ or $u$.

**Definition 8.25.** A family $\{K_u\}_{u\in U}$ of cubes is called an *extension* of another family $\{Q_x\}_{x\in\Omega}$ if for every $x \in \Omega \cap U$ we have

$$Q_x \subset K_x \tag{8.28}$$

and for every $x \in \Omega$ there exists an element $u = u(x) \in U$ such that

$$Q_x \subset K_u. \tag{8.29}$$

**Definition 8.26.** We say that a family $\{K_u\}_{u\in U}$ of cubes is *controlled* by another family $\{Q_x\}$ if the latter contains a subfamily $\{Q_{x_i}\}_{i\in I}$ of finite multiplicity $M_c$ (the index $c$ is for "control") and such that for every $u \in U$ there exists $i \in I$ with

$$K_u \subset \gamma Q_{x_i}, \tag{8.30}$$

where $\gamma$ does not depend on $u$ and $x_i$.

The quantity $\gamma$ will be called the *control constant.* The above condition does not allow the cubes $\{K_u\}_{u\in U}$ to be "too" large.

When talking of a controlled extension of some family $\{Q_x\}$ of cubes, we mean an extension controlled by the same family $\{Q_x\}$. It turns out that, under the assumptions of the Besicovitch theorem, there is a controlled extension that is a WB-covering.

**Theorem 8.27** (On controlled extension). *Any family $\{Q_x\}_{x\in\Omega}$ of cubes with*

$$\sigma = \sup_{x\in\Omega} r(Q_x) < \infty \tag{8.31}$$

*admits a controlled extension $\{K_{u_i}\}_{i\in I}$ that is simultaneously a* WB*-covering of the set $\bigcup_{i\in I} K_{u_i}$.*

*Moreover, all constants involved, that is, $\gamma$ and $M_c$ in Definition* 8.26 *and the constants related to the* WB*-covering in question can be chosen to depend on the dimension only.*

*Remark* 8.28. The control constant $\gamma$ and Whitney constant $\delta$ for the WB-covering $\{K_{u_i}\}_{i\in I}$ constructed in the above theorem are universal in the sense that they do not depend on the dimension and the initial family of cubes (they depend only on the parameter $q$ in the Besicovitch process, which is used in the proof; this parameter must be smaller than $\sqrt{2}-1$, but otherwise the choice of it is free). The multiplicity constants $M$, $M_c$ and engagement constant $\varepsilon$ depend on $q$ and the dimension $n$.

The algorithm of constructing the family $\{K_{u_i}\}_{i\in I}$ consists of three steps.

- By a modification of the Besicovitch $q$-process, starting with the family $\{Q_x\}_{x\in\Omega}$ we construct its almost disjoint Lipschitz subfamily $\{Q_{x_i}\}_{i\in I}$.
- For any point $y$ outside $\bigcup Q_{x_i}$, we consider the cube $U_y$ of radius

$$r(U_y) = \sup_{i\in I} \left(r(Q_{x_i}) - q\,\|y - x_i\|\right)$$

  and then introduce the family $\{U_y\}_{r(U_y)>0}$. To this family, we apply the Besicovitch $q$-process, which yields a family $\left\{U_{y_j}\right\}_{j\in J}$ of Besicovitch cubes.
- The required cubes $\{K_{u_i}\}_{i\in I}$ are constructed with the help of the meshing algorithm (see Lemma 8.23) applied to the family $\{Q_{x_i}\}_{i\in I} \cup \left\{U_{y_j}\right\}_{j\in J}$.

We mention a useful consequence of the controlled extension theorem.

**Definition 8.29.** The *$\alpha$-capacity* $|\{Q_x\}|_\alpha$ of a family $\{Q_x\}_{x\in\Omega}$ of cubes is the quantity

$$|\{Q_x\}|_\alpha = \sup_{\pi=\left\{Q_{x_i}\right\}} \sum_i |Q_{x_i}|^\alpha ,$$

where the supremum is taken over all packings $\pi = \{Q_{x_i}\} \subset \{Q_x\}_{x\in\Omega}$.

We are mainly interested in families with finite $\alpha$-capacity (for some $\alpha$). It turns out that, if $\alpha \geq 1$, then the $\alpha$-capacity does not increase much after controlled extension of a family of cubes.

**Theorem 8.30** (Control on $\alpha$-capacity). *Suppose that $\alpha \geq 1$, and $\left\{K_{u_j}\right\}_{j\in J}$ is a controlled extension of a family $\{Q_x\}_{x\dot{\in}\Omega}$. Then*

$$\left|\left\{K_{u_j}\right\}\right|_\alpha \leq c\,|\{Q_x\}|_\alpha .$$

*Remark* 8.31. If $\left\{K_{u_j}\right\}_{j\in J}$ is the family constructed in Theorem 8.27, then the number $c > 0$ can be chosen to depend only on the dimension $n$.

*Proof.* Take a subset $J_1$ of the index set $J$ such that the family $\left\{K_{u_j}\right\}_{j\in J_1}$ is a packing. By assumption, there is a finite multiplicity subfamily $\{Q_{x_i}\}_{i\in I}$ of the family $\{Q_x\}_{x\in\Omega}$ such that, for every cube $K_{u_j}$, $j \in J_1$, there exists a cube $Q_{x_i}$, $i = i(j) \in I$ with

$$K_{u_j} \subset \gamma Q_{x_i}, \tag{8.32}$$

where $\gamma$ is the control constant. For every $i \in I$, we denote by $F_i$ the set of $j \in J_1$ for which (8.32) is fulfilled with this $i$. Since the family $\{Q_{x_i}\}_{i\in I}$ is of finite multiplicity, it splits into finitely many packings (their number depending only on the multiplicity constant and the dimension $n$), see Proposition 3.5. Thus, the proof will be finished if we establish the estimate

$$\sum_{j\in F_i} |K_{u_j}|^\alpha \le c|Q_{x_i}|^\alpha$$

with a quantity $c$ depending only on the control constant $\gamma$ and the dimension $n$.

But the family $\{K_{u_j}\}_{j\in J_1}$ is itself a packing. Since $\alpha \ge 1$, by (8.32) we have

$$\sum_{j\in F_i} |K_{u_j}|^\alpha \le \left(\sum_{j\in F_i} |K_{x_i}|\right)^\alpha \le |\gamma Q_{x_i}|^\alpha = \gamma^{n\alpha} |Q_{x_i}|^\alpha . \qquad \square$$

It is interesting to know what happens if we drop the condition $\alpha \ge 1$. After the proof of the controlled extension theorem, we shall establish the following statement.

**Theorem 8.32.** *Suppose that $\alpha > 1 - \frac{1}{n}$ and a family $\{Q_x\}_{x\in\Omega}$ of cubes satisfies (8.31). Then the* WB-*covering $\{K_{u_j}\}_{j\in J}$ constructed in the controlled extension theorem obeys the inequality*

$$\left|\{K_{u_j}\}\right|_\alpha \le c|\{Q_x\}|_\alpha ,$$

*where $c > 0$ depends only on the dimension $n$.*

Concluding the discussion, we note that in the case where $\Omega = \mathbb{R}^n$ the statement of the controlled extension theorem becomes somewhat simpler.

**Theorem 8.33.** *Suppose that for every $x \in \mathbb{R}^n$ we are given a cube $Q_x$, and let $\sup_{x\in\mathbb{R}^n} r(Q_x) < \infty$. Then there exists a* WB-*covering $\{K_{x_i}\}_{i\in I}$ of $\mathbb{R}^n$ and a finite multiplicity family $\{Q_{x_j}\}_{j\in J}$ (with multiplicity constant depending only on the dimension) such that*

(a) *$Q_{x_i} \subset K_{x_i}$, $i \in I$;*

(b) *for every cube $Q_x$, there exists $i = i(x) \in I$ satisfying $Q_x \subset K_{x_i}$;*

(c) *for every cube $K_{x_i}$, $i \in I$, there exists $j = j(i) \in J$ satisfying $K_{x_i} \subset \gamma Q_{x_j}$ with a universal control constant $\gamma > 0$.*

The proof of this statement is a consequence of the proof of the controlled extension theorem (to be presented below). In its turn, the latter involves the following modification of the Besicovitch $q$-process.

### 8.2.1 The Besicovitch process with a Lipschitz condition

We fix $q$ with $0 < q < 1$. The standard Besicovitch $q$-process described in Chapter 3 results in a family of cubes, which may fail to be Lipschitz with constant $q$.

However, the construction can be modified so as to give a Lipschitz family. The idea is to choose every new cube in such a way that it satisfies additionally the Lipschitz conditions involving all cubes chosen previously. Here are the details, with repetition of what remains intact from Chapter 3.

We put

$$\sup_{x\in\Omega} r(Q_x) = \sigma < \infty$$

and consider the half-open intervals $\left(\frac{\sigma}{(1+q)^k}, \frac{\sigma}{(1+q)^{k-1}}\right]$ with $k = 1, \dots$. The first step of the new process is the same as it was in Chapter 3, but we describe it again. Take the interval $\left(\frac{\sigma}{1+q}, \sigma\right]$, which corresponds to $k = 1$, and consider all cubes $Q_x$ with radius $r(Q_x)$ in this interval. The set of these cubes will be denoted by $\Omega_1$. In $\Omega_1$, we choose a cube $Q_{x_1}$ arbitrarily and, as before, assume without loss of generality that the origin is placed at its center, that is, $x_1 = 0$. Next, we choose a cube $Q_{x_2} \in \Omega_1$ whose center $x_2$ lies outside $Q_{x_1}$ and is "almost" the nearest to zero, i.e.,

$$\|x_2\| \le 2\inf\{\|x\| : Q_x \in \Omega_1, x \notin Q_{x_1}\}.$$

After that we choose a cube $Q_{x_3} \in \Omega_1$ whose center $x_3$ lies outside the cubes chosen previously and is "almost" the nearest to zero. Continuing in this manner, we construct a subfamily $\{Q_{x_i}\}_{i\in I_1}$ of $\Omega_1$. We observe that this family satisfies the Lipschitz condition with constant $q$ *automatically*. Indeed, the family $\{Q_{x_i}\}_{i\in I_1}$ is almost disjoint because the center of any new-chosen cube lies outside the cubes chosen previously, so that

$$\min(r(Q_{x_i}), r(Q_{x_j})) < \|x_i - x_j\|.$$

Therefore,

$$r(Q_{x_i}) - r(Q_{x_j}) < \sigma - \frac{\sigma}{1+q} = q\frac{\sigma}{1+q} < q\min(r(Q_{x_i}), r(Q_{x_j})) < q\|x_i - x_j\|,$$

as claimed.

Now, since the family constructed above is almost disjoint and Lipschitz, Proposition 8.19 shows that

$$\frac{1}{2+q}Q_{x_i} \cap \frac{1}{2+q}Q_{x_j} = \varnothing$$

for $i \ne j$. Since the volumes of the cubes $\frac{1}{2+q}Q_{x_i}$ are at least $(\frac{1}{2+q})^n 2^n (\frac{\sigma}{1+q})^n$, we see that an arbitrary cube centered at zero contains finitely many centers $x_i$. Thus, if the resulting family $Q_{x_i}$ is infinite, then the distance from $Q_{x_i}$ to zero tends to infinity. Consequently, for every $Q_x \in \Omega_1$ there is a cube $Q_{x_i}$ such that

$$\|x - x_i\| < r(Q_{x_i}).$$

Indeed, this is clear if the family $\{Q_{x_i}\}_{i\in I_1}$ is finite; otherwise, some $x_j$ is much farther away from the origin than $x$, and, by construction, $x$ must be in some $Q_{x_i}$ chosen prior to $Q_{x_j}$. Moreover, since $r(Q_x), r(Q_{x_i}) \in \left(\frac{\sigma}{1+q}, \sigma\right]$, we have

$$r(Q_x) < (1+q)r(Q_{x_i}).$$

This completes the first step of the process.

At the second step, we consider all cubes with radii in $\left(\frac{\sigma}{(1+q)^2}, \frac{\sigma}{1+q}\right]$ (this corresponds to the case of $k = 2$). Should we proceed in the same way as before, we would have been unable to ensure the Lipschitz condition between cubes arising at different steps. Therefore, we slightly modify the selection criterion. Specifically, we demand not only that the center of any new-chosen cube lie outside all cubes chosen previously (including those chosen at the first step), but also that the Lipschitz condition involving this cube and all cubes chosen previously be fulfilled. That is, the choice of $x_k$ is restricted by two conditions:

$$\|x_k - x_i\| > r(Q_{x_i})$$

(if this number is positive) and

$$|r(Q_{x_k}) - r(Q_{x_i})| \leq q\,\|x_k - x_i\|$$

for all cubes $Q_{x_i}$ chosen previously. (Surely, again at each step we choose a cube whose center is "almost" the closest to zero.) This will result in a certain family $\{Q_{x_i}\}_{i\in I_2}$ of cubes. As at the first step, if it is infinite, then the centers of these cubes tend to infinity. Therefore, for every $x$ with $r(Q_x) \in \left(\frac{\sigma}{(1+q)^2}, \frac{\sigma}{1+q}\right]$, there exists $i \in I_1 \cup I_2$ such that at least one of the following inequalities is fulfilled:

$$\|x - x_i\| \leq r(Q_{x_i}), \tag{8.33}$$

$$|r(Q_x) - r(Q_{x_i})| > q\,\|x - x_i\|. \tag{8.34}$$

Since for every $i \in I_1 \cup I_2$ we have $r(Q_{x_i}) > \frac{\sigma}{(1+q)^2}$, the inclusion $r(Q_x) \in \left(\frac{\sigma}{(1+q)^2}, \frac{\sigma}{1+q}\right]$ shows that

$$r(Q_x) \leq \frac{\sigma}{1+q} < (1+q)r(Q_{x_i}). \tag{8.35}$$

We observe also that the union of the families $\{Q_{x_i}\}_{i\in I_1}$ and $\{Q_{x_i}\}_{i\in I_2}$ is almost disjoint and satisfies the Lipschitz condition with constant $q$.

Continuing in the same manner, finally we arrive at an almost disjoint family $\{Q_{x_i}\}_{i\in I}$ of cubes that is also Lipschitz with constant $q$ (the index set $I$ is the union of the sets $I_k$ arising at the $k$th step). These cubes will be called Besicovitch cubes satisfying the $q$-Lipschitz condition.

**Lemma 8.34.** *The above family $\{Q_{x_i}\}_{i\in I}$ of cubes possesses the following properties:*

(a) *it is almost disjoint, that is,* $\min(r(Q_{x_i}), r(Q_{x_j})) < \|x_i - x_j\|$, $i, j \in I$;

(b) *it is Lipschitz with constant* $q$, *that is,* $|r(Q_{x_j}) - r(Q_{x_i})| \leq q\,\|x_j - x_i\|$, $i, j \in I$;

(c) *it has finite overlap with overlap number* $\nu$ *depending only on* $q$ *and the dimension* $n$;

(d) *for every* $x \in \Omega$ *there exists* $i \in I$ *such that either*

$$x \in Q_{x_i} \text{ and } r(Q_x) < (1+q)r(Q_{x_i}), \tag{8.36}$$

*or*

$$x \notin Q_{x_i} \text{ and } r(Q_x) < r(Q_{x_i}) - q\,\|x - x_i\|\,. \tag{8.37}$$

*Proof.* Properties (a) and (b) are clear from the construction. The finite overlap property is a consequence of Lemma 8.21.

We verify (d). Choose $k$ in such a way that $r(Q_x) \in \left(\frac{\sigma}{(1+q)^k}, \frac{\sigma}{(1+q)^{k-1}}\right]$. We prove that either (8.36) or (8.37) is true *for some* cube $Q_{x_i}$ *chosen at the first* $k$ *steps of the process.* We saw that the second condition in (8.36) is fulfilled automatically for such cubes, cf. (8.35). If $x$ belongs to the union of cubes chosen at the first $k$ steps, then (8.36) follows. Suppose that $x$ is outside their union. Then for some of these cubes $Q_{x_i}$ we must have

$$|r(Q_x) - r(Q_{x_i})| > q\,\|x - x_i\|\,. \tag{8.38}$$

If we show that

$$r(Q_x) - r(Q_{x_i}) < 0, \tag{8.39}$$

then, clearly, (8.37) will follow.

Suppose that, to the contrary, $r(Q_x) - r(Q_{x_i}) \geq 0$. Then (8.38) and the inequality $r(Q_x) < (1+q)r(Q_{x_i})$ yield

$$q\,\|x - x_i\| < r(Q_x) - r(Q_{x_i}) < (1+q)r(Q_{x_i}) - r(Q_{x_i}) = qr(Q_{x_i}).$$

That is, $x \in Q_{x_i}$, which is impossible, and statement (d) follows. □

### 8.2.2 Construction of a WB-covering

Suppose that $q < 1$ and the cubes $\{Q_{x_i}\}_{i\in I}$ are the outcome of the Besicovitch $q$-process with Lipschitz condition. Consider the union

$$\Omega_q = \bigcup_{i\in I}(\frac{1}{q}\mathring{Q}_{x_i})$$

($\mathring{E}$ denotes the interior of a set $E$). The set $\frac{1}{q}\mathring{Q}_{x_i}$ can be described as the set of all $y \in \mathbb{R}^n$ such that

$$r(Q_{x_i}) - q\,\|y - x_i\| > 0$$

for some $i$. Therefore, $\Omega_q$ is the set where the following function is positive:

$$G(y) = \sup_{i\in I} \left(r(Q_{x_i}) - q\,\|y - x_i\|\right). \tag{8.40}$$

For $y \in \Omega_q \setminus \bigcup Q_{x_i}$, we introduce the cube $U_y$ of radius

$$r(U_y) = G(y) = \sup_{i\in I} \left(r(Q_{x_i}) - q\,\|y - x_i\|\right). \tag{8.41}$$

The next proposition establishes an important property of the family

$$\{U_y\}_{y\in\Omega_q\setminus\bigcup Q_{x_i}}.$$

**Proposition 8.35.** *For fixed $y \in \Omega_q \setminus \bigcup Q_{x_i}$, the lowest upper bound in* (8.41) *is attained at some $i \in I$:*

$$r(U_y) = r(Q_{x_i}) - q\,\|y - x_i\|. \tag{8.42}$$

*Proof.* We observe that the inequality $r(Q_{x_i}) - q\,\|y - x_i\| > 0$ implies

$$\|y - x_i\| \le \frac{r(Q_{x_i})}{q} \le \frac{\sigma}{q}.$$

Therefore, the supremum in (8.41) can be taken only over the indices $i$ such that $x_i$ belongs to the cube of radius $\|y\| + \frac{\sigma}{q}$ and centered at zero. Next, we choose $k$ with $r(U_y) \in \left(\frac{\sigma}{(1+q)^k}, \frac{\sigma}{(1+q)^{k-1}}\right]$. Then, clearly, the supremum in question can be taken only over $i$ with $r(Q_{x_i}) > \frac{\sigma}{(1+q)^k}$, that is, it suffices to examine the cubes $Q_{x_i}$ constructed at the first $k$ steps of the process. But there are only finitely many such cubes with centers within a given distance from zero (this was established in the course of the construction). So, for every $y$ the supremum can be taken over a finite set of indices $i \in I$, and it is attained. □

Every function $F_i(y) = r(Q_{x_i}) - q\,\|y - x_i\|$ is Lipschitz with constant $q$, consequently, so is the function $G(y) = \sup_{i\in I} F_i(y)$. Thus, the cubes $\{U_y\}_{y\in\Omega_q\setminus\bigcup Q_{x_i}}$ constitute a Lipschitz family with constant $q$. Since also

$$\sup_{y\in\Omega_q\setminus\bigcup Q_{x_i}} r(U_y) \le \sup_{i\in I} r(Q_{x_i}) \le \sigma < \infty,$$

the standard Besicovitch $q$-process applies, yielding a new family $\{U_{y_j}\}_{j\in J}$ of cubes. This latter is Lipschitz with constant $q$, as a subfamily of a Lipschitz family. Now, we consider the union $\{Q_{x_i}\}_{i\in I} \cup \{U_{y_j}\}_{j\in J}$ and show that this family is also Lipschitz. Since, separately, the families $\{Q_{x_i}\}_{i\in I}$ and $\{U_{y_j}\}_{j\in J}$ are Lipschitz, it suffices to prove the inequality

$$\left|r(U_{y_j}) - r(Q_{x_i})\right| \le q\,\|y_j - x_i\|.$$

By (8.41), we have $r(U_{y_j}) \geq r(Q_{x_i}) - q\,\|y_j - x_i\|$, consequently,

$$r(Q_{x_i}) - r(U_{y_j}) \leq q\,\|y_j - x_i\|\,,$$

and it suffices to show that

$$r(U_{y_j}) - r(Q_{x_i}) \leq q\,\|y_j - x_i\|\,.$$

Since the family $\{Q_{x_i}\}_{i\in I}$ is Lipschitz, for every $k \in I$ we obtain

$$r(Q_{x_k}) - r(Q_{x_i}) \leq q\,\|x_k - x_i\| \leq q\,\|y_j - x_k\| + q\,\|y_j - x_i\|\,,$$

whence

$$(r(Q_{x_k}) - q\,\|y_j - x_k\|) - r(Q_{x_i}) \leq q\,\|y_j - x_i\|\,.$$

Therefore, passing to the supremum over $k$, we obtain

$$r(U_{y_j}) - r(Q_{x_i}) = \sup_{k\in I}\,(r(Q_{x_k}) - q\,\|y_j - x_k\|) - r(Q_{x_i}) \leq q\,\|y_j - x_i\|\,.$$

This shows the Lipschitz property for the family $\{Q_{x_i}\}_{i\in I} \cup \{U_{y_j}\}_{j\in J}$.

Since the centers $y_j$ of $U_{y_j}$ are outside the $Q_{x_i}$ and the Besicovitch cubes $U_{y_j}$ are almost disjoint, the entire family $\{Q_{x_i}\}_{i\in I} \cup \{U_{y_j}\}_{j\in J}$ is also almost disjoint. Since $q < 1$, Lemmas 8.21 and 8.24 (about finite overlap and strong engagement) are applicable. They show that, fixing two constants $0 < \varepsilon_1 < \varepsilon_0 < \frac{1}{q} - 1$, we can find numbers $\lambda_k \in [1+\varepsilon_1, 1+\varepsilon_0]$ such that the family $\{\lambda_i Q_{x_i}\}_{i\in I} \cup \{\lambda_j U_{y_j}\}_{j\in J}$ is of finite overlap (hence, of finite multiplicity) and consists of strongly engaged cubes, the relevant constants depending only on $q$ and the dimension $n$.

So, to prove that the family $\{\lambda_i Q_{x_i}\}_{i\in I} \cup \{\lambda_j U_{y_j}\}_{j\in J}$ is a WB-covering, it remains to check the Whitney property for it. To do this, first we invoke Proposition 8.35 to conclude that for every $y \in \Omega_q \setminus \bigcup Q_{x_i}$ there exists $i \in I$ with

$$\frac{1}{q} r(Q_{x_i}) = \frac{1}{q} r(U_y) + \|y - x_i\|\,,$$

whence

$$\frac{1}{q} U_y \subset \frac{1}{q} Q_{x_i}.$$

So, the inequality $1 + \varepsilon_1 \leq \lambda_i \leq 1 + \varepsilon_0 < \frac{1}{q}$ shows that

$$\bigcup_{i\in I, j\in J} (\lambda_i Q_{x_i} \cup \lambda_j U_{y_j}) \subset \bigcup_{i\in I} (\frac{1}{q} \mathring{Q}_{x_i}) = \Omega_q. \tag{8.43}$$

On the other hand, since $\{U_{y_j}\}_{j\in J}$ is a family of Besicovitch cubes constructed starting with the family $\{U_y\}_{y\in\Omega_q\setminus\bigcup Q_{x_i}}$, we have

$$\Omega_q \setminus \bigcup Q_{x_i} \subset \bigcup_{j\in J} U_{y_j}$$

and, consequently,

$$\Omega_q \subset \bigcup_{i\in I, j\in J} (Q_{x_i} \cup U_{y_j}).$$

Since $\lambda_i \geq 1+\varepsilon_1$, this shows that

$$\Omega_q \subset \bigcup_{i\in I, j\in J} \left[ (\frac{1}{1+\varepsilon_1}\lambda_i Q_{x_i}) \cup (\frac{1}{1+\varepsilon_1}\lambda_j U_{y_j}) \right].$$

Together with (8.43), this inclusion shows that the family $\{\lambda_i Q_{x_i}\}_{i\in I} \cup \{\lambda_j U_{y_j}\}_{j\in J}$ has the Whitney property with constant $\delta = \varepsilon_1$.

### 8.2.3 Proof of the controlled extension theorem

We have already proved that the family $\{\lambda_i Q_{x_i}\}_{i\in I} \cup \{\lambda_j U_{y_j}\}_{j\in J}$ is a WB-covering. So, it suffices to show that it is a controlled extension of the initial family $\{Q_x\}_{x\in\Omega}$. We begin with proving that this is an extension. The first condition to be verified (see (8.28)) says that $Q_{x_i} \subset \lambda_i Q_{x_i}$, $i \in I$, and $Q_{y_j} \subset \lambda_j U_{y_j}$ whenever $y_j \in \Omega$. Since $\lambda_i > 1$, the inclusion $Q_{x_i} \subset \lambda_i Q_{x_i}$, $i \in I$, is clear. To prove the inclusion $Q_{y_j} \subset \lambda_j U_{y_j}$, we observe that, since the point $y_j$ is in $\Omega$ and in $\Omega_q \setminus \bigcup Q_{x_i}$, it must satisfy (8.37). But then

$$r(Q_{y_j}) < r(Q_{x_i}) - q\,\|y_j - x_i\| \leq \sup_{i\in I}(r(Q_{x_i}) - q\,\|y_j - x_i\|) = r(U_{y_j})$$

for some $i \in I$. Therefore, $Q_{y_j} \subset U_{y_j} \subset \lambda_j U_{y_j}$. This proves the first condition in the definition of extension.

Now we check the second condition (8.29), which takes the following form: for every $x \in \Omega$, the family $\{\lambda_i Q_{x_i}\}_{i\in I} \cup \{\lambda_j U_{y_j}\}_{j\in J}$ contains a cube $K$ with $Q_x \subset K$. If $x \in \Omega$ satisfies (8.36), there is $i \in I$ such that

$$\|x - x_i\| \leq r(Q_{x_i}) \text{ and } r(Q_x) < (1+q) r(Q_{x_i}).$$

Then

$$Q_x \subset (2+q) Q_{x_i},$$

and the inclusion $Q_x \subset \lambda_i Q_{x_i}$ is fulfilled if $2+q \leq 1+\varepsilon_1$ (recall that $1+\varepsilon_1 \leq \lambda_i$).

Before we proceed, we make a remark.

*Remark* 8.36. Since $1+\varepsilon_1 < 1/q$, the inequality $2+q \leq 1+\varepsilon_1$ imposes a restriction on $q$:

$$2+q < \frac{1}{q},$$

which is equivalent to $q < \sqrt{2} - 1$.

Now, suppose that $x \in \Omega$ does not satisfy (8.36) for any $i$. Then (8.37) is true: there exists $i \in I$ such that

$$x \notin Q_{x_i}, \quad r(Q_x) < r(Q_{x_i}) - q\,\|x - x_i\|\,. \tag{8.44}$$

We show that $x \in \Omega_q \setminus \bigcup Q_{x_i}$. By (8.44), $r(Q_{x_i}) - q\,\|x - x_i\| > 0$ and, therefore, $x \in \frac{1}{q}\mathring{Q}_{x_i} \subset \Omega_q$. Next, should we have $x \in Q_{x_k}$ for some $k$, the failure of (8.36) would imply $r(Q_x) \geq (1+q)r(Q_{x_k})$. By (8.44) this implies

$$q\,\|x_i - x_k\| \leq q\,\|x - x_i\| + q\,\|x - x_k\| < r(Q_{x_i}) - r(Q_x) + qr(Q_{x_k}) \leq r(Q_{x_i}) - r(Q_{x_k}),$$

which contradicts the Lipschitz condition for the family $\{Q_{x_i}\}_{i\in I}$. Thus, indeed, $x \in \Omega_q \setminus \bigcup Q_{x_i}$ and, consequently, the cube $U_x$ is well defined. Now, (8.44) implies

$$r(Q_x) < \sup_{i\in I}\left(r(Q_{x_i}) - q\,\|x - x_i\|\right) = r(U_x),$$

whence

$$Q_x \subset U_x.$$

Since the cubes $\{U_{y_j}\}_{j\in J}$ are constructed with the help of the Besicovitch $q$-process applied to the family $\{U_y\}_{y\in\Omega_q\setminus\bigcup Q_{x_i}}$, from (3.3), see Chapter 3, it follows that there exists a cube $U_{y_j}$ with

$$U_x \subset (2+q)U_{y_j}.$$

Thus, if

$$2 + q \leq 1 + \varepsilon_1,$$

then $Q_x \subset (1+\varepsilon_1)U_{y_j} \subset \lambda_j U_{y_j}$, and the second condition in the definition of extension is verified. The inequality $2 + q \leq 1 + \varepsilon_1$ imposes the same restriction on $q$ as in Remark 8.36.

To complete the proof, we must verify the control property (8.30). We take $\{Q_{x_i}\}_{i\in I}$ for the role of a controlling family. Clearly, this family is of finite multiplicity (the multiplicity constant depending only on the dimension for fixed $q$, see Lemma 8.34, (c)). We show that this family controls $\{\lambda_i Q_{x_i}\}_{i\in I} \cup \{\lambda_j U_{y_j}\}_{j\in J}$ with the constant $\gamma = \frac{2}{q}$, more precisely, $\lambda_i Q_{x_i} \subset \gamma Q_{x_i}$ and for every $j \in J$ there exists $i = i(j)$ such that

$$\lambda_j U_{y_j} \subset \gamma Q_{x_i}. \tag{8.45}$$

But $\lambda_i \leq \frac{1}{q}$, so the inclusions $\lambda_i Q_{x_i} \subset \frac{1}{q}Q_{x_i}$, $i \in I$, are obvious, and we only need to prove (8.45). By (8.42), for every cube $U_{y_j}$ there is a cube $Q_{x_i}$ with

$$r(U_{y_j}) = r(Q_{x_i}) - q\,\|y_j - x_i\|\,.$$

Therefore, $\|y_j - x_i\| \leq \frac{1}{q}r(Q_{x_i})$ and $r(U_{y_j}) \leq r(Q_{x_i})$. Consequently, taking the inequality $\lambda_j \leq \frac{1}{q}$ into account, we obtain

$$\lambda_j U_{y_j} \subset \frac{1}{q}U_{y_j} \subset Q(x_i, \|y_j - x_i\| + r(\frac{1}{q}U_{y_j})) \subset \frac{2}{q}Q_{x_i},$$

as claimed. □

### 8.2.4 Proof of the theorem on the preservation of $\alpha$-capacity for $\alpha \in \left(1 - \frac{1}{n}, 1\right)$

We proceed to the proof of Theorem 8.32. It suffices to dominate the $\alpha$-capacity of the family $\{\lambda_i Q_{x_i}\}_{i\in I} \cup \{\lambda_j U_{y_j}\}_{j\in J}$ by that of the family $\{Q_{x_i}\}_{i\in I}$. Since $\lambda_i$, $\lambda_j \le \frac{1}{q}$, it suffices to dominate the $\alpha$-capacity of the family $\{U_{y_j}\}_{j\in J}$. Consider a packing $\pi = \{U_{y_j}\}$ formed by cubes of this family. By (8.42), for every $j \in J$ there is an index $i = i(j) \in I$ such that

$$r(U_{y_j}) = r(Q_{x_i}) - q\,\|y_j - x_i\|\,. \tag{8.46}$$

The collection of cubes $U_{y_j}$ for which (8.46) is satisfied with a fixed $i$ will be denoted by $F_i$. It suffices to show that

$$\sum_{U_{y_j}\in\pi\cap F_i} |U_{y_j}|^\alpha \le c\,|Q_{x_i}|^\alpha \tag{8.47}$$

with a constant $c$ depending only on the dimension. To do this, we recall that $y_j \notin Q_{x_i}$ (because the centers of the cubes $U_y$ are outside $\bigcup_{i\in I} Q_{x_i}$ by construction). So, we use the positivity of the right-hand side in (8.46) to conclude that $y_j \in (\frac{1}{q}\mathring{Q}_{x_i}) \setminus Q_{x_i}$. We split the set $(\frac{1}{q}\mathring{Q}_{x_i})$ into layers in accordance with the distance from the boundary of the cube $\frac{1}{q}Q_{x_i}$ :

$$S_k = \left(\frac{1}{q}(1 - \frac{1}{2^k})Q_{x_i}\right) \setminus \left(\frac{1}{q}(1 - \frac{1}{2^{k-1}})Q_{x_i}\right), \quad k = 1, 2, \ldots,$$

and estimate the number of points $y_j$ belonging to $S_k$.

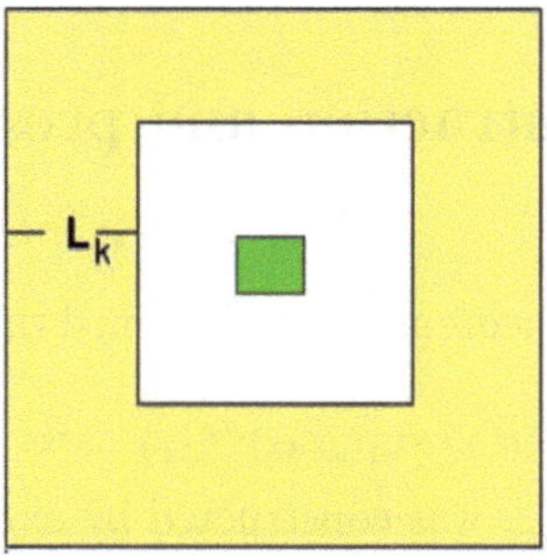


Figure 8.1: The cube $Q_{x_i}$ (colored in green) and the layer $S_k$ (colored in yellow).

We observe that the volume of $S_k$ satisfies

$$|S_k| = \left(\left(\frac{1}{q}\left(1 - \frac{1}{2^k}\right)\right)^n - \left(\frac{1}{q}\left(1 - \frac{1}{2^{k-1}}\right)\right)\right)^n 2^n r(Q_{x_i})^n \le c\frac{1}{2^k} r(Q_{x_i})^n$$

with a constant $c$ depending only on $q$ and $n$. Moreover, the "width" of $S_k$ admits the estimate

$$L_k = \left(\frac{1}{q}\left(1-\frac{1}{2^k}\right) - \frac{1}{q}\left(1-\frac{1}{2^{k-1}}\right)\right) r(Q_{x_i}) = \frac{1}{q}\frac{1}{2^k} r(Q_{x_i}) > \frac{1}{2^k} r(Q_{x_i}).$$

Formula (8.46) shows that if $y_j \in S_k$, then

$$r(U_{y_j}) \geq r(Q_{x_i}) - q\frac{1}{q}\left(1-\frac{1}{2^k}\right) r(Q_{x_i}) = \frac{1}{2^k} r(Q_{x_i}), \tag{8.48}$$

$$r(U_{y_j}) \leq r(Q_{x_i}) - q\frac{1}{q}\left(1-\frac{1}{2^{k-1}}\right) r(Q_{x_i}) = \frac{1}{2^{k-1}} r(Q_{x_i}). \tag{8.49}$$

By (8.48), the cube $U_{y_j}$ cuts a volume of at least $(\frac{1}{2^k} r(Q_{x_i}))^n$ out of $S_k$. Since the cubes $U_{y_j} \subset \pi$ have mutually disjoint interiors, their number cannot exceed the quantity

$$C_k = \frac{|S_k|}{(\frac{1}{2^k} r(Q_{x_i}))^n} \leq c 2^{k(n-1)}.$$

By (8.49) the volume of each of these cubes $U_{y_j}$ does not exceed $2^n(\frac{1}{2^{k-1}} r(Q_{x_i}))^n$. So, their total contribution in (8.47) is at most

$$c 2^{k(n-1)} (\frac{1}{2^{k-1}} r(Q_{x_i}))^{\alpha n}.$$

Summing over $k$, we obtain a convergent geometric series if $\alpha > 1 - \frac{1}{n}$:

$$\sum_{U_{y_j} \in \pi} \left|U_{y_j}\right|^\alpha \leq \sum_{k=1}^{\infty} c 2^{k(n-1)} (\frac{1}{2^{k-1}} r(Q_{x_i}))^{\alpha n} = C \left|Q_{x_i}\right|^\alpha$$

with a constant $C$ depending only on the dimension $n$ and the parameter $q$. □

## 8.3 Controlled contraction and preservation of the $\alpha$-capacity

In the controlled extension theorem, it was assumed that the initial family $\{Q_x\}_{x \in \Omega}$ of cubes satisfies

$$\sigma = \sup{}_{x \in \Omega} r(Q_x) < \infty$$

and a WB-covering $\{K_{u_j}\}_{j \in J}$ was constructed by expansion of some cubes in the initial family. In the controlled contraction theorem, we shall assume that

$$\sigma = \inf{}_{x \in \Omega} r(Q_x) > 0$$

and construct a WB-covering $\{K_{u_j}\}_{j \in J}$ by contraction of some initial cubes. Thus, in a sense, we shall deal with a statement "dual" to the controlled extension theorem.

We start with necessary definitions.

**Definition 8.37.** A family $\{K_u\}_{u\in U}$ of cubes is called a *contraction* of another family $\{Q_x\}_{x\in\Omega}$ (of cubes) if

$$K_x \subset Q_x \tag{8.50}$$

whenever $x \in U \cap \Omega$, and for every $x \in \Omega$ there exists $u = u(x) \in U$ such that

$$x \in K_u \subset Q_x. \tag{8.51}$$

**Definition 8.38.** A contraction $\{K_u\}_{u\in U}$ of a family $\{Q_x\}_{x\in\Omega}$ is said to be *controlled with constant* $\gamma$ if there is a finite multiplicity subfamily $\{Q_{x_i}\}_{i\in I}$ of $\{Q_x\}_{x\in\Omega}$ with multiplicity constant $M_c$ ($c$ is again for "control") and with the property that for every $u \in U$ there exists $i = i(x) \in I$ such that

$$Q_{x_i} \subset \gamma K_u.$$

Thus, the cubes $\{K_u\}_{u\in U}$ are not permitted to be "too" small.

**Theorem 8.39** (On controlled contraction). *Suppose we are given a family $\{Q_x\}_{x\in\Omega}$ of cubes satisfying*

$$\sigma = \inf{}_{x\in\Omega} r(Q_x) > 0. \tag{8.52}$$

*Then there is a controlled contraction $\{K_{u_i}\}_{i\in I}$ of this family that is simultaneously a* WB*-covering. As in Theorem* 8.27*, all relevant constants can be chosen to depend on the dimension only.*

*Remark* 8.40. More precisely, in the above theorem the control constant $\gamma$ and the Whitney constant $\delta$ can be chosen universal, whereas the multiplicity and engagement constants $M$ and $\varepsilon$ as well as the constant $M_c$ depend only on the dimension $n$.

The algorithm for constructing the family $\{K_{u_i}\}_{i\in I}$ resembles that used in the controlled extension theorem, and is based on a somewhat similar (though different) modification of the Besicovitch process. This algorithm consists of three steps.

- At the first step, we apply a newly modified Besicovitch $q$-process to the family $\{Q'_x = \mu Q_x\}_{x\in\Omega}$, where $\mu \le 1$ is a positive constant. This results in a Lipschitz almost disjoint family $\{Q'_{x_i}\}_{i\in I}$.
- For $y \in \mathbb{R}^n \setminus \bigcup Q'_{x_i}$, we introduce the cube $U_y$ of radius

$$r(U_y) = \sup_{i\in I} \left( r(Q'_{x_i}) + q \left\| y - x_i \right\| \right).$$

  To the family $\{U_y\}_{y\in\mathbb{R}^n\setminus\bigcup Q'_{x_i}}$, we again apply the version of the Besicovitch $q$-process used at the first step. This will give a Lipschitz and almost disjoint subfamily $\{U_{y_j}\}_{j\in J}$.

- The cubes $\{K_{u_i}\}_{i\in I}$ are obtained by application of the meshing algorithm (Lemma 8.23) to the family $\{Q'_{x_i}\}_{i\in I} \cup \{U_{y_j}\}_{j\in J}$. The parameters $q$ and $\mu$ are chosen so as to ensure that the resulting WB-covering is a controlled contraction of the initial family of cubes.

In fact, the construction makes it possible to prove the following statement about preservation of the $\alpha$-capacity.

**Theorem 8.41.** *Suppose that $\alpha < 0$ and a family $\{Q_x\}_{x\in\Omega}$ of cubes obeys the condition $\sigma = \inf_{x\in\Omega} r(Q_x) > 0$. Then the WB-covering $\{K_{u_j}\}_{j\in J}$ constructed in the above theorem has the following property:*

$$\left|\{K_{u_j}\}\right|_\alpha \leq c\left|\{Q_x\}\right|_\alpha ,$$

*where $c > 0$ is a constant depending only on the dimension $n$.*

### 8.3.1 Besicovitch $q$-process with a Lipschitz condition for controlled contraction

We shall need the following version of the Besicovitch $q$-process.

Let $\{Q'_x\}_{x\in\Omega}$ be a family of cubes such that

$$\sigma' = \inf_{x\in\Omega} r(Q'_x) > 0. \tag{8.53}$$

We split the interval $[\sigma', \infty)$ into half-open intervals $\left[\sigma'(1+q)^{k-1}, \sigma'(1+q)^k\right)$, $k = 1, 2, \ldots$. First, we consider the cubes $Q'_x$ with radii in the first interval, that is, those satisfying $r(Q') \in [\sigma', \sigma'(1+q))$, and choose cubes $\{Q'_{x_i}\}_{i\in I_1}$ from this family as in the standard Besicovitch $q$-process. Since the center of each new-chosen cube lies outside all cubes chosen previously, we have

$$r(Q'_{x_i}) - r(Q'_{x_j}) \leq \sigma'(1+q) - \sigma' = q\sigma' \leq q \min(r(Q'_{x_i}), r(Q'_{x_j})) < q\,\|x_i - x_j\|$$

and, consequently, the family $\{Q'_{x_i}\}_{i\in I_1}$ satisfies the Lipschitz condition with constant $q$. After that, we consider the cubes $Q'_x$ with radii in the next interval $\left[\sigma'(1+q), \sigma'(1+q)^2\right)$. As in the case of controlled extension, we modify the Besicovitch process by requiring that every new-chosen cube satisfy the Lipschitz condition with constant $q$ relative to all cubes chosen previously *and* have center outside all these cubes. Then we do likewise with the cubes whose radii fall within the next interval $\left[\sigma'(1+q)^2, \sigma'(1+q)^3\right)$, and so on. As a result, we arrive at a family $\{Q'_{x_i}\}_{i\in I}$ of cubes.

**Lemma 8.42.** *The family $\{Q'_{x_i}\}_{i\in I}$ constructed in this way has the following properties:*

- *it is almost disjoint, i.e.,*

$$\min(r(Q'_{x_i}), r(Q'_{x_j})) < \|x_i - x_j\| ; \tag{8.54}$$

- *it satisfies the Lipschitz condition with constant* $q$:

$$\left|r(Q'_{x_i}) - r(Q'_{x_j})\right| \le q \,\|x_i - x_j\|\,; \tag{8.55}$$

- *for every* $x \in \Omega$ *there is an index* $i \in I$ *such that either*

$$x \in r(Q'_{x_i}) \text{ and } r(Q'_x) > \frac{1}{1+q} r(Q'_{x_i}) \tag{8.56}$$

*or*

$$x \notin r(Q'_{x_i}) \text{ and } r(Q'_x) > q\,\|x - x_i\| + r(Q'_{x_i}). \tag{8.57}$$

*Proof.* The first two properties are clear from the construction. The third is proved as follows. Let $x \in \Omega$. Choose $k$ with $r(Q'_x) \in \big[\sigma'(1+q)^{k-1}, \sigma'(1+q)^k\big)$. If $x$ is contained in the union of the cubes $Q'_{x_i}$ constructed at the first $k$ steps, then (8.56) is fulfilled, because these cubes satisfy

$$r(Q'_{x_i}) < \sigma'(1+q)^k.$$

But if $x$ is not in the union of these cubes, then, as in the previous variation of this theme, for some cube $Q'_{x_i}$ chosen at the first $k$ steps we must have

$$\left|r(Q'_x) - r(Q'_{x_i})\right| > q\,\|x - x_i\|\,. \tag{8.58}$$

As before, slight concentration (the same as previously!) is needed to realize this. Again, we observe that if there are infinitely many cubes chosen at some step, then their centers tend to infinity. Then we recall that, at each stage, we do not choose "at random" among "admissible" cubes but require that the center be "almost" the closest to the origin, etc.

It suffices to show that $r(Q'_x) - r(Q'_{x_i}) \ge 0$ (then (8.57) will follow from (8.58)). Suppose the contrary, then (8.58) implies

$$r(Q'_{x_i}) - r(Q'_x) > q\,\|x_i - x\|\,.$$

Since also $r(Q'_{x_i}) < \sigma'(1+q)^k \le (1+q) r(Q'_x)$, we see that

$$r(Q'_{x_i}) - \frac{1}{1+q} r(Q'_{x_i}) > q\,\|x_i - x\|\,,$$

i.e., $\frac{1}{1+q} r(Q'_{x_i}) > \|x_i - x\|$ and $x \in Q'_{x_i}$, which is impossible. □

*Remark* 8.43. If the family $\{Q'_{x_i}\}_{i\in I}$ is Lipschitz with constant $q$, i.e.,

$$|r(Q'_u) - r(Q'_v)| \le q\,\|u - v\| \text{ for all } u, v \in \Omega,$$

then the inequality in (8.57) is never fulfilled. Therefore, (8.56) is true for all $x \in \Omega$. In particular,

$$\Omega \subset \bigcup_{i\in I} Q'_{x_i}. \tag{8.59}$$

### 8.3.2 Construction of a WB-covering

Suppose we are given a family $\{Q_x\}_{x\in\Omega}$ of cubes with $\sigma = \inf_{x\in\Omega} r(Q_x) > 0$. Consider the family $\{Q'_x = \mu Q_x\}_{x\in\Omega}$, where the constant $\mu > 0$ will be chosen later. We still have

$$\sigma' = \inf_{x\in\Omega} r(Q'_x) = \mu\sigma > 0.$$

So, the above version of the Besicovitch process applies, yielding a new family $\{Q'_{x_i}\}_{i\in I}$ of cubes. On the space $\mathbb{R}^n$, we consider the function

$$G(y) = \inf_{i\in I}(r(Q'_{x_i}) + q\,\|x_i - y\|).$$

For every $y \in \mathbb{R}^n \setminus \bigcup_{i\in I} Q'_{x_i}$, we introduce the cube $U_y$ centered at $y$ and of radius

$$r(U_y) = G(y) = \inf_{i\in I}(r(Q'_{x_i}) + q\,\|x_i - y\|). \tag{8.60}$$

It is important that this infimum is attained.

**Proposition 8.44.** *For every $y \in \mathbb{R}^n \setminus \bigcup_{i\in I} Q'_{x_i}$ there is $i \in I$ such that*

$$r(U_y) = r(Q'_{x_i}) + q\,\|x_i - y\|\,. \tag{8.61}$$

*Proof.* It suffices to take the infimum in (8.60) over $i \in I$ such that $\|x_i - y\| \le \frac{1}{q} r(U_y)$. This means that $x_i$ must be in the cube of radius $\|y\| + \frac{1}{q} r(U_y)$ and centered at zero. Since the family $\{Q'_{x_i}\}_{i\in I}$ is Lipschitz and almost disjoint, Proposition 8.19 shows that the cubes $\frac{1}{2+q} Q'_{x_i}$ do not intersect. Their radii being at least $\frac{\sigma'}{2+q} > 0$, by volume considerations there are only finitely many such cubes. Consequently, there exists an index $i \in I$ such that (8.61) is fulfilled. □

Since the functions $F_i(y) = r(Q'_{x_i}) + q\,\|x_i - y\|$ are Lipschitz with constant $q$, the function $G(y) = \inf_{i\in I} F_i(y)$ has the same property. Therefore, the family $\{U_y\}_{\mathbb{R}^n\setminus\bigcup Q'_{x_i}}$ is Lipschitz with constant $q$. By (8.61), it is also clear that

$$\inf_{y\in\mathbb{R}^n\setminus\bigcup Q'_{x_i}} r(U_y) \ge \sigma' > 0.$$

Therefore, the above version of the Besicovitch $q$-process applies to the family $\{U_y\}_{\mathbb{R}^n\setminus\bigcup Q'_{x_i}}$, yielding a Lipschitz and almost disjoint family $\{U_{y_j}\}_{j\in J}$ of cubes. Moreover, by Remark 8.43, we have (8.59), that is,

$$\mathbb{R}^n \setminus \bigcup Q'_{x_i} \subset \bigcup_{j\in J} U_{y_j}$$

and, consequently,

$$\mathbb{R}^n = \bigcup_{i\in I} Q'_{x_i} \cup \bigcup_{j\in J} U_{y_j}.$$

Now, we show that the union $\{Q'_{x_i}\}_{i\in I}\cup\{U_{y_j}\}_{j\in J}$ is an almost disjoint family that is also Lipschitz with constant $q$. Separately, the families $\{Q'_{x_i}\}_{i\in I}$ and $\{U_{y_j}\}_{j\in J}$ are almost disjoint; since the centers $y_j$ are outside $\bigcup_{i\in I} Q'_{x_i}$, their union is also almost disjoint. Next, separately the two families are Lipschitz with constant $q$; so, to prove the Lipschitz property for the union, it suffices to verify the inequality

$$\left|r(Q'_{x_i}) - r(U_{y_j})\right| \le q\,\|x_i - y_j\|$$

for arbitrary $i \in I$ and $j \in J$. Since $r(U_{y_j}) = \inf_{i\in I}(r(Q'_{x_i}) + q\,\|x_i - y_j\|)$, we see that $r(U_{y_j}) - r(Q'_{x_i}) \le q\,\|x_i - y_j\|$. Thus, it suffices to prove that

$$r(Q'_{x_i}) - r(U_{y_j}) \le q\,\|x_i - y_j\|\,. \tag{8.62}$$

The family $\{Q'_{x_i}\}_{i\in I}$ being Lipschitz, for every $k \in I$ we have

$$r(Q'_{x_i}) - r(Q'_{x_k}) \le q\,\|x_i - x_k\| \le q\,\|y_j - x_i\| + q\,\|y_j - x_k\|\,.$$

Therefore,

$$r(Q'_{x_i}) - q\,\|y_j - x_i\| \le r(Q'_{x_k}) + q\,\|y_j - x_k\|\,,$$

and

$$r(Q'_{x_i}) - q\,\|y_j - x_i\| \le \inf_{k\in I}(r(Q'_{x_k}) + q\,\|y_j - x_k\|) = r(U_{y_j}),$$

implying (8.62).

So, the family $\{Q'_{x_i}\}_{i\in I} \cup \{U_{y_j}\}_{j\in J}$ is Lipschitz with constant $q$ and almost disjoint. Therefore, Lemmas 8.21 and 8.24 (about finite overlap and strong engagement) are applicable. They say that for given $0 < \varepsilon_1 < \varepsilon_0 < \frac{1}{q} - 1$ there exist numbers

$$\lambda_i, \lambda_j \in [1+\varepsilon_1, 1+\varepsilon_0]\,, i \in I, j \in J$$

such that the family $\{\lambda_i Q'_{x_i}\}_{i\in I} \cup \{\lambda_j U_{y_j}\}_{j\in J}$ is strongly engaged and of finite overlap (*a fortiori*, of finite multiplicity). Since

$$\bigcup_{i\in I} Q'_{x_i} \cup \bigcup_{j\in J} U_{y_j} = \mathbb{R}^n$$

and $\lambda_i, \lambda_j \ge 1+\varepsilon_1$, we see that the family $\{\lambda_i Q'_{x_i}\}_{i\in I} \cup \{\lambda_j U_{y_j}\}_{j\in J}$ satisfies the Whitney condition with $\delta = \varepsilon_1$ and, therefore, is a WB-covering for $\mathbb{R}^n$.

### 8.3.3 Proof of the contraction theorem

First, we prove that the family $\{\lambda_i Q'_{x_i}\}_{i\in I} \cup \{\lambda_j U_{y_j}\}_{j\in J}$ is a contraction of $\{Q_x\}_{x\in\Omega}$.

We show that the first condition (8.50) in the definition of a contraction is fulfilled if

$$\mu \le \frac{1}{1+\varepsilon_0}. \tag{8.63}$$

Indeed, $\lambda_i Q'_{x_i} = \lambda_i \mu Q_{x_i}$, and since $\lambda_i \le 1+\varepsilon_0$, by (8.63) we obtain

$$\lambda_i Q'_{x_i} \subset Q_{x_i},$$

and we have verified the required inclusion (8.50) for the cubes of the form $\lambda_i Q'_{x_i}$.

We treat the remaining cubes. If $y_j \in \Omega$ and $y_j \in \mathbb{R}^n \setminus \bigcup Q'_{x_i}$, then (8.57) is fulfilled, i.e., there exists $i \in I$ with

$$r(Q'_{y_j}) > r(Q'_{x_i}) + q\,\|x_i - y_j\|\,.$$

Therefore,

$$r(Q'_{y_j}) > \inf{}_{i\in I}(r(Q'_{x_i}) + q\,\|x_i - y_j\|) = r(U_{y_j})$$

and, since $\lambda_j \le 1+\varepsilon_0$, we have

$$\lambda_j U_{y_j} \subset \lambda_j Q'_{y_j} \subset (1+\varepsilon_0)\mu Q_{y_j}.$$

Now, (8.63) implies $\lambda_j U_{y_j} \subset Q_{y_j}$, and (8.50) is true also for the cubes $\lambda_j U_{y_j}$.

Now, we show that the second condition (8.51) in the definition of contraction is fulfilled provided that

$$\mu \le \min\left(\frac{1}{(2+\varepsilon_0)(1+q)}, \frac{1-q}{2+\varepsilon_0}\right) = \frac{1-q}{2+\varepsilon_0}. \tag{8.64}$$

Indeed, suppose $x \in \Omega$ satisfies (8.56), that is, there exists $i \in I$ with

$$x \in Q'_{x_i} \text{ and } r(Q'_{x_i}) < r(Q'_x)(1+q).$$

But $x \in \lambda_i Q'_{x_i}$ for this $i$ (because $\lambda_i \ge 1$). Next,

$$\begin{aligned}\lambda_i Q'_{x_i} &\subset Q(x, r(Q'_{x_i}) + \lambda_i r(Q'_{x_i})) \subset Q(x,(2+\varepsilon_0)r(Q'_{x_i}))\\ &\subset (2+\varepsilon_0)(1+q)Q'_x = (2+\varepsilon_0)(1+q)\mu Q_x.\end{aligned}$$

By (8.64), we see that

$$\lambda_i Q'_{x_i} \subset Q_x,$$

that is, we have verified the required condition in the case where $x \in \Omega$ satisfies (8.56).

But if $x \in \Omega$ does not satisfy (8.56), then (8.57) is true for it, i.e., there exists $i \in I$ with

$$r(Q'_x) > r(Q'_{x_i}) + q\,\|x - x_i\|\,. \tag{8.65}$$

Two cases are possible: $x \in \bigcup_k Q'_{x_k}$ or $x \in \Omega \setminus \bigcup_k Q'_{x_k}$. We consider them separately. Suppose $x \in Q'_{x_k}$. Since the family $\{Q'_{x_i}\}_{i\in I}$ is Lipschitz, we have

$$\begin{aligned}r(Q'_{x_k}) &\le r(Q'_{x_i}) + q\,\|x_k - x_i\| \le r(Q'_{x_i}) + q\,\|x_k - x\| + q\,\|x - x_i\|\\ &\le r(Q'_{x_i}) + q r(Q'_{x_k}) + q\,\|x - x_i\|\,.\end{aligned}$$

Therefore,

$$r(Q'_{x_k}) \le \frac{1}{1-q}(r(Q'_{x_i}) + q\,\|x - x_i\|) < \frac{1}{1-q} r(Q'_x).$$

It follows that

$$\begin{aligned}\lambda_k Q'_{x_k} &\subset Q(x, r(Q'_{x_k}) + \frac{\lambda_k}{1-q} r(Q'_x)) \\ &\subset \frac{1}{1-q} Q'_x + \lambda_k \frac{1}{1-q} Q'_x \subset \frac{2+\varepsilon_0}{1-q} Q'_x = \frac{2+\varepsilon_0}{1-q}\mu Q_x.\end{aligned}$$

By (8.64), the required inclusion

$$\lambda_k Q'_{x_k} \subset Q_x,$$

is proved in the case in question.

Now, we consider the case where $x \in \Omega \setminus \bigcup_k Q'_{x_k}$. Formula (8.65) shows that

$$r(Q'_x) > \inf{}_{i\in I}(r(Q'_{x_i}) + q\,\|x_i - x\|) = r(U_x). \tag{8.66}$$

Since the family $\{U_y\}_{\mathbb{R}^n \setminus \bigcup Q'_{x_i}}$ is Lipschitz and the family $\{U_{y_j}\}_{j\in J}$ is obtained from it by the modified Besicovitch process, by Remark 8.43 and formula (8.56) there is $j \in J$ with (see (8.66))

$$x \in U_{y_j} \text{ and } \frac{1}{1+q} r(U_{y_j}) \le r(U_x) \le r(Q'_x).$$

Thus, $x \in U_{y_j}$ and for this $j \in J$ we have

$$\begin{aligned}\lambda_j U_{y_j} &\subset Q(x, r(U_{y_j}) + \lambda_j r(U_{y_j})) \subset Q(x, (2+\varepsilon_0) r(U_{y_j})) \\ &\subset (2+\varepsilon_0)(1+q) Q'_x = (2+\varepsilon_0)(1+q)\mu Q_x.\end{aligned}$$

Then condition (8.64) implies

$$\lambda_j U_{y_j} \subset Q_x,$$

which completes the proof of (8.51).

Now, we explore the control property. We take $\left\{Q_{x_i} = \frac{1}{\mu} Q'_{x_i}\right\}_{i\in I}$ for the role of a controlling family. The family $\{Q'_{x_i}\}_{i\in I}$ is almost disjoint and Lipschitz with constant $q$. By the finite overlap lemma (see Lemma 8.21), the family $\left\{Q_{x_i} = \frac{1}{\mu} Q'_{x_i}\right\}_{i\in I}$ is of finite multiplicity provided $\frac{1}{\mu} < \frac{1}{q}$, i.e.,

$$q < \mu. \tag{8.67}$$

*Remark* 8.45. Taken together, conditions (8.64) and (8.67) impose a restriction on $\mu$ :

$$q < \mu \le \frac{1-q}{2+\varepsilon_0}.$$

Such a $\mu$ exists if $3 + \varepsilon_0 < \frac{1}{q}$, that is, when $q < \frac{1}{3}$.

Now, we show that the family $\{Q_{x_i}\}_{i\in I}$ controls the family $\{\lambda_i Q'_{x_i}\}_{i\in I} \cup \{\lambda_j U_{y_j}\}_{j\in J}$. Indeed, the inclusion

$$Q_{x_i} \subset \gamma\lambda_i Q'_{x_i} = \gamma\lambda_i\mu Q_{x_i}$$

is equivalent to the inequality

$$\gamma\lambda_i\mu \geq 1,$$

which is fulfilled for $q < \mu$ provided

$$\gamma \geq \frac{1}{q}.$$

The inclusion $Q_{x_i} \subset \gamma\lambda_j U_{y_j}$, where $i \in I$ depends on $j$, is a consequence of (8.61). Indeed, by that formula, there exists $i = i(j) \in I$ with

$$r(U_{y_j}) = r(Q'_{x_i}) + q\,\|x_i - y_j\|,$$

whence we obtain

$$r(Q'_{x_i}) < r(U_{y_j}) \text{ and } \|x_i - y_j\| < \frac{1}{q} r(U_{y_j}).$$

Combination of this with (8.67) yields

$$Q_{x_i} = \frac{1}{\mu}Q'_{x_i} \subset \frac{1}{q}Q'_{x_i} \subset Q\left(y_j, \|x_i - y_j\| + \frac{1}{q}r(Q'_{x_i})\right) \subset \frac{1}{q}U_{y_j} + \frac{1}{q}U_{y_j} \subset \frac{2}{q}U_{y_j}.$$

Thus, we can take $\frac{2}{q}$ for the role of the control constant $\gamma$. □

## 8.4 Preservation of the $\alpha$-capacity ($\alpha < 0$)

We estimate the $\alpha$-capacity of the family $\{\lambda_i Q'_{x_i}\}_{i\in I} \cup \{\lambda_j U_{y_j}\}_{j\in J}$ in terms of that of the family $\{Q_{x_i}\}_{i\in I}$. First, recall that the latter family is of finite multiplicity, so it splits in finitely many packings (the number of which only depends on the dimension and the multiplicity constant). This shows that

$$\sum_{i\in I} |Q_{x_i}|^\alpha \leq c\left|\{Q_x\}_{x\in\Omega}\right|_\alpha.$$

The relation $q < \mu < 1$ implies $Q'_{x_i} \subset Q_{x_i} \subset \frac{1}{q}Q'_{x_i}$. Since $\alpha < 0$, we obtain

$$\sum_{i\in I} \left|Q'_{x_i}\right|^\alpha \leq cq^{n\alpha}\left|\{Q_x\}_{x\in\Omega}\right|_\alpha. \tag{8.68}$$

Combined with the inequality $\lambda_i \geq 1$, $i \in I$, this gives the estimate

$$\left|\{\lambda_i Q'_{x_i}\}_{i\in I}\right|_\alpha \leq cq^{n\alpha}\left|\{Q_x\}_{x\in\Omega}\right|_\alpha.$$

Next, again referring to the inequality $\lambda_j \geq 1$, $j \in J$, we see that the $\alpha$-capacity of the family $\{\lambda_j U_{y_j}\}_{j\in J}$ does not exceed that of the family $\{U_{y_j}\}_{j\in J}$. To estimate the latter quantity, we use the fact (see (8.61)) that $y_j \in \mathbb{R}^n \setminus \bigcup_{i\in I} Q'_{x_i}$ and for every $y_j$ there exists an index $i = i(j) \in I$ with

$$r(U_{y_j}) = r(Q'_{x_i}) + q\,\|x_i - y_j\|\,. \tag{8.69}$$

The set of cubes $U_{y_j}$ corresponding to one and the same index $i$ in (8.69) will be denoted by $F_i$. So, to finish the proof, it suffices to verify the inequality

$$\sum_{U_{y_j}\in F_i} \left|U_{y_j}\right|^\alpha \leq c\left|Q'_{x_i}\right|^\alpha \tag{8.70}$$

with a constant $c > 0$ depending only on the dimension $n$. Since the family $\{U_{y_j}\}_{j\in J}$ is of finite multiplicity, it splits into finitely many packings, their number depending only on the dimension. So, we may assume that summation on the left in (8.70) is over the intersection of $F_i$ with one of these packings. We split $\mathbb{R}^n \setminus Q'_{x_i}$ into the layers

$$S_k = 2^k Q'_{x_i} \setminus 2^{k-1}Q'_{x_i}, k = 1, 2, \ldots$$

of "width" $2^{k-1}r(Q'_{x_i})$ and of volume

$$|S_k| = (2^n - 1)2^{kn} r(Q'_{x_i})^n.$$

Since $y_j \in \mathbb{R}^n \setminus \bigcup_{i\in I} Q'_{x_i}$, it follows that $y_j$ is in some $S_k$. By (8.69), for this $k$ we have

$$r(U_{y_j}) \geq q 2^{k-1} r(Q'_{x_i}).$$

Consequently, for $q < 1/3$, the cube $U_{y_j}$ cuts out of $S_k$ a volume of at least

$$V = \left(q 2^{k-1} r(Q'_{x_i})\right)^n$$

Since we stay within the packing $\pi$, the number of cubes $U_{y_j}$ in question is at most

$$\frac{|S_k|}{V} = \frac{(2^n-1)2^{kn}}{q^n 2^{(k-1)n}} = \frac{2^n(2^n-1)}{q^n},$$

which does not depend on $k$. Next, for these $U_{y_j}$ we have

$$r(U_{y_j}) \leq (1 + q2^k) r(Q'_{x_i}) \leq 2^{k+1} r(Q'_{x_i})$$

(see (8.69)), and their contribution to the left-hand side of (8.70) does not exceed

$$c(n, \alpha, q) 2^{kn\alpha} r(Q'_{x_i})^{n\alpha}.$$

Summing over $k$, for $\alpha < 0$ we obtain a convergent geometric series:

$$\sum_{U_{y_j}\in\pi} \left|U_{y_j}\right|^\alpha \leq \sum_{k=1}^{\infty} c(n, \alpha, q) 2^{kn\alpha} r(Q'_{x_i})^{n\alpha} \leq c\left|Q'_{x_i}\right|^\alpha$$

($c$ is a positive constant depending only on the dimension $n$). This proves inequality (8.70) and, with it, the theorem.

# Notes and remarks

The results of this chapter were obtained by Kruglyak. The covering theorems presented here develop the original results in [K1], [K0] in the following way: in those papers it was assumed that the initial covering has bounded $\alpha$-capacity, and the notion of a controlled covering was not used. Accordingly, the arguments here differ somewhat from those in [K1], [K0].

Presumably, the restriction $\alpha \notin [0, 1 - \frac{1}{n}]$ in the results on the preservation of the $\alpha$-capacity is essential. See a counterexample for $\alpha = 0$ in [K1].

We want to give more emphasis to the (dis)similarity between the results of this chapter and the classical theorems by Whitney and Besicovitch.

a) In the Whitney theorem, the cubes are "movable", there is much freedom in their choice, the only important restriction on their position is that they must lie inside a given open set $\Omega$ whose complement is nonempty. These cubes can be used to construct a smooth partition of unity.

   The cubes in controlled covering theorems share the same features. However, the restriction on their position acquires a different form: they must be controlled by another family of cubes given beforehand. In fact, we explained at the beginning (see Theorem 8.12) that the Whitney theorem corresponds to the particular case of the controlled extension theorem when the initial family is Lipschitz with constant $q < 1$.

b) In the Besicovitch theorem, the cubes are "unmovable" and cannot be used for construction of a smooth partition of unity with due properties. However, the proofs of controlled covering theorems are modifications of the arguments leading to Besicovich's result.

# Chapter 9

# Construction of near-minimizers

In this chapter we present long-promised explicit constructions of near-minimizers for couples in which one space is $L^p$, and the other is either a Sobolev or a Morrey–Campanato space.

We start with recalling some basic definitions. Let $(X_0, X_1)$ be a couple of Banach or quasi-Banach spaces, and let $x \in X_0$. We remind the reader that an element $x_t$ is called a near-minimizer for the distance functional if

$$\|u_t\|_{X_1} \leq ct \text{ and } \|x - u_t\|_{X_0} \leq c \operatorname{dist}_{X_0}(x, B_{X_1}(\frac{t}{c})),$$

where

$$\operatorname{dist}_{X_0}(x, B_{X_1}(t)) = \inf_{\|u\|_{X_1} \leq t} \|x - u\|_{X_0}$$

is the distance in the metric of $X_0$ from $x$ to the ball $B_{X_1}(t)$ of $X_1$ centered at zero and of radius $t > 0$. We are interested in finding near-minimizers for which the constant $c$ does not depend on $x$ and $t$.

As has already been said (and as we shall see below), in many cases, instead of the distance functional, it is more convenient to construct near-minimizers for the $L_{p,q}$-functional

$$L_{p,q}(t, x; X_0, X_1) = \inf_{u \in X_1} \|x - u\|_{X_0}^p + t \|u\|_{X_1}^q ,$$

where $p, q$ are some positive constants (they are chosen in a way specific for every couple $(X_0, X_1)$). Recall that $x_t$ is a near-minimizer for the $L_{p,q}$-functional if

$$\|x - x_t\|_{X_0}^p + t \|x_t\|_{X_1}^q \leq cL_{p,q}(t, x; X_0, X_1).$$

We denote by $L^p(Q)$ the space $L^p$ against Lebesgue measure on a cube $Q$. The symbol $\mathcal{P}_k$ is reserved for the subspace of all algebraic polynomials of degree *strictly smaller* than $k$. In particular, $\mathcal{P}_0 = \{0\}$, and $\mathcal{P}_1$ consists of constants. If a function

$f$ defined on $\mathbb{R}^n$ belongs locally to $L^p$, $0 < p < \infty$, then $f_Q$ denotes a polynomial of degree strictly smaller than $k$ that provides the best approximation to $f$ on $Q$ in the $L^p$-metric, that is

$$\inf_{P\in\mathcal{P}_k}\left(\int_Q |f-P|^p\right)^{\frac{1}{p}} = \left(\int_Q |f-f_Q|^p\right)^{\frac{1}{p}}.$$

(Usually, the number $k$ will be clear from the context.) Such a polynomial always exists because the space $\mathcal{P}_k$ is finite-dimensional.

Positive estimational constants are denoted by $c$; the value of $c$ may vary even within one line, but always it only depends on the dimension $n$ and, maybe, on some other parameters unessential at the moment.

The lemma proved in the next section will often be used in the sequel.

## 9.1 Estimates for derivatives of approximants

Suppose that a family $\{K_i\}_{i\in I}$ of cubes is a WB-covering with Whitney, multiplicity, and engagement constants depending only on the dimension $n$. We recall that, by Lemma 3.13, the overlap number of this covering is also controlled in terms of $n$. Let $f$ be a function defined on $\mathbb{R}^n$ and belonging to $L^p$ locally. Let $P_i, i \in I$, be polynomials of degree strictly smaller than $k$ that nearly realize the best approximation on the cubes $K_i, i \in I$, that is,

$$\|f - P_i\|_{L^p(K_i)} \le c \inf_{g\in\mathcal{P}_k} \|f-g\|_{L^p(K_i)},$$

where $c \ge 1$ is a constant independent of the function and of the cubes (but we allow it to depend on the dimension $n$).

We are interested in estimates for derivatives of the function

$$\tilde{f} = \sum_{i\in I} P_i\psi_i,$$

where $\{\psi_i\}_{i\in I}$ is a smooth partition of unity adjusted to the WB-covering $\{K_i\}_{i\in I}$.

**Lemma 9.1.** *Let $k_1 + k_2 + \cdots + k_n = k$, then for every cube $K_{i_0}$ we have*

$$\left\|\frac{\partial^k}{\partial^{k_1}x_1\partial^{k_2}x_2\ldots\partial^{k_n}x_n}\tilde{f}\right\|_{L^\infty(K_{i_0})} \le c \sum_{j:K_j\cap K_{i_0}\neq\varnothing} \frac{1}{|K_i|^{\frac{k}{n}}}\left(\frac{1}{|K_i|}\int_{K_i}|f-f_{K_i}|^p\right)^{\frac{1}{p}},$$

*where $c > 0$ depends only on $k$ and the dimension $n$.*

*Proof.* All derivatives of order $k$ of the $P_i$ vanish, whence

$$\frac{\partial^k}{\partial^{k_1}x_1\partial^{k_2}x_2\ldots\partial^{k_n}x_n}\tilde{f} = \frac{\partial^k}{\partial^{k_1}x_1\partial^{k_2}x_2\ldots\partial^{k_n}x_n}\sum_{i\in I}P_i\psi_i$$

$$= \sum_{i\in I}\sum_{l_1,\ldots,l_n} c_{l_1,\ldots,l_n}\frac{\partial^l}{\partial^{l_1}x_1\partial^{l_2}x_2\ldots\partial^{l_n}x_n}P_i \cdot \frac{\partial^{k-l}}{\partial^{k_1-l_1}x_1\partial^{k_2-l_2}x_2\ldots\partial^{k_n-l_n}x_n}\psi_i.$$

Here the internal summation is taken over all indices $l_j \leq k_j$ such that $l = l_1 + \cdots + l_n < k$. Now, we multiply the relation $\sum_{i\in I} \psi_i = \chi_{\cup K_i}$ by $P_{i_0}$ and differentiate the resulting formula as above. Since $k - l \geq 1$, this results in

$$\sum_{i\in I} \sum_{l_1,\dots,l_n} c_{l_1,\dots,l_n} \frac{\partial^l}{\partial^{l_1} x_1 \partial^{l_2} x_2 \dots \partial^{l_n} x_n} P_{i_0} \cdot \frac{\partial^{k-l}}{\partial^{k_1-l_1} x_1 \partial^{k_2-l_2} x_2 \dots \partial^{k_n-l_n} x_n} \psi_i = 0$$

on $K_{i_0}$ (summation is over the same collections of $l_j$'s as above), and we obtain

$$\begin{aligned} &\frac{\partial^k}{\partial^{k_1} x_1 \partial^{k_2} x_2 \dots \partial^{k_n} x_n} \tilde{f} \\ &= \sum_{i\in I} \sum_{l_1,\dots,l_n = l<k} c_{l_1,\dots,l_n} \frac{\partial^l}{\partial^{l_1} x_1 \dots \partial^{l_n} x_n} (P_i - P_{i_0}) \cdot \frac{\partial^{k-l}}{\partial^{k_1-l_1} x_1 \dots \partial^{k_n-l_n} x_n} \psi_i. \end{aligned}$$

Since the family $\{K_i\}_{i\in I}$ is a WB-covering, it has finite overlap, so only finitely many summands on the right may be nonzero on $K_{i_0}$. Moreover, by Lemma 3.13, their number is controlled in terms of the dimension only in our setting. We estimate one of them (note that $K_i \cap K_{i_0} \neq \varnothing$ for it). The Markov inequality (see Theorem 0.21) shows that

$$\left\| \frac{\partial^l}{\partial^{l_1} x_1 \partial^{l_2} x_2 \dots \partial^{l_n} x_n} (P_i - P_{i_0}) \right\|_{L^\infty(K_{i_0})} \leq c \frac{1}{|K_{i_0}|^{\frac{l}{n}}} \|(P_i - P_{i_0})\|_{L^\infty(K_{i_0})}.$$

By the strong engagement property, we see that the $L^\infty$-norm of the polynomial on the right is equivalent to the $L^p$-norm of the same polynomial on $K_i \cap K_{i_0}$ (the corresponding constants depending only on $k$ and the dimension). Also (see Theorem 0.22), a polynomial on $K_{i_0}$ can be estimated in terms of its values on $K_i \cap K_{i_0}$. Therefore,

$$\begin{aligned} &\|(P_i - P_{i_0})\|_{L^\infty(K_{i_0})} \leq c \|P_i - P_{i_0}\|_{L^\infty(K_i \cap K_{i_0})} \\ &\leq c \left( \frac{1}{|K_i \cap K_{i_0}|} \int_{K_i \cap K_{i_0}} |P_i - P_{i_0}|^p \right)^{\frac{1}{p}} \\ &\leq c \left[ \left( \frac{1}{|K_i|} \int_{K_i} |f - P_i|^p \right)^{\frac{1}{p}} + \left( \frac{1}{|K_{i_0}|} \int_{K_{i_0}} |f - P_{i_0}|^p \right)^{\frac{1}{p}} \right]. \end{aligned}$$

Combined with the strong engagement property of the family $\{K_i\}_{i\in I}$, this yields

$$\begin{aligned}
&\left\|\frac{\partial^l}{\partial^{l_1}x_1\dots\partial^{l_n}x_n}(P_i-P_{i_0})\right\|_{L^\infty(K_{i_0})}\\
&\le c\frac{1}{|K_{i_0}|^{\frac{l}{n}}}\left[\left(\frac{1}{|K_i|}\int_{K_i}|f-P_i|^p\right)^{\frac1p}+\left(\frac{1}{|K_{i_0}|}\int_{K_{i_0}}|f-P_{i_0}|^p\right)^{\frac1p}\right]\\
&\le c\frac{1}{|K_{i_0}|^{\frac{l}{n}+\frac1p}}\left[\left(\int_{K_i}|f-P_i|^p\right)^{\frac1p}+\left(\int_{K_{i_0}}|f-P_{i_0}|^p\right)^{\frac1p}\right].
\end{aligned}$$

But since $\{\psi_i\}_{i\in I}$ is a smooth partition of unity adjusted to the WB-covering $\{K_i\}_{i\in I}$, by Theorem 3.19 we see that

$$\left\|\frac{\partial^{k-l}}{\partial^{k_1-l_1}x_1\dots\partial^{k_n-l_n}x_n}\psi_i\right\|_{L^\infty(K_{i_0})}\le c\frac{1}{|K_i|^{\frac{k-l}{n}}}\le c\frac{1}{|K_{i_0}|^{\frac{k-l}{n}}}.$$

Therefore,

$$\begin{aligned}
&\left\|\frac{\partial^l}{\partial^{l_1}x_1\dots\partial^{l_n}x_n}(P_i-P_{i_0})\cdot\frac{\partial^{k-l}}{\partial^{k_1-l_1}x_1\dots\partial^{k_n-l_n}x_n}\psi_i\right\|_{L^\infty(K_{i_0})}\\
&\le\frac{c}{|K_{i_0}|^{\frac{k}{n}+\frac1p}}\left[\left(\int_{K_i}|f-P_i|^p\right)^{\frac1p}+\left(\int_{K_{i_0}}|f-P_{i_0}|^p\right)^{\frac1p}\right]\\
&\le c\left[\frac{1}{|K_i|^{\frac{k}{n}}}\left(\frac{1}{|K_i|}\int_{K_i}|f-P_i|^p\right)^{\frac1p}+\frac{1}{|K_{i_0}|^{\frac{k}{n}}}\left(\frac{1}{|K_{i_0}|}\int_{K_{i_0}}|f-P_{i_0}|^p\right)^{\frac1p}\right].
\end{aligned}$$

Since $\|f-P_i\|_{L^p(K_i)}\le c\left(\int_{K_i}|f-f_{K_i}|^p\right)^{\frac1p}$ (as previously, $f_{K_i}$ stands for an appropriate polynomial of best approximation), summing over $i$ we arrive at the required statement (we have again used the finite overlap property here). □

## 9.2 Near-minimizers for Sobolev spaces: the couples $(L^p,\dot W_q^k)$

We recall the formula for the standard seminorm on $\dot W_q^k$:

$$\|g\|_{\dot W_q^k}=\sup_{k_1+\dots+k_n=k}\left\|\frac{\partial^k}{\partial^{k_1}x_1\partial^{k_2}x_2\dots\partial^{k_n}x_n}g(x)\right\|_{L^q}.$$

The indices $p$ and $q$ will be subject to the conditions $0 < p < \infty$ and $1 \le q < \infty$. We shall construct near-minimizers for the $L_{p,q}$-functional of the couple $(L^p, \dot{W}^k_q)$ under the assumption

$$\frac{k}{n} + \frac{1}{p} - \frac{1}{q} > 0. \tag{9.1}$$

As was mentioned at the very beginning of this book, this condition is well known in embedding theorems. Also, it guarantees the validity of the Sobolev–Poincaré inequality (see Theorem 4.2). We reproduce the statement. It should be kept in mind that the constant $c$ in it depends also on $p$ and $q$ (and the same applies to various estimation constants in the subsections that follow): these two parameters are viewed as fixed in each particular statement.

**Theorem.** *Suppose $g \in \dot{W}^k_q$, and let $g_Q \in \mathcal{P}_k$ be a polynomial of the best approximation for $g$ on a cube $Q$ in the $L^p$-metric. There exists a constant $c = c(n, k)$ such that*

$$\left(\frac{1}{|Q|}\int_Q |g(x) - g_Q|^p\,dx\right)^{\frac{1}{p}}$$
$$\le c\,|Q|^{\frac{k}{n}} \sup_{k_1+\dots+k_n=k} \left(\frac{1}{|Q|}\int_Q \left|\frac{\partial^k}{\partial^{k_1}x_1\dots\partial^{k_n}x_n} g(x)\right|^q dx\right)^{\frac{1}{q}}.$$

We mention an obvious consequence of this inequality, which looks slightly stronger. It will also be called the Sobolev–Poincaré inequality. Namely, for every family $\pi = \{Q_{x_i}\}_{i\in I}$ of mutually nonintersecting cubes we have

$$\left(\sum_{i\in I}\left(\frac{\left(\frac{1}{|Q_{x_i}|}\int_{Q_{x_i}} |g(x) - g_{Q_{x_i}}|^p\,dx\right)^{\frac{1}{p}}}{|Q_{x_i}|^{\frac{k}{n}}}\right)^q |Q_{x_i}|\right)^{\frac{1}{q}} \le c\,\|g\|_{\dot{W}^k_q}. \tag{9.2}$$

This inequality remains true for the limit exponent, i.e., when $k/n+1/p-1/q = 0$, provided $p < \infty$.

### 9.2.1 Near-minimizers for the couple $(L^p, \dot{W}^k_p)$

We start with the case where $q = p$, which stands somewhat apart.

We remind the reader that the $K$-functional of a function $f \in L^p$ corresponding to the couple $(L^p, \dot{W}^k_p)$ is defined for $t > 0$ as follows:

$$K(t, f; L^p, \dot{W}^k_p) = \inf_{g\in \dot{W}^k_p} (\|f - g\|_{L^p} + t\,\|g\|_{\dot{W}^k_p}).$$

To construct a near-minimizer, we split $\mathbb{R}^n$ into congruent cubes $Q_i$ with edge length $t$ (thus, of volume $t^n$) and put $K_i = \frac{3}{2}Q_i$. Then (see Example 8.3) the family $\{K_i\}_{i\in I}$ is a WB-covering of $\mathbb{R}^n$ with Whitney, multiplicity, and engagement constants depending only on $n$. Let $\{\psi_i\}$ be a smooth partition of unity adjusted to the WB-covering $\{K_i\}_{i\in I}$. We put

$$f_t = \sum_{i\in I} P_i\psi_i, \tag{9.3}$$

where the $P_i$ are polynomials of degree strictly smaller than $k$ that provide nearly best approximation to $f \in L^p$ on the cubes $K_i$ in the $L^p$-metric.

**Theorem 9.2.** *Suppose $f \in L^p$ and $p < \infty$. Then the function $f_t$ given by (9.3) is a near-minimizer for the $K$-functional of $f$ relative to the couple $(L^p, \dot{W}_p^k)$ at the point $t^k$, that is,*

$$\|f - f_t\|_{L^p} + t^k \|f_t\|_{\dot{W}_p^k} \le cK(t^k, f; L^p, \dot{W}_p^k).$$

*Moreover, for the cubes $K_i$ defined above, we have*

$$K(t^k, f; L^p, \dot{W}_p^k) \approx \left(\sum_{i\in I} \int_{K_i} |f - f_{K_i}|^p\right)^{\frac{1}{p}}.$$

*All constants depend on the dimension $n$ only.*

*Proof.* Lemma 9.1 and the finite overlap property of the family $\{K_i\}$ show that for every cube $K_{i_0}$ we have

$$\begin{aligned}\left\|\frac{\partial^k}{\partial^{k_1}x_1\ldots\partial^{k_n}x_n} f_t\right\|^p_{L^\infty(K_{i_0})} &\le \left(c \sum_{i:K_i\cap K_{i_0}\neq\varnothing} \frac{1}{|K_i|^{\frac{k}{n}}} \left(\frac{1}{|K_i|}\int_{K_i} |f - P_i|^p\right)^{\frac{1}{p}}\right)^p \\ &\le c \sum_{i:K_i\cap K_{i_0}\neq\varnothing} \frac{1}{|K_i|^{\frac{kp}{n}+1}} \int_{K_i} |f - P_i|^p .\end{aligned}$$

Using the strong engagement property (and again the finite overlap property) for the family $\{K_i\}_{i\in I}$, we obtain

$$\begin{aligned}\left\|\frac{\partial^k}{\partial^{k_1}x_1\partial^{k_2}x_2\ldots\partial^{k_n}x_n} f_t\right\|^p_{L^p} &\le c\sum_{i\in I} \frac{1}{|K_i|^{\frac{kp}{n}}} \int_{K_i} |f - P_i|^p \\ &\le c\frac{1}{t^{kp}} \sum_{i\in I} \int_{K_i} |f - f_{K_i}|^p .\end{aligned}$$

Therefore,

$$t^k \|f_t\|_{\dot{W}_p^k} \le c\left(\sum_{i\in I} \int_{K_i} |f - f_{K_i}|^p\right)^{\frac{1}{p}}.$$

Furthermore, since the functions $\{\psi_i\}_{i\in I}$ form a partition of unity of $\mathbb{R}^n$ adjusted to the family $\{K_i\}_{i\in I}$ with finite overlap, we see that

$$\|f - f_t\|_{L^p}^p = \left\| \sum_{i\in I} f\psi_i - \sum_{i\in I} P_i\psi_i \right\|_{L^p}^p \leq c \sum_{i\in I} \|(f - P_i)\psi_i\|_{L^p}^p$$
$$\leq c \sum_{i\in I} \int_{K_i} |f - P_i|^p \leq c \sum_{i\in I} \int_{K_i} |f - f_{K_i}|^p .$$

Therefore,

$$\|f - f_t\|_{L^p} + t^k \|f_t\|_{\dot{W}_p^k} \leq c \left( \sum_{i\in I} \int_{K_i} |f - f_{K_i}|^p \right)^{\frac{1}{p}} .$$

Thus, to prove the theorem, it suffices to find a constant $\varepsilon > 0$ such that

$$K(t^k, f; L^p, \dot{W}_p^k) = \inf_{g\in \dot{W}_p^k} (\|f - g\|_{L_p} + t^k \|g\|_{\dot{W}_p^k}) \geq \varepsilon \left( \sum_{i\in I} \int_{K_i} |f - f_{K_i}|^p \right)^{\frac{1}{p}} . \tag{9.4}$$

Suppose a function $g \in \dot{W}_p^k$ is chosen in such a way that

$$\|f - g\|_{L_p} + t^k \|g\|_{\dot{W}_p^k} \leq 2K(t^k, f; L^p, \dot{W}_p^k).$$

If $\|g\|_{\dot{W}_p^k} \geq \frac{2\varepsilon}{t^k} \left( \sum_{i\in I} \int_{K_i} |f - f_{K_i}|^p \right)^{\frac{1}{p}}$, then inequality (9.4) is true. So, we consider the opposite case, where

$$\|g\|_{\dot{W}_p^k} \leq \frac{2\varepsilon}{t^k} \left( \sum_{i\in I} \int_{K_i} |f - f_{K_i}|^p \right)^{\frac{1}{p}} . \tag{9.5}$$

The family $\{K_i\}_{i\in I}$ splits into finitely many packings (whose number is estimated from above in terms of the dimension $n$). Consequently, for some packing $\pi \subset \{K_i\}_{i\in I}$ we have

$$\left( \sum_{i\in I} \int_{K_i} |f - f_{K_i}|^p \right)^{\frac{1}{p}} \leq c \left( \sum_{K_i\in \pi} \int_{K_i} |f - f_{K_i}|^p \right)^{\frac{1}{p}} . \tag{9.6}$$

Clearly, an analog of (9.5) is fulfilled for this packing $\pi$ :

$$\|g\|_{\dot{W}_p^k} \leq \frac{\varepsilon c}{t^k} \left( \sum_{K_i\in \pi} \int_{K_i} |f - f_{K_i}|^p \right)^{\frac{1}{p}} .$$

Therefore, by the Sobolev–Poincaré inequality (9.2) with $q = p$ and $|K_i| = \left(\frac{3}{2}\right)^n t^n$, the function $g$ satisfies

$$\left(\sum_{K_i \in \pi} \int_{K_i} |g(x) - g_{K_i}|^p \, dx\right)^{\frac{1}{p}} \leq ct^k \, \|g\|_{\dot{W}_p^k} \leq c\varepsilon \left(\sum_{K_i \in \pi} \int_{K_i} |f - f_{K_i}|^p\right)^{\frac{1}{p}}.$$

Now, the triangle inequality $\|f\|_{L^p} \leq c_p(\|f - g\|_{L^p} + \|g\|_{L^p})$ (with constant) implies

$$\|f - g\|_{L^p} \geq \frac{1}{c_p} \|f\|_{L^p} - \|g\|_{L^p} .$$

Thus, for every function $h$ we have

$$\|f - g\|_{L^p} = \|(f - h) - (g - h)\|_{L^p} \geq \frac{1}{c_p} \|f - h\|_{L^p} - \|g - h\|_{L^p} .$$

Consequently,

$$\begin{aligned}
&\|f - g\|_{L^p} \geq \|f - g\|_{L^p(\cup K_i)} \\
&\geq \frac{1}{c_p} \left(\sum_{K_i \in \pi} \int_{K_i} |f(x) - g_{K_i}|^p \, dx\right)^{\frac{1}{p}} - \left(\sum_{K_i \in \pi} \int_{K_i} |g(x) - g_{K_i}|^p \, dx\right)^{\frac{1}{p}} \\
&\geq \frac{1}{c_p} \left(\sum_{K_i \in \pi} \int_{K_i} |f(x) - f_{K_i}|^p \, dx\right)^{\frac{1}{p}} - c\varepsilon \left(\sum_{K_i \in \pi} \int_{K_i} |f(x) - f_{K_i}|^p \, dx\right)^{\frac{1}{p}} \\
&\geq (\frac{1}{c_p} - c\varepsilon) \left(\sum_{K_i \in \pi} \int_{K_i} |f(x) - f_{K_i}|^p \, dx.\right)^{\frac{1}{p}} .
\end{aligned}$$

Taking (9.6) into account, we obtain

$$\|f - g\|_{L^p} \geq (\frac{1}{c_p} - c\varepsilon)c \left(\sum_{i \in I} \int_{K_i} |f - f_{K_i}|^p\right)^{\frac{1}{p}} .$$

Therefore, in the case (9.5) we have

$$K(t^k, f; L^p, \dot{W}_p^k) \geq \frac{1}{2} \|f - g\|_{L^p} \geq (\frac{1}{c_p} - c\varepsilon)c \left(\sum_{i \in I} \int_{K_i} |f - f_{K_i}|^p\right)^{\frac{1}{p}} ;$$

thus, (9.4) is true if $\varepsilon$ is sufficiently small. □

### 9.2.2 Near-minimizers for the couple $(L^p, \dot{W}_q^k)$ when $q \neq p$

Below we construct near-minimizers $f_t$ for the $L_{p,q}$-functional of $f \in L^p$ with respect to the couple $(L^p, \dot{W}_q^k)$ at the point $t^{p-q}$, that is, $f_t \in \dot{W}_q^k$ will satisfy

$$\|f - f_t\|_{L^p}^p + t^{p-q} \|f_t\|_{\dot{W}_q^k}^q \leq cL_{p,q}(t^{p-q}, f; L^p, \dot{W}_q^k) \tag{9.7}$$

with a constant $c > 0$ depending only on the dimension $n$. We remind the reader that

$$L_{p,q}(t, f; L^p, \dot{W}_q^k) = \inf(\|f - g\|_{L^p}^p + t\|g\|_{\dot{W}_q^k}^q),$$

therefore in fact we have equivalence in (9.7).

Though the proofs and the results are quite similar in the cases of $q > p$ and $q < p$, the constructions themselves of the minimizers are different. We present both.

**Near-minimizers for the couple $(L^p, \dot{W}_q^k)$ if $q < p$ and $p \neq \infty$**

Suppose $f \in L^p$ and $t > 0$, and let (9.1) be fulfilled. A near-minimizer for the couple $(L^p, \dot{W}_q^k)$ is constructed as described below.

We define the parameter $\lambda$ by the formula

$$\lambda = 1 + \frac{k}{n(\frac{1}{p} - \frac{1}{q})} = \left(\frac{k}{n} + \frac{1}{p} - \frac{1}{q}\right) \frac{1}{\frac{1}{p} - \frac{1}{q}} = \left(\frac{k}{n} + \frac{1}{p} - \frac{1}{q}\right) \frac{qp}{q - p}. \tag{9.8}$$

Since $q < p$, by (9.1) and the last expression in (9.8) we have $\lambda < 0$. For every $x \in \mathbb{R}^n$, consider the following function of $r > 0$:

$$\varphi_x(r) = \left(\frac{1}{|Q(x,r)|^\lambda} \int_{Q(x,r)} |f(y) - f_{Q(x,r)}(y)|^p \, dy\right)^{\frac{1}{p}}. \tag{9.9}$$

By the absolute continuity of a Lebesgue integral, this function is continuous and monotone nondecreasing (because $\lambda$ is negative). Putting[1]

$$\Omega = \left\{x \in \mathbb{R}^n : \sup_{r>0} \varphi_x(r) > t\right\}, \tag{9.10}$$

we see that for every $x \in \Omega$ there exists a number $r_x > 0$ with

$$\varphi_x(r_x) = t \text{ and } \sup_{r \leq r_x} \varphi_x(r) \leq t. \tag{9.11}$$

We set

$$Q_x = Q(x, r_x)$$

[1]Note that $\Omega = \mathbb{R}^n$ because $\lambda < 0$. A similar set will have nonempty complement in the case of $q > p$.

and consider the family $\{Q_x\}_{x\in\Omega}$ of these cubes.

By (9.11), we have

$$\left(\frac{1}{|Q_x|^\lambda}\int_{Q_x}|f(x)-f_{Q_x}(x)|^p\,dx\right)^{\frac{1}{p}}=t. \tag{9.12}$$

Since $\lambda<0$, it follows that

$$|Q_x|\geq\left(\frac{t}{\|f\|_{L^p}}\right)^{-\frac{p}{\lambda}}.$$

Therefore, our family $\{Q_x\}_{x\in\Omega}$ satisfies the assumption (8.52) of the controlled contraction theorem. We apply that theorem to obtain a family $\{K_i\}_{i\in I}$ that forms a WB-covering of the entire space $\mathbb{R}^n$. Let $\{\psi_i\}_{i\in I}$ be a smooth partition of unity adjusted to this WB-covering, and let $P_i$, $i\in I$, be polynomials of degree at most $k-1$ that provide an almost optimal approximation of $f$ on the cubes $K_i$:

$$\left(\int_{K_i}|f(x)-P_i(x)|^p\,dx\right)^{\frac{1}{p}}\leq c\inf_{P\in\mathcal{P}_k}\left(\int_{K_i}|f(x)-P(x)|^p\,dx\right)^{\frac{1}{p}},$$

where $c>0$ is a constant depending only on the dimension $n$.

The required near-minimizer is defined by the formula

$$f_t=\sum_{i\in I}P_i\psi_i. \tag{9.13}$$

We isolate some properties of the cubes $\{K_i\}_{i\in I}$, which will be used in the proofs.

First, we observe that the first condition (8.50) in the definition of a contraction shows that

$$\left(\frac{1}{|K_i|^\lambda}\int_{K_i}|f(x)-P_i(x)|^p\,dx\right)^{\frac{1}{p}}\leq ct \tag{9.14}$$

for all $i\in I$, with a constant $c>0$ depending on the dimension only. Moreover, Theorem 8.41 (about the preservation of the $\alpha$-capacity for negative $\alpha$) and the fact that the family $\{K_i\}_{i\in I}$ splits into a controllable finte number of packings (see Proposition 3.5) imply the inequality

$$\sum_{i\in I}|K_i|^\lambda\leq c\left|\{Q_x\}_{x\in\Omega}\right|_\lambda. \tag{9.15}$$

Here the right-hand side is finite because, by (9.12), for every packing $\pi\subset\{Q_x\}_{x\in\Omega}$ we obtain

$$\sum_{Q_{x_i}\in\pi}|Q_{x_i}|^\lambda=\frac{1}{t^p}\sum_{Q_{x_i}\in\pi}\int_{Q_{x_i}}|f(x)-f_{Q_{x_i}}(x)|^p\,dx\leq\frac{1}{t^p}\|f\|_{L^p}^p.$$

The proofs themselves will be postponed till we shall have described near-minimizers in the case where $q>p$.

**Near-minimizers for the couple $(L^p, \dot{W}_q^k)$ when $q > p$**

We introduce the parameter $\lambda$ by formula (9.8) and consider the function $\varphi_x(r)$ defined by (9.9), see Figure 1. Now $\lambda > 0$, and the behavior of $\varphi_x(r)$ is somewhat more complicated than for $q < p$. However, from the continuity of the Lebesgue integral it follows that the function $\varphi_x(r)$ is continuous in $r$ and (since $\lambda > 0$) tends to zero as $r \to \infty$. Therefore, for every $x \in \Omega$ (where, again, $\Omega$ is defined by (9.10)), there exists a number $r_x > 0$ with

$$\varphi_x(r_x) = t \quad \text{and} \quad \sup_{r \geq r_x} \varphi_x(r) \leq t. \tag{9.16}$$

We put

$$Q_x = Q(x, r_x)$$

a

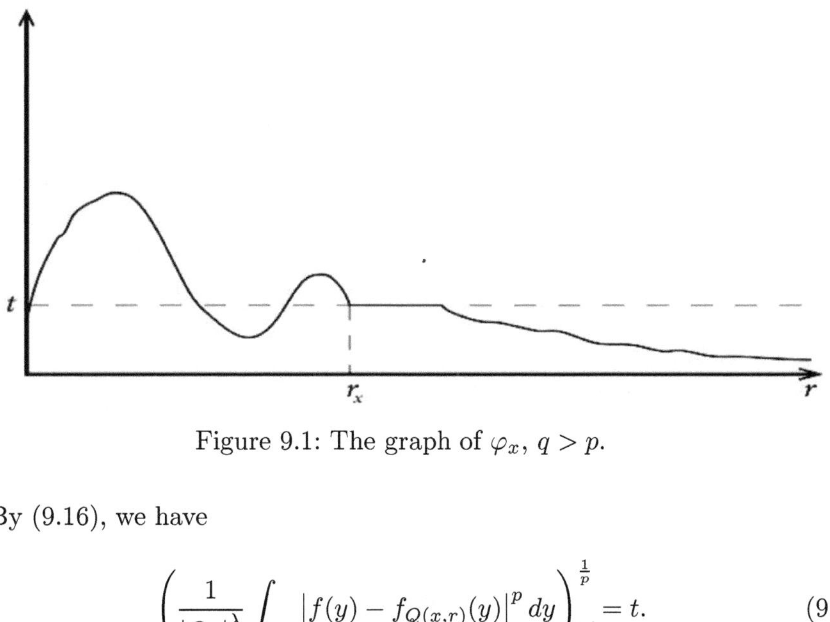


Figure 9.1: The graph of $\varphi_x$, $q > p$.

By (9.16), we have

$$\left( \frac{1}{|Q_x|^\lambda} \int_{Q_x} \left| f(y) - f_{Q(x,r)}(y) \right|^p dy \right)^{\frac{1}{p}} = t. \tag{9.17}$$

Now, for every packing $\pi \subset \{Q_x\}_{x \in \Omega}$, formula (9.17) implies the inequality

$$\sum_{Q_{x_i} \in \pi} |Q_{x_i}|^\lambda = \frac{1}{t^p} \sum_{Q_{x_i} \in \pi} \int_{Q_{x_i}} \left| f(x) - f_{Q_{x_i}}(x) \right|^p dx \leq \frac{1}{t^p} \|f\|_{L^p}^p .$$

Thus, the family $\{Q_x\}_{x \in \Omega}$ has bounded $\lambda$-capacity. We observe that $\lambda > 1$ by (9.8). Now, we supplement the family $\{Q_x\}_{x \in \Omega}$ with cubes with centers lying in $\mathbb{R}^n \setminus \Omega$ in such a way that the $\lambda$-capacity of the resulting family increases at most twice. For this, we split $\mathbb{R}^n$ into cubes $Q_k$, $k = 1, 2, \dots$, of unit volume ($|Q_k| = 1$) and for $x \in Q_k \setminus \Omega$ put $Q_x = Q(x; \varepsilon^k)$. Let $\pi = \{Q_{x_i}\}$ be a packing included in the family $\{Q_x\}_{\mathbb{R}^n \setminus \Omega}$. Volume estimates show that each $Q_k$ contains at most $\frac{1}{\varepsilon^{kn}}$ points $x_i$. Since $\lambda > 1$, for $\varepsilon < 1$ we have

$$\sum_i |Q_{x_i}|^\lambda \le 2^{n\lambda} \sum_{k=1}^{\infty} \varepsilon^{kn\lambda} \frac{1}{\varepsilon^{kn}} = 2^{n\lambda} \sum_{k=1}^{\infty} \varepsilon^{kn(\lambda-1)} = 2^{n\lambda} \frac{\varepsilon^{n(\lambda-1)}}{1 - \varepsilon^{n(\lambda-1)}}.$$

Therefore, if $\varepsilon$ is sufficiently small, then $\left|\{Q_x\}_{x \in \mathbb{R}^n \setminus \Omega}\right|_\lambda \le \left|\{Q_x\}_{x \in \Omega}\right|_\lambda$ and, consequently,

$$\left|\{Q_x\}_{x \in \mathbb{R}^n}\right|_\lambda \le 2 \left|\{Q_x\}_{x \in \Omega}\right|_\lambda \le 2 \frac{1}{t^p} \|f\|_{L^p}^p. \tag{9.18}$$

Since $\lambda > 0$, it follows that the family $\{Q_x\}_{\mathbb{R}^n}$ satisfies the assumption

$$\sup_{x \in \mathbb{R}^n} |Q_x| < \infty$$

of the controlled extension theorem. Applying that theorem, we obtain a family $\{K_i\}_{i \in I}$ of cubes, which is a WB-covering of the entire space $\mathbb{R}^n$ (because the initial family is defined on $\mathbb{R}^n$, see the second condition (8.29) in the definition of an extension). Let $\{\psi_i\}_{i \in I}$ be a smooth partition of unity adjusted to this WB-covering, and let $P_i$, $i \in I$, be polynomials of degree strictly smaller than $k$ that provide an almost optimal approximation to $f$ on the cubes $K_i$, that is

$$\left( \int_{K_i} |f(x) - P_i(x)|^p \, dx \right)^{\frac{1}{p}} \le c \inf_{P \in \mathcal{P}_k} \left( \int_{K_i} |f(x) - P(x)|^p \, dx \right)^{\frac{1}{p}},$$

where the constant $c > 0$ depends only on the dimension $n$.

We define a function $f_t$ by the formula

$$f_t = \sum_{i \in I} P_i \psi_i. \tag{9.19}$$

In the next remark, we show that counterparts of properties (9.14) and (9.15) are satisfied.

*Remark* 9.3. The construction of the initial family of cubes and the first condition (8.28) in the definition of an extension show that

$$\left( \frac{1}{|K_i|^\lambda} \int_{K_i} |f(x) - P_i(x)|^p \, dx \right)^{\frac{1}{p}} \le ct \tag{9.20}$$

for all $i \in I$. Next, Theorem 8.30 on preservation of the $\alpha$-capacity, combined again with Proposition 3.5, implies

$$\sum_{i \in I} |K_i|^\lambda \le c \left|\{Q_x\}_{x \in \Omega}\right|_\lambda. \tag{9.21}$$

### 9.2.3 Statement and proof of the main result

We have finished the preparations. We recall that $p$ and $q$ obey (9.1). Also, we assume that $0 < p < \infty$, $1 \le q < \infty$, and $q \neq p$.

**Theorem 9.4.** *Given $f \in L^p$ and $t > 0$, the function $f_t$ defined in (9.13) or (9.19) is a near-minimizer for the $L_{p,q}$-functional of $f$ at the point $t^{p-q}$ with respect to the couple $(L^p, \dot{W}_q^k)$, that is*

$$\|f - f_t\|_{L^p}^p + t^{p-q}\|f_t\|_{\dot{W}_q^k}^q \le cL_{p,q}(t^{p-q}, f; L^p, \dot{W}_q^k)$$

*with a constant $c > 0$ depending only on the dimension $n$.*

For the proof, we need a lemma. Let $\{Q\}_{f,t}$ be the family of all cubes satisfying

$$\left(\frac{1}{|Q|^\lambda}\int_Q |f(x) - f_Q(x)|^p\,dx\right)^{\frac{1}{p}} \ge t. \tag{9.22}$$

**Lemma 9.5.** *There is a positive number $\varepsilon$ depending only on the dimension such that*

$$L_{p,q}(t^{p-q}, f; L^p, \dot{W}_q^k) \ge \varepsilon t^p \left|\{Q\}_{f,t}\right|_\lambda. \tag{9.23}$$

*Proof.* Choose a function $g \in \dot{W}_q^k$ in such a way that

$$\|f - g\|_{L^p}^p + t^{p-q}\|g\|_{\dot{W}_q^k}^q \le 2L_{p,q}(t^{p-q}, f; L^p, \dot{W}_q^k).$$

Inequality (9.23) is clear if

$$\frac{1}{2}t^{p-q}\|g\|_{\dot{W}_q^k}^q \ge \varepsilon t^p \left|\{Q\}_{f,t}\right|_\lambda.$$

So, it suffices to consider the case where

$$\|g\|_{\dot{W}_q^k}^q \le 2\varepsilon t^q \left|\{Q\}_{f,t}\right|_\lambda.$$

By the definition of the quantity $\left|\{Q\}_{f,t}\right|_\lambda$, there exists a packing $\pi \subset \{Q\}_{f,t}$ such that

$$\sum_{Q_i \in \pi} |Q_i|^\lambda \ge \frac{1}{2}\left|\{Q\}_{f,t}\right|_\lambda.$$

By the Sobolev–Poincaré inequality (9.2), for this packing $\pi$ we have

$$\sum_{Q_i \in \pi}\left(\frac{\left(\frac{1}{|Q_i|}\int_{Q_i}|g(x) - g_{Q_i}(x)|^p\,dx\right)^{\frac{1}{p}}}{|Q_i|^{\frac{k}{n}}}\right)^q |Q_i| \le c\|g\|_{\dot{W}_q^k}^q$$

$$\le c\varepsilon t^q \left|\{Q\}_{f,t}\right|_\lambda \le c\varepsilon t^q \sum_{Q_i \in \pi}|Q_i|^\lambda.$$

The definition of $\lambda$ (see (9.8)) allows us to rewrite this in the form

$$\sum_{Q_i\in\pi}\left(\frac{\left(\int_{Q_i}|g(x)-g_{Q_i}(x)|^p\,dx\right)^{\frac{1}{p}}}{|Q_i|^{\frac{\lambda}{p}}}\right)^q |Q_i|^\lambda \le c\varepsilon t^q\sum_{Q_i\in\pi}|Q_i|^\lambda . \tag{9.24}$$

Denote by $\pi_0$ the set of all cubes $Q_i$ in $\pi$ for which

$$\frac{\left(\int_{Q_i}|g(x)-g_{Q_i}(x)|^p\,dx\right)^{\frac{1}{p}}}{|Q_i|^{\frac{\lambda}{p}}} \ge (2c\varepsilon)^{\frac{1}{q}}t.$$

By (9.24) we obtain

$$\sum_{Q_i\in\pi_0}|Q_i|^\lambda \le \frac{1}{2}\sum_{Q_i\in\pi}|Q_i|^\lambda .$$

The triangle inequality with a constant $c_p$, namely, $\|f\|_{L^p}\le c_p(\|f-g\|_{L^p}+\|g\|_{L^p})$, shows that

$$\|f-g\|_{L^p}\ge\frac{1}{c_p}\|f\|_{L^p}-\|g\|_{L^p}.$$

At the same time, for $Q_i\in\pi\setminus\pi_0$ we have

$$\left(\int_{Q_i}|g(x)-g_{Q_i}(x)|^p\,dx\right)^{\frac{1}{p}}\le(2c\varepsilon)^{\frac{1}{q}}t\,|Q_i|^{\frac{\lambda}{p}}.$$

Consequently,

$$\begin{aligned}
&\|f-g\|_{L^p}\\
&\ge\frac{1}{c_p}\left(\sum_{Q_i\in\pi\setminus\pi_0}\int_{Q_i}|f(x)-g_{Q_i}(x)|^p\,dx\right)^{\frac{1}{p}}-\left(\sum_{Q_i\in\pi\setminus\pi_0}\int_{Q_i}|g(x)-g_{Q_i}(x)|^p\,dx\right)^{\frac{1}{p}}\\
&\ge\frac{1}{c_p}\left(\sum_{Q_i\in\pi\setminus\pi_0}\int_{Q_i}|f(x)-f_{Q_i}(x)|^p\,dx\right)^{\frac{1}{p}}-\left(\sum_{Q_i\in\pi\setminus\pi_0}\int_{Q_i}|g(x)-g_{Q_i}(x)|^p\,dx\right)^{\frac{1}{p}}\\
&\ge\frac{1}{c_p}t\left(\sum_{Q_i\in\pi\setminus\pi_0}|Q_i|^\lambda\right)^{\frac{1}{p}}-(2c\varepsilon)^{\frac{1}{q}}t\left(\sum_{Q_i\in\pi\setminus\pi_0}|Q_i|^\lambda\right)^{\frac{1}{p}}\\
&\ge\left(\frac{1}{c_p}-(2c\varepsilon)^{\frac{1}{q}}\right)t\frac{1}{2^{\frac{1}{p}}}\left(\sum_{Q_i\in\pi}|Q_i|^\lambda\right)^{\frac{1}{p}}\ge\left(\frac{1}{c_p}-(2c\varepsilon)^{\frac{1}{q}}\right)\frac{1}{2^{2\frac{1}{p}}}t\left|\{Q\}_{f,t}\right|_\lambda^{\frac{1}{p}}.
\end{aligned}$$

Clearly, this implies the inequality

$$\|f-g\|_{L^p}^p\ge 2\varepsilon t^p\left|\{Q\}_{f,t}\right|_\lambda$$

if $\varepsilon$ is sufficiently small. This proves (9.23). □

We pass to the proof of the theorem.

*Proof.* Since $\{Q_x\}_{x\in\Omega} \subset \{Q\}_{f,t}$, it suffices to prove that

$$\|f - f_t\|_{L^p}^p + t^{p-q} \|f_t\|_{\dot{W}_q^k}^q \le ct^p \left|\{Q_x\}_\Omega\right|_\lambda .$$

We use Lemma 9.1, the strong engagement and finite multiplicity properties of the family $\{K_i\}_{i\in I}$, and inequality (9.14) for $q < p$ (respectively, (9.20) for $q > p$) to conclude that

$$\begin{aligned}\left\|\frac{\partial^k}{\partial^{k_1}x_1\partial^{k_2}x_2\dots\partial^{k_n}x_n} f_t\right\|_{L^\infty(K_{i_0})} &\le c \sum_{i:K_i\cap K_{i_0}\neq\varnothing} \frac{1}{|K_i|^{\frac{k}{n}}} \left(\frac{1}{|K_i|}\int_{K_i} |f - f_{K_i}|^p\right)^{\frac{1}{p}} \\ &\le c \sum_{i:K_i\cap K_{i_0}\neq\varnothing} \frac{t}{|K_i|^{\frac{k}{n}+\frac{1}{p}}} |K_i|^{\frac{\lambda}{p}} \le ct \left|K_{i_0}\right|^{\frac{\lambda}{p}-\frac{k}{n}-\frac{1}{p}}\end{aligned}$$

for every $K_{i_0}$. Since $\lambda = (\frac{k}{n} + \frac{1}{p} - \frac{1}{q})\frac{pq}{q-p}$ (see (9.8)), we have

$$\frac{\lambda}{p}(q-p) = \left(\frac{k}{n} + \frac{1}{p} - \frac{1}{q}\right) q$$

and therefore

$$\left(\frac{\lambda}{p} - \frac{k}{n} - \frac{1}{p}\right) q + 1 = \lambda.$$

Consequently,

$$\left\|\frac{\partial^k}{\partial^{k_1}x_1\dots\partial^{k_n}x_n} f_t\right\|_{L^q} \le ct \left(\sum_{i\in I} \left(|K_i|^{\frac{\lambda}{p}-\frac{k}{n}-\frac{1}{p}}\right)^q |K_i|\right)^{\frac{1}{q}} = ct \left(\sum_{i\in I} |K_i|^\lambda\right)^{\frac{1}{q}}.$$

Taking (9.15) (respectively, (9.21)) into account, we obtain

$$\left\|\frac{\partial^k}{\partial^{k_1}x_1\partial^{k_2}x_2\dots\partial^{k_n}x_n} f_t\right\|_{L^q} \le ct \left(\left|\{Q_x\}_{x\in\Omega}\right|_\lambda\right)^{\frac{1}{q}}.$$

Therefore,

$$\|f_t\|_{\dot{W}_q^k} \le ct \left(\left|\{Q_x\}_{x\in\Omega}\right|_\lambda\right)^{\frac{1}{q}}.$$

Since also

$$\|f - f_t\|_{L^p}^p \le \sum_{i\in I} \int_{K_i} |f(x) - P_i|^p \, dx \le ct^p \sum_{i\in I} |K_i|^\lambda \le ct^p \left|\{Q_x\}_{x\in\Omega}\right|_\lambda$$

(see (9.14) and (9.20)), finally we arrive at

$$\|f - f_t\|_{L^p}^p + t^{p-q} \|f_t\|_{\dot{W}_q^k}^q \le ct^p \left|\{Q_x\}_{x\in\Omega}\right|_\lambda .$$

□

The following statement was established in the course of the above proof.

**Theorem 9.6.** *Let* $\lambda = 1 + \frac{k}{n(\frac{1}{p}-\frac{1}{q})}$, *and let, as before,* $\{Q\}_{f,t}$ *be the family of all cubes satisfying* (9.22). *Then*

$$L_{p,q}(t^{p-q}, f; L^p, \dot{W}_q^k) \approx t^p \left|\{Q\}_{f,t}\right|_\lambda$$

(*and even* $L_{p,q}(t^{p-q}, f; L^p, \dot{W}_q^k) \approx t^p \left|\{Q_x\}_{x\in\Omega}\right|_\lambda$) *with equivalence constants depending only on the dimension* $n$.

## 9.3 Near-minimizers for Morrey–Campanato spaces: the couples $(L^p, \dot{C}_p^{\alpha,k})$

We start with a brief reminder. Some useful information about the (homogeneous) Morrey–Campanato spaces $\dot{C}_p^{\alpha,k}$ was presented in Chapters 1 and 4. For the reader's convenience, we recall the formula for the seminorm on $\dot{C}_p^{\alpha,k}$:

$$\|f\|_{\dot{C}_p^{\alpha,k}} = \sup_Q \frac{1}{|Q|^{\frac{\alpha}{n}}} \left(\frac{1}{|Q|}\int_Q |f(x) - f_Q|^p \, dx\right)^{\frac{1}{p}}, \alpha \in \left(-\frac{n}{p}, k\right]. \tag{9.25}$$

Here, as in the preceding section, $f_Q$ stands for a polynomial of degree strictly smaller than $k$ that provides the best approximation to $f$ on $Q$ in the $L^p$-metric:

$$\left(\int_Q |f(x) - f_Q|^p \, dx\right)^{\frac{1}{p}} = \inf_{P\in\mathcal{P}_k} \left(\int_Q |f(x) - P(x)|^p \, dx\right)^{\frac{1}{p}}.$$

As a particular case, we find here the Morrey spaces $\mathcal{L}^\sigma(L^p)$ determined by the norm

$$\|f\|_{\mathcal{L}^\sigma(L^p)} = \sup_Q |Q|^{\frac{\sigma}{n}} \left(\frac{1}{|Q|}\int_Q |f(x)|^p \, dx\right)^{\frac{1}{p}}, \sigma \in \left[0, \frac{n}{p}\right). \tag{9.26}$$

(specifically, $\mathcal{L}^\sigma(L^p) = \dot{C}_p^{-\sigma,0}$). However, we recall that even more important and well-known spaces can be recognized under the disguise of $\dot{C}_p^{\alpha,k}$. For instance, the classes $\operatorname{Lip}\alpha$ are among them,

$$\|f\|_{\operatorname{Lip}\alpha} = \sup_{x\neq y} \frac{|f(x) - f(y)|}{|x-y|^\alpha}, \quad 0 < \alpha \le 1.$$

Specifically,

$$\dot{C}_p^{\alpha,1} = \operatorname{Lip}\alpha,$$

see Chapters 1 and 4. The norm of $\dot{C}_p^{\alpha,1}$ is called a *Campanato norm* on $\operatorname{Lip}\alpha$. Also, $\dot{C}_p^{0,0} = L^\infty$ and $\dot{C}_p^{0,1} = \mathrm{BMO}$.

Below, some estimational constants claimed to depend on the dimension only, depend in fact also on $p$: we treat this parameter as fixed. We remind the reader that, for $\alpha \geq 0$, only the seminorm on $\dot{C}_p^{\alpha,k}$ rather than the class itself depends on $p$ (see Chapter 1). It turns out to be "proper" to choose always that very value of $p$ which pertains to the space $L^p$ involved.

### 9.3.1 Algorithm for constructing near-minimizers

As has already been said, for the couple $(L^p, \dot{C}_p^{\alpha,k})$ it is convenient to study the distance functional rather than an $L$-functional. However, the procedures will be similar to those for the couple $(L^p, \dot{W}_q^k)$ in the case where $q > p$. We describe the algorithm in detail.

Suppose $f \in L^p$ ($p < \infty$) and $t > 0$. We define the parameter $\lambda$ by

$$\lambda = 1 + \frac{\alpha p}{n}, \quad \alpha \in \left(-\frac{n}{p}, k\right]. \tag{9.27}$$

It should be noted that, for $\alpha = k$, this formula is a limit case (as $q \to \infty$) of the parameter $\lambda$ for the couple $(L^p, \dot{W}_q^k)$ (see (9.8)).

Since $\alpha \in \left(-\frac{n}{p}, k\right]$, the number $\lambda$ is positive. Consider the function $\varphi_x(r)$ defined by (9.9), that is,

$$\begin{aligned} \varphi_x(r) &= \left(\frac{1}{|Q(x,r)|^\lambda} \int_{Q(x,r)} \left|f(x) - f_{Q(x,r)}\right|^p dx\right)^{\frac{1}{p}} \\ &= \frac{1}{|Q(x,r)|^{\frac{\alpha}{n}}} \left(\frac{1}{|Q(x,r)|} \int_{Q(x,r)} \left|f(x) - f_{Q(x,r)}\right|^p dx\right)^{\frac{1}{p}}. \end{aligned} \tag{9.28}$$

This is a continuous function of $r$ and, since $\lambda > 0$, it tends to zero as $r \to \infty$. Consider the set

$$\Omega = \left\{x \in \mathbb{R}^n : \sup_{r>0} \varphi_x(r) > t\right\}. \tag{9.29}$$

For every $x \in \Omega$, there is a positive number $r_x$ with

$$\varphi_x(r_x) = t \quad \text{and} \quad \sup_{r \geq r_x} \varphi_x(r) \leq t. \tag{9.30}$$

We consider the family of cubes

$$Q_x = Q(x, r_x), \; x \in \Omega.$$

By (9.30) we have

$$\left(\frac{1}{|Q_x|^\lambda} \int_{Q_x} |f(x) - f_{Q_x}(x)|^p \, dx\right)^{\frac{1}{p}} = t. \tag{9.31}$$

Hence, for every packing $\pi \subset \{Q_x\}_{x\in\Omega}$ we have

$$\sum_{Q_{x_i}\in\pi} |Q_{x_i}|^\lambda = \frac{1}{t^p}\sum_{Q_{x_i}\in\pi}\int_{Q_{x_i}} \left|f(x) - f_{Q_{x_i}}(x)\right|^p dx \le \frac{1}{t^p}\left\|f\right\|_{L^p}^p .$$

Thus, the $\lambda$-capacity of the family $\{Q_x\}_{x\in\Omega}$ is finite.

Two cases are possible: $\lambda > 1$ and $\lambda \le 1$.

If $\lambda > 1$, then, as in the case of the couple $(L^p, \dot{W}_q^k)$ with $q > p$, we supplement the family $\{Q_x\}_{x\in\Omega}$ by cubes centered at the points of $\mathbb{R}^n \backslash \Omega$ in such a way that the $\lambda$-capacity of the resulting family $\{Q_x\}_{x\in\mathbb{R}^n}$ should not be more than twice as large as initially. Next, we apply the controlled extension theorem to obtain a family $\{K_i\}_{i\in I}$ that is a WB-covering of the entire space $\mathbb{R}^n$. Take a smooth partition $\{\psi_i\}_{i\in I}$ of unity adjusted to this WB-covering, and let $P_i$ $(i \in I)$ be polynomials of degree at most $k-1$ that provide an almost optimal approximation to $f$ on the cubes $K_i$ in the $L^p$-metric. Put

$$f_t = \sum_{i\in I} P_i\psi_i. \tag{9.32}$$

In the case where $\lambda \le 1$, it is impossible to supplement the family $\{Q_x\}_{x\in\Omega}$ by cubes centered in $\mathbb{R}^n \setminus \Omega$ in such a way that the $\lambda$-capacity do not increase too much. Therefore, we apply the controlled extension theorem directly, obtaining a family $\{K_i\}_{i\in I}$ of cubes that is a WB-covering of the set $\cup K_i$. Take a smooth partition $\{\psi_i\}_{i\in I}$ of unity corresponding to this WB-covering, and polynomials $P_i$, $i \in I$, of degree at most $k-1$ that provide an almost optimal approximation to $f$ on the cubes $K_i$ in the $L^p$-metric, and put

$$f_t = \sum_{i\in I} P_i\psi_i + f\,\chi_{\mathbb{R}^n\setminus\cup K_i}. \tag{9.33}$$

Thus, if $\lambda \le 1$, the formula for $f_t$ involves the supplementary term $f\,\chi_{\mathbb{R}^n\setminus\cup K_i}$.

### 9.3.2 Statement and the proof of the main result

We shall show that the functions $f_t$ constructed above are near-minimizers for the distance functional

$$\operatorname{dist}_{L^p}(f, B_{\dot{C}_p^{\alpha,k}}(t)) = \inf_{\|g\|_{\dot{C}_p^{\alpha,k}}\le t} \|f-g\|_{L^p} .$$

**Theorem 9.7.** *Suppose $f \in L^p$ $(p<\infty)$ and $t > 0$. Define $f_t$ by (9.32) for $\alpha \in (0,k]$ and by (9.33) for $\alpha \in \left(-\frac{n}{p}, 0\right]$. Then $f_t$ is a near-minimizer for the distance functional of the couple $(L^p, \dot{C}_p^{\alpha,k})$, that is*

$$\|f_t\|_{\dot{C}_p^{\alpha,k}} \le ct \text{ and } \|f-f_t\|_{L^p} \le c\operatorname{dist}_{L^p}(f, B_{\dot{C}_p^{\alpha,k}}(\frac{t}{c}))$$

*with a constant $c > 0$ depending on the dimension $n$.*

For the proof, we need a lemma. Let $\{Q\}_{f,t}$ denote the family of all cubes satisfying

$$\left(\frac{1}{|Q|^\lambda}\int_Q |f(x)-f_Q(x)|^p\,dx\right)^{\frac{1}{p}} \ge t. \tag{9.34}$$

**Lemma 9.8.** *We have*

$$\left|\{Q\}_{f,t}\right|_\lambda \le \frac{c}{t^p}\left(\mathrm{dist}_{L^p}(f, B_{\dot{C}_p^{\alpha,k}}(\frac{t}{c}))\right)^p$$

*with a constant* $c>0$ *depending only on the dimension* $n$.

*Proof.* Let $\pi=\{Q_j\}$ be a packing formed by cubes of the family $\{Q\}_{f,t}$, and let $g\in B_{\dot{C}_p^{\alpha,k}}(\frac{t}{c})$, where $c>0$ is a sufficiently large number to be chosen later. Since $\lambda = 1+\frac{\alpha p}{n}$, by (9.25) we have

$$\left(\frac{1}{|Q_j|^\lambda}\int_{Q_j}|g(x)-g_{Q_j}(x)|^p\,dx\right)^{\frac{1}{p}} = \frac{1}{|Q_j|^{\frac{1}{p}+\frac{\alpha}{n}}}\left(\int_{Q_j}|g(x)-g_{Q_j}(x)|^p\,dx\right)^{\frac{1}{p}} \le \frac{t}{c}.$$

Therefore,

$$\begin{aligned}
&\|f-g\|_{L^p}\\
&\ge \frac{1}{c_p}\left(\sum_j\int_{Q_j}|f(x)-g_{Q_j}(x)|^p\,dx\right)^{\frac{1}{p}} - \left(\sum_j\int_{Q_j}|g(x)-g_{Q_j}(x)|^p\,dx\right)^{\frac{1}{p}}\\
&\ge \frac{1}{c_p}\left(\sum_j\int_{Q_j}|f(x)-f_{Q_j}(x)|^p\,dx\right)^{\frac{1}{p}} - \frac{t}{c}\left(\sum_j|Q_j|^\lambda\right)^{\frac{1}{p}}\\
&\ge \frac{1}{c_p}t\left(\sum_j|Q_j|^\lambda\right)^{\frac{1}{p}} - \frac{t}{c}\left(\sum_j|Q_j|^\lambda\right)^{\frac{1}{p}} \ge \gamma t\left(\sum_j|Q_j|^\lambda\right)^{\frac{1}{p}}.
\end{aligned}$$

The last inequality is true if the constant $c$ is sufficiently large. □

We observe that for all cubes $\{Q_x\}_{x\in\Omega}$ we have

$$\frac{1}{|Q_x|^{\frac{1}{p}+\frac{\alpha}{n}}}\left(\int_{Q_x}|f(y)-f_{Q_x}(y)|^p\,dy\right)^{\frac{1}{p}} = t$$

and, consequently, the above lemma implies the estimate

$$t\left(\left|\{Q_x\}_{x\in\Omega}\right|_\lambda\right)^{\frac{1}{p}} \le c\,\mathrm{dist}_{L^p}(f, B_{\dot{C}_p^{\alpha,k}}(\frac{t}{c})).$$

Thus, the theorem follows from the next statement.

**Lemma 9.9.** *The functions $f_t$ constructed above satisfy*

$$\|f_t\|_{\dot{C}_p^{\alpha,k}} \le ct \text{ and } \|f - f_t\|_{L^p} \le ct\,|\{Q_x\}_\Omega|_\lambda^{\frac{1}{p}}.$$

*Proof.* First, we estimate the quantity $\|f - f_t\|_{L^p}$. Since

$$f - f_t = \sum_{i\in I}(f - P_i)\psi_i$$

and the family $\{K_i\}_{i\in I}$ is of finite multiplicity, we have

$$\|f - f_t\|_{L^p}^p \le c\sum_{i\in I}\int_{K_i}|f(x) - P_i(x)|^p\,dx \le c\sum_{i\in I}\int_{K_i}|f(x) - f_{K_i}(x)|^p\,dx.$$

To estimate the right-hand side, we use the fact that the family $\{K_i\}_{i\in I}$ is controlled, with a universal constant $\gamma \ge 1$, by a finite multiplicity subfamily $\{Q_{x_j}\}_{j\in J}$ of the initial family of cubes (see Definition 8.26). Moreover, the multiplicity depends only on the dimension $n$. Therefore, for every cube $K_i$ there exists a cube $\gamma Q_{x_j}$ with $K_i \subset \gamma Q_{x_j}$. Since the family $\{K_i\}_{i\in I}$ is of finite multiplicity, for every $j \in J$ we have

$$\sum_{i:K_i\subset\gamma Q_{x_j}}\int_{K_i}|f(x) - f_{K_i}(x)|^p\,dx \le c\int_{\gamma Q_{x_j}}\left|f(x) - f_{\gamma Q_{x_j}}(x)\right|^p dx$$

and, therefore,

$$\|f - f_t\|_{L^p}^p \le c\sum_{j\in J}\int_{\gamma Q_{x_j}}\left|f(x) - f_{\gamma Q_{x_j}}(x)\right|^p dx.$$

The construction of the initial family $\{Q_x\}$ of cubes (see (9.30)) shows that

$$\varphi_x(r(\gamma Q_{x_j})) = \left(\frac{1}{\left|\gamma Q_{x_j}\right|^\lambda}\int_{\gamma Q_{x_j}}\left|f(x) - f_{\gamma Q_{x_j}}(x)\right|^p\right)^{\frac{1}{p}} \le t.$$

Thus,

$$\|f - f_t\|_{L^p}^p \le ct^p\sum_{j\in J}\left|Q_{x_j}\right|^\lambda.$$

Furthermore, since the family $\{Q_{x_j}\}_{j\in J}$ is of finite multiplicity, we obtain

$$\|f - f_t\|_{L^p} \le ct\left|\{Q_x\}_{x\in\Omega}\right|_\lambda^{\frac{1}{p}}.$$

Now we prove that $\|f_t\|_{\dot{C}_p^{\alpha,k}} \le ct$. Take an arbitrary cube $Q$. We must show that

$$\frac{1}{|Q|^{\frac{1}{p}+\frac{\alpha}{n}}}\left(\int_Q|f_t(x) - (f_t)_Q(x)|^p\,dx\right)^{\frac{1}{p}} \le ct. \tag{9.35}$$

Suppose first that $Q$ is included in some cube $K_{i_0}$. In this case the function $f_t$ is infinitely differentiable on $K_{i_0}$, and by Lemma 9.1 and the finite multiplicity and engagement properties of the family $\{K_i\}_{i\in I}$, we have

$$\left\|\frac{\partial^k}{\partial^{k_1}x_1\partial^{k_2}x_2\dots\partial^{k_n}x_n}f_t\right\|_{L^\infty(K_{i_0})} \le c \sum_{i:K_i\cap K_{i_0}\neq\varnothing} \frac{1}{|K_i|^{\frac{k}{n}}}\left(\frac{1}{|K_i|}\int_{K_i}|f-f_{K_i}|^p\right)^{\frac{1}{p}}$$

$$\le ct \sum_{i:K_i\cap K_{i_0}\neq\varnothing} \frac{1}{|K_i|^{\frac{k}{n}+\frac{1}{p}}}|K_i|^{\frac{\lambda}{p}} = ct \sum_{i:K_i\cap K_{i_0}\neq\varnothing}\frac{1}{|K_i|^{\frac{k-\alpha}{n}}} \le ct\frac{1}{|K_{i_0}|^{\frac{k-\alpha}{n}}}.$$

Therefore, replacing $(f_t)_Q(x)$ with a Taylor polynomial of degree $k-1$ for $f_t$ relative to the center of $Q$, we have (see Theorem 0.20)

$$\frac{1}{|Q|^{\frac{1}{p}+\frac{\alpha}{n}}}\left(\int_Q |f_t(x)-(f_t)_Q(x)|^p\,dx\right)^{\frac{1}{p}}$$

$$\le \frac{c}{|Q|^{\frac{1}{p}+\frac{\alpha}{n}}}\sup_{k_1+\dots+k_n=k}\left\|\frac{\partial^k}{\partial^{k_1}x_1\dots\partial^{k_n}x_n}f_t\right\|_{L^\infty(K_{i_0})}|Q|^{\frac{k}{n}+\frac{1}{p}} \le c\frac{|Q|^{\frac{k-\alpha}{n}}}{|K_{i_0}|^{\frac{k-\alpha}{n}}}t \le ct.$$

If $Q$ is not included in any of the cubes $K_i$, the proof of (9.35) is more involved. First, we show that for such $Q$'s we have

$$\frac{1}{|Q|^{\frac{1}{p}+\frac{\alpha}{n}}}\left(\int_Q |f(x)-f_Q(x)|^p\,dx\right)^{\frac{1}{p}} \le t. \tag{9.36}$$

Indeed, otherwise we can find a cube $Q_x$ in the initial family $\{Q_x\}_\Omega$ such that $Q\subset Q_x$. Then the controlled extension theorem shows that (8.29) is fulfilled, and, therefore, there is a cube $K_i$ containing $Q_x$ and, consequently, $Q$. This contradiction proves (9.36).

Now, (9.36) implies

$$\frac{1}{|Q|^{\frac{1}{p}+\frac{\alpha}{n}}}\left(\int_Q |f_t(x)-(f_t)_Q(x)|^p\,dx\right)^{\frac{1}{p}} \le \frac{1}{|Q|^{\frac{1}{p}+\frac{\alpha}{n}}}\left(\int_Q |f_t(x)-f_Q(x)|^p\,dx\right)^{\frac{1}{p}}$$

$$\le \frac{c_p}{|Q|^{\frac{1}{p}+\frac{\alpha}{n}}}\left(\int_Q |f(x)-f_t(x)|^p\,dx\right)^{\frac{1}{p}} + \frac{c_p}{|Q|^{\frac{1}{p}+\frac{\alpha}{n}}}\left(\int_Q |f(x)-f_Q(x)|^p\,dx\right)^{\frac{1}{p}}$$

$$\le \frac{c_p}{|Q|^{\frac{1}{p}+\frac{\alpha}{n}}}\left(\int_Q |f(x)-f_t(x)|^p\,dx\right)^{\frac{1}{p}} + ct.$$

Thus, it suffices to estimate the quantity

$$\frac{1}{|Q|^{\frac{1}{p}+\frac{\alpha}{n}}}\left(\int_Q |f(x)-f_t(x)|^p\,dx\right)^{\frac{1}{p}}.$$

Formulas (9.32)-(9.33) for $f_t$ and the finite multiplicity property of the family $\{K_i\}_{i\in I}$ imply

$$\frac{1}{|Q|^{\frac{1}{p}+\frac{\alpha}{n}}}\left(\int_Q |f(x)-f_t(x)|^p\,dx\right)^{\frac{1}{p}} \tag{9.37}$$
$$\leq \frac{c}{|Q|^{\frac{1}{p}+\frac{\alpha}{n}}}\left(\sum_{i:K_i\cap Q\neq\varnothing}\int_{K_i}|f(x)-f_{K_i}(x)|^p\,dx\right)^{\frac{1}{p}}.$$

In order to estimate the right-hand side, we prove that, if none of the $K_i$ includes $Q$, then there is $\gamma$ (depending only on the dimension $n$) such that

$$\bigcup_{i:K_i\cap Q\neq\varnothing} K_i\subset\gamma Q. \tag{9.38}$$

Indeed, take a point $x\in K_i\cap Q\neq\varnothing$. By the Whitney property, there exists a cube $K_j$ with $x\in\frac{1}{1+\delta}K_j$. Since $Q$ is not a subset of $K_j$, it follows that $2r(Q)\geq\frac{\delta}{1+\delta}r(K_j)$. Now, the strong engagement property implies the estimate $r(K_i)\leq cr(K_j)$ with a constant $c>0$ depending only on the dimension $n$. Therefore,

$$r(K_i)\leq cr(Q),$$

and (9.38) is fulfilled.

With (9.38) at hand, we can use the finite multiplicity property of the family $\{K_i\}_{i\in I}$ to estimate the right-hand side of (9.37):

$$\frac{c}{|Q|^{\frac{1}{p}+\frac{\alpha}{n}}}\left(\sum_{i:K_i\cap Q\neq\varnothing}\int_{K_i}|f(x)-f_{K_i}(x)|^p\,dx\right)^{\frac{1}{p}}$$
$$\leq\frac{c}{|Q|^{\frac{1}{p}+\frac{\alpha}{n}}}\left(\int_{\gamma Q}|f(x)-f_{\gamma Q}(x)|^p\,dx\right)^{\frac{1}{p}}\leq\frac{c}{|Q|^{\frac{1}{p}+\frac{\alpha}{n}}}t\,|Q|^{\frac{1}{p}+\frac{\alpha}{n}}=ct$$

Thus, we have proved (9.35) also in the case where $Q$ is not a subset of any $K_i$. This finishes the proof of the theorem. □

As in (9.34), we denote by $\{Q\}_{f,t}$ the family of all cubes satisfying

$$\frac{1}{|Q|^{\frac{1}{p}+\frac{\alpha}{n}}}\left(\int_Q|f(x)-f_Q(x)|^p\,dx\right)^{\frac{1}{p}}\geq t.$$

**Theorem 9.10.** *Let $-\frac{n}{p}<\alpha\leq k$. There exist constants $c_1$ and $c_2>0$ (depending only on the dimension $n$ if $p$ is fixed) such that*

$$c_1\operatorname{dist}_{L^p}\left(f,B_{\dot C_p^{\alpha,k}}\left(\frac{t}{c_1}\right)\right)\leq t\left(\left|\{Q\}_{f,t}\right|_\lambda\right)^{\frac{1}{p}}\leq c_2\operatorname{dist}_{L^p}\left(f,B_{\dot C_p^{\alpha,k}}\left(\frac{t}{c_2}\right)\right), \tag{9.39}$$

*where $\lambda=1+\frac{\alpha p}{n}$.*

*Proof.* Lemma 9.8 implies the second inequality in (9.39). By Lemma 9.9,

$$\|f_t\|_{\dot{C}_p^{\alpha,k}} \leq ct \text{ and } \|f - f_t\|_{L^p} \leq ct\,|\{Q_x\}_\Omega|_\lambda^{\frac{1}{p}}.$$

Therefore,

$$\operatorname{dist}_{L^p}(f, B_{\dot{C}_p^{\alpha,k}}(ct)) \leq ct\,|\{Q_x\}_\Omega|_\lambda^{\frac{1}{p}}.$$

Since the family $\{Q_x\}_\Omega$ is a part of $\{Q\}_{f,t}$, we see that

$$\operatorname{dist}_{L^p}(f, B_{\dot{C}_p^{\alpha,k}}(ct)) \leq ct\left|\{Q\}_{f,t}\right|_\lambda^{\frac{1}{p}}.$$

It remains to take $\frac{1}{c}$ for the role of $c_1$. □

## Notes and remarks

The results of this chapter were obtained by Kruglyak and were published (partly) in [K1] and [K0]. For this book, the exposition was revised.

Note that we did not consider the couples ($L^\infty$, a Sobolev space) in this chapter. It would be interesting and important to cover this case. This task may well happen to be feasible, see [CK]. In fact, in that paper an *exact* minimizer was constructed, unlike what we do here. So, the general question of constructing exact minimizers in certain situations treated in this chapter may also be reasonable. This is important in some applications, in particular in image processing.

# Chapter 10

# Stability of near-minimizers

Near-minimizers constructed in the preceding chapter have the form

$$f_t = \sum_{i \in I} P_i \psi_i + f\, \chi_{\mathbb{R}^n \setminus \cup K_i},$$

where $\{K_i\}_{i \in I}$ is a WB-covering, $\{\psi_i\}_{i \in I}$ is a smooth partition of unity adjusted to it, and the $\{P_i\}_{i \in I}$ are polynomials of degree strictly smaller than $k$ that provide a nearly optimal approximation of $f$ on the cubes $K_i$.

We shall show that, for many couples, if the polynomials $\{P_i\}_{i \in I}$ are chosen in a special way, the near-minimizers $f_t$ turn out to be stable under the action of certain operators that may be discontinuous. Though we have already encountered this sort of stability, we repeat the explanations of what this means, by using the couple $(L^p, \dot{W}_q^k)$, $q > p$, as an example. In this case, for any $f \in L^p$ and $t > 0$ we have constructed a function $f_t \in \dot{W}_p^k$ of the above type and such that

$$\|f - f_t\|_{L^p}^p + t^{p-q} \|f_t\|_{\dot{W}_q^k}^q \le c L_{p,q}(t^{p-q}, f; L^p, \dot{W}_q^k),$$

where

$$L_{p,q}(t^{p-q}, f; L^p, \dot{W}_q^k) = \inf_{g \in \dot{W}_q^k} (\|f - g\|_{L^p}^p + t^{p-q} \|g\|_{\dot{W}_q^k}^q).$$

If a linear operator $T$ is bounded both on $L^p$ and on $\dot{W}_q^k$, then

$$L(t^{p-q}, Tf; L^p, \dot{W}_q^k) \le \|Tf - Tf_t\|_{L^p}^p + t^{p-q} \|Tf_t\|_{\dot{W}_p^k}^q \le c(T) L(t^{p-q}, f; L^p, \dot{W}_q^k). \tag{10.1}$$

Now, (10.1) can be rewritten in the following equivalent way:

$$\|Tf - Tf_t\|_{L^p}^p + t^{p-q} \|Tf_t\|_{\dot{W}_p^k}^q \le c(T) \left( L(t^{p-q}, f; L^p, \dot{W}_q^k) + L(t^{p-q}, Tf; L^p, \dot{W}_q^k) \right). \tag{10.2}$$

However, (10.2) may be true for some unbounded operators. This inequality says that $Tf_t$ approximates $Tf$ "fairly well". This is especially clear if $Tf = f$ because then (10.2) takes the form

$$\|f - Tf_t\|_{L^p} + t^{p-q} \|Tf_t\|^q_{\dot{W}^k_q} \leq c(T) L(t^{p-q}, f; L^1, \dot{W}^k_q)$$

and says that $Tf_t$ remains a near-minimizer for $f$.

In §1 we describe the choice of approximation polynomials that leads to stable minimizers. In §2 we formulate the stability theorems and present their applications. The proofs are given in §3.

## 10.1 Construction of approximating polynomials

Stability results lean upon a special choice of the polynomials $P_i$ that were mentioned in the preceding subsection. Here we describe the construction.

Suppose $Q$ is a cube in $\mathbb{R}^n$ and $\psi$ is a function satisfying

$$0 \leq \psi \leq 1, \quad \operatorname{supp} \psi \subset Q \tag{10.3}$$

and

$$\psi(x) \geq c > 0 \text{ for } x \in \frac{1}{4}Q, \tag{10.4}$$

where $c$ depends on the dimension only.

In what follows, $L^p(\psi)$ stands for the $L^p$-space with respect to the measure $\psi(x)dx$, and $L^p(Q)$ stands for the $L^p$-space with respect to the measure $\chi_Q dx$, where $\chi_Q$ is the characteristic function of $Q$.

In the space $\mathcal{P}_k$ of polynomials of degree strictly smaller than $k$, there is a natural basis consisting of the monomials

$$x^m = x_1^{m_1} x_2^{m_2} \dots x_n^{m_n}, \; m = (m_1, m_2, \dots, m_n),$$

of degree strictly smaller than $k$, i.e., with $|m| = m_1 + m_2 + \dots + m_n < k$.

We order these monomials in accordance with the growth of the degree $|m|$ (within the same degree, the order is inessential, but we fix some). We orthogonalize the functions $x^m$ in $L^2(\psi)$ (without normalization!) by the Gram-Schmidt process. The resulting orthogonal polynomials will be denoted by $x^m_\psi$.

We need the following lemma.

**Lemma 10.1.** *With constants depending only on $k$ and the dimension, we have*

$$\left\|x^m_\psi\right\|_{L^2(\psi)} \approx |Q|^{\frac{1}{2}+\frac{|m|}{n}}, \quad \left\|x^m_\psi\right\|_{L^\infty(Q)} \approx |Q|^{\frac{|m|}{n}}. \tag{10.5}$$

*Proof.* First, we observe that on $P_k$ we have

$$\|\cdot\|_{L^2(\psi)} \approx \|\cdot\|_{L^2(Q)} \approx |Q|^{\frac{1}{2}} \|\cdot\|_{L^\infty(Q)} \tag{10.6}$$

with constants independent of $Q$ and $\psi$. Indeed, denoting by $x_i^0$ the $i$th coordinate of the center of $Q$ and by $r(Q)$ the radius, we make the change of variables

$$y_i = \frac{x_i - x_i^0}{r(Q)} \quad (i = 1, \dots, n), \tag{10.7}$$

which takes the space $\mathcal{P}_k$ onto itself. This reduces the proof of (10.6) to the case of the cube $Q_0 = [-1,1]^n$. However, by (10.3)-(10.4), the norms

$$\|\cdot\|_{L^2(Q_0)}, \quad \|\cdot\|_{L^2(\frac{1}{4}Q_0)}, \quad \text{and} \quad \|\cdot\|_{L^\infty(Q_0)}$$

are equivalent on the finite-dimensional space $\mathcal{P}_k$ with constants depending only on $k$. This proves (10.6).

We denote by $V_m$ the subspace of $\mathcal{P}_k$ spanned by the monomials that precede $x^m$ (strictly) in the order described before the lemma. The properties of the orthogonalization process and the equivalence (10.6) imply

$$\left\|x_\psi^m\right\|_{L^2(\psi)} = \operatorname{dist}_{L^2(\psi)}(x^m, V_m) \approx \operatorname{dist}_{L^2(Q)}(x^m, V_m) \approx |Q|^{\frac{1}{2}} \operatorname{dist}_{L^\infty(Q)}(x^m, V_m) \tag{10.8}$$

with equivalence constants depending only on $k$ and the dimension $n$. (As usual, we denote by $\operatorname{dist}_X(x^m, V_m)$ the distance from the monomial $x^m$ to the subspace $V_m$ in the metric of $X$.)

Now, under the change of variables (10.7), the monomial $x^m$ is taken to the sum of the monomial $r(Q)^{|m|}y^m$ and a polynomial belonging to $V_m$. Therefore,

$$\operatorname{dist}_{L^\infty(Q)}(x^m, V_m) = r(Q)^{|m|} \operatorname{dist}_{L^\infty(Q_0)}(y^m, V_m).$$

Since the number of the monomials to be taken into account is finite (moreover, depends only on $k$ and $n$) and the quantities $\operatorname{dist}_{L^\infty(Q_0)}(y^m, V_m)$ are nonzero, we obtain

$$0 < \min_{|m|<k} \operatorname{dist}_{L^\infty(Q_0)}(y^m, V_m) \quad \text{and} \quad \max_{|m|<k} \operatorname{dist}_{L^\infty(Q_0)}(y^m, V_m) < \infty.$$

(In fact, these minimum and maximum are controlled in terms of $k$ and $n$.) Therefore, (10.8) and (10.6) imply the claim. □

The polynomials $x_\psi^m$ constructed above form a basis in $P_k$ (because they are obtained by $L^2(\psi)$-orthogonalization of the basis of monomials) and can be used to exhibit almost optimal approximants on $Q$ among polynomials of degree strictly smaller than $k$. To describe the procedure, we introduce the orthogonal projection of $L^2(\psi)$ onto $\mathcal{P}_k$. It is given by the formula

$$P_\psi(h) = \sum_{|m|<k} c_m x_\psi^m, \text{ where } c_m = \frac{\int (h x_\psi^m)\psi}{\int (x_\psi^m)^2 \psi}. \tag{10.9}$$

We establish some important properties of $P_\psi$.

**Lemma 10.2.** *Suppose $h \in L^p(Q)$ and $1 \le p < \infty$. Then*

$$\|P_\psi h\|_{L^p(Q)} \le c\,\|h\|_{L^p(Q)}\,, \tag{10.10}$$

$$\int_Q |h - P_\psi h|^p \le c \inf_{v \in \mathbb{P}_k} \int_Q |h - v|^p \tag{10.11}$$

*and*

$$\int (h - P_\psi h)\,\psi = 0, \tag{10.12}$$

*where $c > 0$ is a constant depending only on $k$ and the dimension $n$.*

*Proof.* To verify (10.10), it suffices to estimate one term in (10.9). Lemma 10.1, the properties of $\psi$, and the Hölder inequality show that

$$\int_Q \left( \left| \frac{\int (h x_\psi^m)\psi}{\int (x_\psi^m)^2 \psi} \right| |x_\psi^m| \right)^p \le c \int_Q \left( \frac{\int |h|\,\psi \cdot |Q|^{\frac{|m|}{n}}}{|Q|^{\frac{2|m|}{n}+1}} |Q|^{\frac{|m|}{n}} \right)^p$$
$$\le c|Q|^{1-p} \left( \int_Q |h| \right)^p \le c\,\|h\|_{L^p(Q)}^p\,.$$

To prove (10.11), we observe that $P_\psi$ is identical on $P_k$. Therefore, by (10.10), for every $v \in \mathcal{P}_k$ we have

$$\left( \int_Q |h - P_\psi h|^p \right)^{\frac{1}{p}} \le \left( \int_Q |h - v|^p \right)^{\frac{1}{p}} + \left( \int_Q |v - P_\psi h|^p \right)^{\frac{1}{p}}$$
$$= \left( \int_Q |h - v|^p \right)^{\frac{1}{p}} + \left( \int_Q |P_\psi (v - h)|^p \right)^{\frac{1}{p}} \le c \left( \int_Q |h - v|^p \right)^{\frac{1}{p}}.$$

Since $v \in \mathcal{P}_k$ is arbitrary, inequality (10.11) follows.

We prove (10.12). Under orthogonalization, the first monomial $x_\psi^0 \equiv 1$ does not change, and the other monomials turn into functions orthogonal to it, so we have

$$\int x_\psi^m \psi = \int x_\psi^m x_\psi^0 \psi = 0, \quad |m| > 0.$$

Therefore,

$$\int (P_\psi h)\psi = \int \left( \sum_{|m|<k} c_m x_\psi^m \right) \psi = \int c_0 \psi = \int \left( \frac{\int (h x_\psi^0)\psi}{\int (x_\psi^0)^2 \psi} \right) \psi = \int h\psi.$$

This implies (10.12). □

It has already been mentioned that, given $f \in L^p(\mathbb{R}^n)$ and $t > 0$, for fairly many couples $(L^p, X)$, a near-minimizer $f_t$ can be found in the form

$$f_t = \sum_{i \in I} P_i \psi_i + f \chi_{\mathbb{R}^n \setminus \cup K_i}\,,$$

where $\{K_i\}_{i\in I}$ is a certain WB-covering, $\{\psi_i\}_{i\in I}$ is a smooth partition of unity adjusted to it, and $\{P_i\}_{i\in I}$ is a family of polynomials of degree strictly smaller than $k$ that approximate $f$ on the cubes $K_i$ in a roughly optimal way. By (10.11), we can take $P_i$ to be equal to $P_{\psi_i}(f)$, so that the function

$$f_t = \sum_{i\in I} P_{\psi_i}(f)\psi_i + f\chi_{\mathbb{R}^n\setminus\cup K_i}$$

can play the role of a near-minimizer. Precisely this choice will be meant in the sequel.

Consider the linear operator

$$P_{t,f}(h) = \sum_{i\in I} P_{\psi_i}(h)\psi_i + h\chi_{\mathbb{R}^n\setminus\cup K_i}. \tag{10.13}$$

Clearly,

$$P_{t,f}(f) = f_t. \tag{10.14}$$

We shall use the following properties of the operator $P_{t,f}$ (compare with similar properties in the classical case, see Chapters 1 and 3; note also that all these particular cases fit into the abstract linearization pattern discussed in §7 of Chapter 5).

**Lemma 10.3** (On operator estimates). *We have*

$$\|P_{t,f}\|_{L^p\to L^p} \le c. \tag{10.15}$$

a) *If $q > p$, then*

$$\|I - P_{t,f}\|_{\dot W^k_q\to L^p} \le c\left(\frac{L_{p,q}(t^{p-q}, f; L^p, \dot W^k_q)}{t^p}\right)^{\frac{1}{p}-\frac{1}{q}}. \tag{10.16}$$

b) *If $q = p$, then*

$$\|I - P_{t,f}\|_{\dot W^k_p\to L^p} \le ct^k. \tag{10.17}$$

c) *If $\alpha \in \left(-\frac{1}{p}, k\right]$, then*

$$\|I - P_{t,f}\|_{\dot C^{\alpha,k}_p\to L^p} \le c\frac{1}{t}\operatorname{dist}_{L_1}\left(f, B_{\dot C^{\alpha,k}_p}\left(\frac{t}{c}\right)\right). \tag{10.18}$$

*The positive constant $c$ in the above formulas depends only on $k, p, q$, and the dimension $n$.*

*Proof.* Since the family of cubes $\{K_i\}_{i\in I}$ has finite multiplicity, from (10.10) we deduce that

$$\begin{aligned}\int_{\mathbb{R}^n} |P_{t,f}(h)|^p &\le c\sum_{i\in I}\int_{\cup K_i} |P_{\psi_i}(h)\psi_i|^p + \int_{\mathbb{R}^n\setminus\cup K_i} |h|^p \\ &\le c\sum_{i\in I}\int_{K_i} |P_{\psi_i}(h)|^p + \int_{\mathbb{R}^n\setminus\cup K_i} |h|^p \\ &\le c\sum_{i\in I}\int_{K_i} |h|^p + \int_{\mathbb{R}^n\setminus\cup K_i} |h|^p \le c\,\|h\|_{L^p}\,,\end{aligned}$$

i.e., (10.15) is true. We prove a). Again by finite multiplicity, using (10.11) we obtain

$$\begin{aligned}\int_{\mathbb{R}^n} |h-P_{t,f}(h)|^p &\le c\sum_{i\in I}\int_{K_i} |h-P_{\psi_i}(h)|^p \le c\sum_{i\in I}\int_{K_i}|h-h_{K_i}|^p \\ &\le c\sum_{i\in I}\left(\frac{1}{|K_i|^{1+\frac{kp}{n}-\frac{p}{q}}}\int_{K_i}|h-h_{K_i}|^p\right)\cdot |K_i|^{1+\frac{kp}{n}-\frac{p}{q}}.\end{aligned}$$

Since $q>p$, application of the Hölder inequality to the right-hand side yields

$$\begin{aligned}&\int_{\mathbb{R}^n} |h-P_{t,f}(h)|^p \\ &\le c\left(\sum_i\left(\frac{1}{|K_i|^{1+\frac{kp}{n}-\frac{p}{q}}}\int_{K_i}|h-h_{K_i}|^p\right)^{\frac{q}{p}}\right)^{\frac{p}{q}}\left(\sum_{i\in I}|K_i|^{(1+\frac{kp}{n}-\frac{p}{q})\frac{1}{1-\frac{p}{q}}}\right)^{1-\frac{p}{q}} \\ &= c\left(\left(\sum_i \frac{(\frac{1}{|K_i|}\int_{K_i}|h-h_{K_i}|^p)^{\frac{1}{p}}}{|K_i|^{\frac{k}{n}}}\right)^q |K_i|\right)^{\frac{p}{q}}\left(\sum_{i\in I}|K_i|^{(\frac{1}{p}+\frac{k}{n}-\frac{1}{q})\frac{pq}{q-p}}\right)^{1-\frac{p}{q}}.\end{aligned}$$

Yet another reference to finite multiplicity combined with the Sobolev–Poincaré inequality (9.2) yields the following estimate for the first factor:

$$\left(\left(\sum_i \frac{(\frac{1}{|K_i|}\int_{K_i}|h-h_{K_i}|^p)^{\frac{1}{p}}}{|K_i|^{\frac{k}{n}}}\right)^q |K_i|\right)^{\frac{p}{q}} \le c\,\|h\|^p_{\dot W^k_q}\,.$$

To estimate the second factor, we use Theorem 8.30 (preservation of the $\alpha$-capacity under extension) and Theorem 9.6 (about the $L$-functional of the couple $(L^p,\dot W^k_q)$). In the latter statement, the parameter $\lambda$ is equal to $(\frac{1}{p}+\frac{k}{n}-\frac{1}{q})\frac{pq}{q-p}$ (see (9.8)). Consequently,

$$\left(\sum_{i\in I}|K_i|^{(1+\frac{k}{n}-\frac{1}{q})\frac{pq}{q-p}}\right)^{1-\frac{p}{q}} \le c\left|\{Q_x\}_{x\in\Omega}\right|_\lambda^{1-\frac{p}{q}} \le c\left(\frac{L(t^{p-q},f;L^p,W^k_q)}{t^p}\right)^{1-\frac{p}{q}}.$$

Therefore,

$$\left(\int_{\mathbb{R}^n} |h - P_{t,f}(h)|^p\right)^{\frac{1}{p}} \le c\left(\frac{L(t^{p-q}, f; L^p, \dot{W}_q^k)}{t^p}\right)^{\frac{1}{p}-\frac{1}{q}} \|h\|_{\dot{W}_q^k},$$

as required. Now, we prove b). By the Sobolev–Poincaré inequality and the fact that $|K_i| \approx t^n$ in the case in question, we obtain

$$\int_{\mathbb{R}^n} |h - P_{t,f}(h)|^p \le c\sum_{i\in I}\int_{K_i} |h - h_{K_i}|^p \le c\sum_{i\in I} |K_i|^{\frac{kp}{n}} \|h\|^p_{\dot{W}_p^k(K_i)} \le ct^{kp} \|h\|^p_{\dot{W}_p^k},$$

that is, (10.17) is fulfilled. We verify c). If $h \in \dot{C}_p^{\alpha,k}$, then the definition of the seminorm on $\dot{C}_p^{\alpha,k}$ shows that

$$\inf_{P\in\mathcal{P}_k}\int_{K_i} |h - P|^p \le c|K_i|^{1+\frac{\alpha p}{n}} \|h\|^p_{\dot{C}_p^{\alpha,k}}.$$

Consequently,

$$\int_{\mathbb{R}^n} |h - P_{t,f}(h)|^p \le c(\sum_{i\in I} |K_i|^{1+\frac{\alpha p}{n}}) \|h\|^p_{\dot{C}_p^{\alpha,k}}.$$

By Theorem 8.32 (preservation of the $\alpha$-capacity under controlled extension), we obtain

$$(\sum_{i\in I} |K_i|^{1+\frac{\alpha p}{n}})^{\frac{1}{p}} \le c(\left|\{Q_x\}_{x\in\Omega}\right|_{1+\frac{\alpha p}{n}})^{\frac{1}{p}} \le c\frac{1}{t}\operatorname{dist}_{L_p}\left(f, B_{\dot{C}_p^{\alpha,k}}\left(\frac{t}{c}\right)\right).$$

This implies (10.18). It should be noted that the restriction $\alpha > -\frac{1}{p}$ stems from the condition $1 + \frac{\alpha p}{n} > 1 - \frac{1}{n}$ of Theorem 8.32 (preservation of the $\alpha$-capacity under controlled extension), whereas the restriction $\alpha \le k$ is imposed because the space $\dot{C}_p^{\alpha,k}$ is defined only for $\alpha \le k$. □

## 10.2 Stability theorems: statements and applications

### 10.2.1 Statements

In what follows, $P_{t,f}$ is the linear operator defined as in the preceding section, i.e.,

$$P_{t,f}(h) = \sum_{i\in I} P_{\psi_i}(h)\psi_i + h\chi_{\mathbb{R}^n\setminus\cup K_i}, P_{\psi_i}(h) = \sum_{|m|<k}\left(\frac{\int(hx^m_{\psi_i})\psi_i}{\int(x^m_{\psi_i})^2\psi_i}\right)x^m_{\psi_i},$$

where $\{\psi_i\}_{i\in I}$ is a smooth partition of unity corresponding to some WB-covering $\{K_i\}_{i\in I}$. As has been shown above, for many couples a near-minimizer (for a duly chosen functional) has the form

$$f_t = P_{t,f}(f)$$

if the WB-covering $\{K_i\}_{i\in I}$ is chosen appropriately.

Next, in what follows $T$ will be either a multiplier with symbol $m$ satisfying (2.15), that is

$$|D^\alpha m(\xi)| \le c_\alpha |\xi|^{-|\alpha|}, \ \xi \in \mathbb{R}^n,$$

for all multiindices $\alpha$, or an operator related to a wavelet system as in Theorem 4.11, that is,

$$Tf = \sum_{j,m} t_{j,m} \langle f, \psi_{j,m} \rangle \psi_{j,m},$$

where $\{t_{j,m}\}$ is a bounded numerical sequence and $\{\psi_{j,m}\}$ is a system of infinitely smooth wavelets with compactly supported Fourier transforms. We saw in Chapters 2 and 4 that then $T$ is a singular integral operator that takes the Morrey-Campanato space $\dot{C}_p^{\alpha,k}$ to itself if $k > \alpha$, and the Sobolev space $\dot{W}_q^k$ to itself if $1 < q < \infty$. Our main results in this chapter say that in many cases the near-minimizer exhibited above is $T$-stable. We give the details, starting with a fairly easy case.

**Theorem 10.4.** *Suppose* $1 < p < \infty$, $f \in L^p$, *and* $t > 0$. *Let* $f_t = P_{t,f}(f)$ *be the above explicit near-minimizer for the* $K$*-functional* $K(t^k, f; L^p, \dot{W}_p^k)$. *Then this near-minimizer is* $T$*-stable, i.e., along with the inequality*

$$\|f - f_t\|_{L^p} + t^k \|f_t\|_{\dot{W}_p^k} \le cK(t^k, f; L^p, \dot{W}_p^k),$$

*which describes the property of being a near-minimizer, we also have*

$$\|Tf - Tf_t\|_{L^p} + t^k \|Tf_t\|_{\dot{W}_p^k} \le c(T) \left( K(t^k, f; L^p, \dot{W}_p^k) + K(t^k, Tf; L^p, \dot{W}_p^k) \right).$$

This is nontrivial and at the same time does not require any proof, because this follows directly from the (*nontrivial!*) fact that a singular integral operator of the form considered here is bounded on $L^p$ and on $\dot{W}_p^k$. Moreover, we see that the term $K(t^k, Tf; L^p, \dot{W}_p^k)$ on the right in the last display is redundant, because it is dominated by the first; we write it only for uniformity reasons. Also, in fact here the particular form of a near-minimizer is irrelevant.

The next result pertains to the couple $(L^p, \dot{W}_q^k)$ with $p \ne q$, $q > 1$. In accordance with the results proved in the preceding chapter for this case, now we treat the $L_{p,q}$-functional.

**Theorem 10.5.** *Suppose* $1 \le p < \infty$, $p \ne q$, $q > 1$, $f \in L^p$, *and* $t > 0$. *Let* $f_t = P_{t,f}(f)$ *be the above explicit near-minimizer for the* $L_{p,q}$*-functional* $L_{p,q}(t^{p-q}, f; L^p, \dot{W}_q^k)$. *If* $Tf \in L^p$, *then this near-minimizer is* $T$*-stable, i.e., along with the inequality*

$$\|f - f_t\|_{L^p}^p + t^{p-q} \|f_t\|_{\dot{W}_q^k}^q \le cL(t^{p-q}, f; L^p, \dot{W}_q^k)$$

*we also have*

$$\|Tf - Tf_t\|_{L^p}^p + t^{p-q} \|Tf_t\|_{\dot{W}_p^k}^q \le c(T) \left( L(t^{p-q}, f; L^p, \dot{W}_q^k) + L(t^{p-q}, Tf; L^p, \dot{W}_q^k) \right).$$

If $p > 1$, then, again, this is nontrivial but immediate, because then $T$ operates both on $L^p$ and on $\dot{W}_q^k$. All said above concerning Theorem 10.4 can be repeated in this case; moreover, the condition $Tf \in L^p$ also becomes redundant. However, the theorem is true also for $p = 1$, and this will be the subject of our work in the next section. If $p = 1$, then neither the second term on the right in the last display, nor the condition $Tf \in L^p$ is redundant.

A similar result holds true in the case of couples of the form $(L^p, \dot{C}_p^{\alpha,k})$, but this time for the distance functional.

**Theorem 10.6.** *Suppose* $1 \le p < \infty$, $f \in L^p$, *and* $t > 0$. *Let* $f_t = P_{t,f}(f)$ *be the above explicit near-minimizer for the distance functional of* $f$ *at* $t$ *for the couple* $(L^p, \dot{C}_p^{\alpha,k})$ *with* $k > 0$ *and* $\alpha \in \left(-\frac{1}{p}, k\right)$. *If* $Tf \in L^p$, *then this near-minimizer is* $T$*-stable, i.e., along with the inequalities*

$$\|f_t\|_{\dot{C}_p^{\alpha,k}} \le ct \text{ and } \|f - f_t\|_{L^p} \le c\,\mathrm{dist}_{L^p}(f, B_{\dot{C}_p^{\alpha,k}}(\frac{t}{c})),$$

*it also satisfies*

$$\|Tf - Tf_t\|_{L^p} \le c(T)\left(\mathrm{dist}_{L^p}(f, B_{\dot{C}_p^{\alpha,k}}(\frac{t}{c})) + \mathrm{dist}_{L^p}(Tf, B_{\dot{C}_p^{\alpha,k}}(\frac{t}{c}))\right).$$

Again, this is nontrivial but immediate if $p > 1$, the condition $Tf \in L^p$ is redundant in this case, etc. In the next section we shall deal with the proof for $p = 1$. We emphasize once again that, for nonnegative $\alpha$, $\dot{C}_p^{\alpha,k}$ does not depend on $p$ (see Chapter 1) and depends only slightly on $k$ (see Chapter 4), and it is merely a renorming, convenient under the circumstances, of a space defined in terms of a smoothness modulus (also see Chapter 4).

### 10.2.2 Applications

#### Shift in smoothness

In this section we present an explicit formula for a near-minimizer for a couple of two Sobolev spaces. The same method applies to a couple consisting of a Sobolev space and a Morrey–Campanato space (the required modifications of the arguments below are easy, and we do not dwell on them).

For two nonnegative integers $m$ and $l$ with $m > l$, consider the Sobolev spaces $\dot{W}_p^l$ and $\dot{W}_q^m$, where $p$ and $q$ are finite and

$$\frac{m-l}{n} + \frac{1}{p} - \frac{1}{q} > 0.$$

We introduce the operator $J$ sending any function $f$ to the collection of all derivatives of $f$ of order $l$. Let $N$ be the number of these derivatives, and let $E$ be an $N$-dimensional space (say, Euclidean or unitary, depending on the scalar field; in principle, a particular norm on $E$ is irrelevant). Then $J$ is an isomorphic embedding of $\dot{W}_p^l$ into $L^p(E)$ and of $\dot{W}_q^m$ into $\dot{W}_q^{m-l}(E)$ (the definition of a Sobolev space of $E$-valued functions is obvious, and we do not write it out). We put $k = m - l$.

Surely, the theory of near-minimizers exposed in this book can be extended without any effort and nearly without changes to spaces of $E$-valued functions. Given $f \in L^p(E)$ and $x \in \mathbb{R}^n$, as in (9.9) we consider the function

$$\varphi_x(r) = \left( \frac{1}{|Q(x,r)|^\lambda} \int_{Q(x,r)} \left| f(y) - f_{Q(x,r)}(y) \right|^p dx \right)^{\frac{1}{p}}.$$

The difference with the scalar case is that this time $f_{Q(y,r)}$ is an *E-valued* polynomial yielding an optimal approximation ... (etc.). The symbol $|\cdot|$ in the integrand now stands for the norm in $E$. Then we define the set $\Omega$ and the cubes $\{Q_x\}_{x\in\Omega}$ and, precisely as in the scalar case (Subsections 7.2.2 and 7.7.3), construct a controlled extension of this family, which will be a WB-covering, etc.

Next, we interpret the operator $P_{f,t}$ (see (10.13)) as an operator sending $E$-valued functions to $E$-valued functions. To do this, we should first understand formula (10.9) in an appropriate way. Namely, the function $h$ in it is $E$-valued now, but (surely!) $\psi$ is scalar-valued and the $x_\psi^m$ are *precisely the same* scalar functions as before. After that, the definition of $P_{f,t}$ becomes clear. Consequently, if $f \in L^p(E)$, then the ($E$-valued) function $f_t$ given by (10.13)–(10.14) is a near-minimizer for $L_{p,q}(t^{p-q}, f; L^p(E), \dot{W}_q^k(E))$ if $q \neq p$ and for $K(t^k, f; L^p(E), \dot{W}_q^k(E))$ for $p = q$; see Lemma 10.3. Moreover, (10.15)–(10.18) remain true in the vector case.[1]

Now, we introduce a projection $\mathcal{Q}$ of the space $L^2(E)$ onto its subspace $J(\dot{W}_2^l(E))$. This is merely a more complicated analog of the projection onto the gradient vector fields, which was discussed thoroughly in §2 of Chapter 2. We label the coordinates of $E$ in a natural way, i.e., by the multiindices $s = (s_1, s_2, \dots, s_n)$ with $|s| = s_1 + \dots + s_n = l$. Let $A$ be the set of such multiindices. For $\alpha \in A$, we put $a_\alpha(\xi) = (2\pi i\xi_1)^{\alpha_1} \dots (2\pi i \xi_n)^{\alpha_n}$ and define $\mathcal{Q}$ in terms of Fourier transforms as follows:

$$[\mathcal{F}(\mathcal{Q}(\{f_\alpha\}_{\alpha\in A}))](\xi) = \left\{ \frac{a_\beta(\xi) \sum_{\alpha\in A} \overline{a_\alpha(\xi)} \hat{f}_\alpha(\xi)}{V(\xi)} \right\}_{\beta\in A}, \tag{10.19}$$

where $V(\xi) = \sum_{\gamma\in A} |a_\gamma(\xi)|^2$. (On the right, the symbol $\mathcal{F}$ means taking the Fourier transform componentwise.) This is a projection indeed, because after the second application of $\mathcal{Q}$ we obtain the vector-valued function whose Fourier transform is equal to

$$\left\{ a_\delta(\xi) \sum_{\beta\in A} \frac{\overline{a_\beta(\xi)} a_\beta(\xi) \sum_{\alpha\in A} \overline{a_\alpha(\xi)} \hat{f}_\alpha}{V(\xi)} \frac{1}{V(\xi)} \right\}_{\delta\in A},$$

[1] Clearly, a similar statement holds true for the distance functional for the couple $(L^p(E), \dot{C}_p^{\alpha,k}(E))$.

and, clearly, this reduces to the right-hand side of (10.19).

Now, we see that the operator matrix of $\mathcal{Q}$ consists of Fourier multipliers whose symbols are homogeneous of degree 0 and infinitely differentiable everywhere except the origin. Hence, these symbols satisfy (2.15), and we see that the (vector versions of) Theorems 10.4 and 10.5 apply. Now, we are ready to state the result.

**Theorem 10.7.** *Under the above assumptions and in the above notation, suppose that either* $1 \le p < \infty$ *and* $p \ne q$ *or* $1 < p = q$. *Then the function* $J^{-1}\mathcal{Q}P_{f,t}J(f)$ *is a near-minimizer for* $L_{p,q}(t^{p-q}, f; \dot{W}_p^l, \dot{W}_q^m)$ *if* $p \ne q$ *and for* $K(t^k, f; \dot{W}_p^l(E), \dot{W}_q^m(E))$ *if* $p = q$.

Here $J$ is the operator defined above: we saw that it is an isomorphic embedding of $\dot{W}_p^l$ into $L^p(E)$ and of $\dot{W}_q^m$ into $\dot{W}_q^{m-l}(E)$. The symbol $J^{-1}$ denotes the inverse of $J$ when the latter is viewed as an operator onto its image; thus, $J^{-1}$ is defined on the set of $E$-valued functions that remain fixed under $\mathcal{Q}$.

*Proof.* This follows immediately from the counterpart of Theorems 10.4 and 10.5 for $E$-valued functions. The role of $T$ in those statements is played by $\mathcal{Q}$. (Since $\mathcal{Q}$ is a projection and we are only interested in $E$-valued functions $f$ satisfying $\mathcal{Q}f = f$, the final estimates in those theorems simplify, as it has already been discussed.) In the next section, we prove the scalar statements only, but formulating and proving the vector versions presents no problem whatsoever. See Subsection 3.2 in Chapter 3 for somewhat nasty but detailed explanations in a similar setting; we do not repeat them here. □

Now, Theorem 0.2 in the Introduction is an immediate consequence of Theorem 10.7. Theorem 0.7 also follows from the counterpart of Theorem 10.7 for Morrey–Campanato spaces mentioned above.

### A property of wavelet expansions

In this subsection, we work in dimension 1. Let $\{\psi_j\}$ be a system of infinitely smooth wavelets whose Fourier transforms are compactly supported (as described in §5, Chapter 4), and let $k + 1/p + 1/q > 0$, $1 \le p < q$.

**Theorem 10.8.** *For* $f \in L^p(\mathbb{R})$, *the functional* $L_{p,q}(t^{p-q}, f; L^p(\mathbb{R}), \dot{W}_q^k(\mathbb{R}))$ *admits a near-minimizer* $g$ *with the property that* $\langle g, \psi_j\rangle = 0$ *whenever* $\langle f, \psi_j\rangle = 0$.

*Proof.* Let $A = \{j : \langle f, \psi_j\rangle \ne 0\}$. We introduce the orthogonal projection $P$ onto the closed linear span of $\{\psi_j : j \in A\}$ in $L^2(\mathbb{R})$. As we saw in Chapter 4, $P$ is a singular integral operator satisfying all the assumptions imposed on "$T$" in the stability theorems of the preceding section. This shows that, for $p = 1$, the function $g = Pf_t$ has the required properties. The same choice of $g$ works for $p > 1$ simply because then $P$ is bounded on $L^p(\mathbb{R})$. □

A similar statement can be proved for the distance functional of the couple $(L^1(\mathbb{R}), \mathrm{Lip}_\alpha(\mathbb{R}))$. We do not repeat the arguments. Clearly, the Lipschitz class can be replaced with other Morrey–Campanato spaces. Next, if we invoke the elementary stability theorem proved in Chapter 1 (Theorem 1.18), in the same way we obtain the following.

**Theorem 10.9.** *Suppose $f \in L^1(\mathbb{R})$, $t > 0$, and $1 < q < \infty$. Then there is a function $g \in L^q(\mathbb{R})$ such that*

$$\|g\|_{L^q} \le Ct, \quad \|f - g\|_{L^1} \le C \operatorname{dist}_{L^1}(f, B_{L^p}(t)), \tag{10.20}$$

*and $\langle g, \psi_j \rangle = 0$ whenever $\langle f, \psi_j \rangle = 0$.*

It should be noted that similar properties are not enjoyed by many classical series expansions. For example, they are not true for expansions in the trigonometric system. We do not enter in a more detailed discussion of that; see a hint in [KK].

## 10.3 Proof of Theorems 10.4–10.6

We must only prove Theorems 10.5 and 10.6 for $p = 1$. In fact, we discuss more general statements in this setting. Namely, we proceed as in stability theorems of Chapter 1, where we did not assume that $T$ is a singular integral operator of special type, but simply isolated the properties of $T$ required in the proofs.

To do the same now, we recall several definitions.

**Definition 10.10.** An operator $T$ is said to be long-range regular if for every $f \in L^1$ that is supported on a cube $Q$ and has mean value zero, we have

$$\int_{\mathbb{R}^n \setminus 2Q} |Tf| \le c \int |f|$$

with a constant $c > 0$ independent of $Q$ and $f$.

**Definition 10.11.** An operator $T$ is said to be weakly $L^1$-bounded if there exists $k \in \mathbb{N}$ such that for every function $\varphi \in C^\infty$ supported on a cube $Q$ and every $f \in L^1$ such that $Tf$ is integrable over $Q$, we have

$$\left| \int T(f)(s)\varphi(s)ds \right| \le C \|f\|_{L^1} \left(\|\varphi\|_{L^\infty} + \sum_{j \le k} (\operatorname{diam} Q)^j \left\|\nabla^j \varphi\right\|_{L^\infty}\right) \tag{10.21}$$

with a constant $c(T) > 0$ independent of $Q$ and $\varphi$.

See Chapters 2 and 4 for the explanations of the fact that the two properties are enjoyed if both $T$ and $T^*$ are singular integral operators.[2] In the theorems

[2] Note that, in distinction from Theorem 3.21, here we do not need strong long range regularity. This is because now we are using finer covering theorems than we did previously.

that follow, it will also be required that $T$ act in a certain space of smooth functions. As was explained in Chapter 4, this is so if $T^*$ is also a singular integral operator sending 1 (more generally, the polynomials of power strictly less than a fixed number) to zero. So, in practice, the theorems below are applicable under the assumption that both $T$ and $T^*$ are singular integrals with some additional properties "in both variables" (see Chapter 4 for more information). In particular, they imply Theorems 10.5 and 10.6 for $p = 1$.

Now we are ready to state the results.

**Theorem 10.12.** *Suppose $f \in L^1$ and $t > 0$. Denote by $f_t = P_{t,f}(f)$ the specific near-minimizer constructed in §1 for the $L_{1,q}$-functional $L_{1,q}(t^{1-q}, f; L^1, \dot{W}_q^k)$ with $q > 1$. Suppose that an operator $T$ takes the space $\dot{W}_q^k$ boundedly into itself and is weakly bounded on $L^1$. Suppose also that it is long-range regular. If $Tf \in L^1$, then the above near-minimizer is $T$-stable, that is, along with the inequality*

$$\|f - f_t\|_{L^1} + t^{1-q}\, \|f_t\|^q_{\dot{W}_q^k} \leq cL(t^{1-q}, f; L^1, \dot{W}_q^k),$$

*it also satisfies*

$$\|Tf - Tf_t\|_{L^1} + t^{1-q}\, \|Tf_t\|^q_{\dot{W}_q^k} \leq c(T) \left( L(t^{1-q}, f; L^1, \dot{W}_q^k) + L(t^{1-q}, Tf; L^1, \dot{W}_q^k) \right).$$

A similar result holds true in the case of couples of the form $(L^1, \dot{C}_1^{\alpha,k})$, but this time for the distance functional.

**Theorem 10.13.** *Suppose $f \in L^1$ and $t > 0$. Denote by $f_t = P_{t,f}(f)$ the specific near-minimizer constructed in §1 for the distance functional of the couple $(L^1, \dot{C}_1^{\alpha,k})$ at the function $f$ and the point $t$. Here $k > 0$ and $\alpha \in (-1, k)$.*

*Suppose that an operator $T$ takes the space $\dot{C}_1^{\alpha,k}$ boundedly to itself and is weakly bounded on $L^1$. Suppose also that it is long-range regular. If $Tf \in L^1$, then the above near-minimizer is $T$-stable, that is, along with the inequality*

$$\|f_t\|_{\dot{C}_1^{\alpha,k}} \leq ct \text{ and } \|f - f_t\|_{L^1} \leq c\, \mathrm{dist}_{L^1}(f, B_{\dot{C}_1^{\alpha,k}}(\frac{t}{c})),$$

*it also satisfies*

$$\|Tf - Tf_t\|_{L^1} \leq c(T) \left( \mathrm{dist}_{L^1}(f, B_{\dot{C}_1^{\alpha,k}}(\frac{t}{c})) + \mathrm{dist}_{L^1}(Tf, B_{\dot{C}_1^{\alpha,k}}(\frac{t}{c})) \right).$$

The difficult part of the proof of both theorems is contained in the following lemma.

**Lemma 10.14** (Main lemma). *Under the assumptions of Theorems* 10.12 *and* 10.13, *we have*

$$\|P_{f,t}T(f - f_t)\|_{L^p} \leq c \sum_i \int_{K_i} |f - f_{K_i}|\,.$$

(Here $P_{f,t}$ is the operator given by (10.13) and $c$ does not depend on $f$ and $t$.)

The proof of this lemma will be postponed. Now we establish the theorems under the assumption that the lemma is true. The arguments in both cases are similar and, more or less, develop along the lines of the proof of Theorem 3.21. We present them for completeness. Note that, once again, the linearization pattern discussed in §7 of Chapter 5 is involved here implicitly. First, we establish Theorem 10.12.

*Proof.* Since $T$ is bounded on $\dot{W}_q^k$, we have

$$t^{1-q}\|Tf_t\|^q_{\dot{W}_q^k} \leq c(T)t^{1-q}\|f_t\|^q_{\dot{W}_q^k} \leq c(T)L_{1,q}(t^{1-q}, f; L^1, \dot{W}_q^k),$$

and it only remains to estimate the quantity $\|Tf - Tf_t\|_{L^1}$. We proceed as follows: take $u \in \dot{W}_q^k$ with

$$\|Tf - u\|_{L^1} + t^{1-q}\|u\|^q_{\dot{W}_q^k} \leq 2L_{1,q}(t^{1-q}, Tf; L^1, \dot{W}_q^k).$$

Then

$$\begin{aligned}\|Tf - Tf_t\|_{L^1} &= \|T(I - P_{f,t})f\|_{L^1} \\ &\leq \|P_{f,t}T(I - P_{f,t})f\|_{L^1} + \|(I - P_{f,t})T(I - P_{f,t})f\|_{L^1} \\ &\leq \|P_{f,t}T(I - P_{f,t})f\|_{L^1} + \|(I - P_{f,t})(Tf - u)\|_{L^1} \\ &\quad + \|(I - P_{f,t})TP_{f,t}f\|_{L^1} + \|(I - P_{f,t})u\|_{L^1}.\end{aligned}$$

The operator $I - P_{f,t}$ is bounded on $L^1$ and takes $\dot{W}_q^k$ to $L^1$ with the norm estimates (10.15) and (10.16). Therefore,

$$\|(I - P_{f,t})(Tf - u)\|_{L^1} \leq c\|(Tf - u)\|_{L^1} \leq cL_{1,q}(t^{1-q}, Tf; L^1, \dot{W}_q^k)$$

and

$$\begin{aligned}\|(I - P_{f,t})u\|_{L^1} &\leq c\left(\frac{L_{1,q}(t^{1-q}, f; L^1, \dot{W}_q^k)}{t}\right)^{1-\frac{1}{q}}\|u\|_{\dot{W}_q^k} \\ &\leq c\left(\frac{L_{1,q}(t^{1-q}, f; L^1, \dot{W}_q^k)}{t}\right)^{1-\frac{1}{q}}\left(\frac{L_{1,q}(t^{1-q}, f; L^1, \dot{W}_q^k)}{t^{1-q}}\right)^{\frac{1}{q}} \\ &\leq cL_{1,q}(t^{1-q}, Tf; L^1, \dot{W}_q^k).\end{aligned}$$

Similarly,

$$
\begin{aligned}
\|(I-P_{f,t})TP_{f,t}f\|_{L^1} &\le c\left(\frac{L_{1,q}(t^{1-q},f;L^1,\dot{W}^k_q)}{t}\right)^{1-\frac{1}{q}}\|TP_{f,t}f\|_{\dot{W}^k_q}\\
&\le c(T)\left(\frac{L_{1,q}(t^{1-q},f;L^1,\dot{W}^k_q)}{t}\right)^{1-\frac{1}{q}}\|f_t\|_{\dot{W}^k_q}\\
&\le c(T)\left(\frac{L_{1,q}(t^{1-q},f;L^1,\dot{W}^k_q)}{t}\right)^{1-\frac{1}{q}}\cdot\left(\frac{L_{1,q}(t^{1-q},f;L^1,\dot{W}^k_q)}{t^{1-q}}\right)^{\frac{1}{q}}\\
&\le c(T)L_{1,q}(t^{1-q},Tf;L^1,\dot{W}^k_q).
\end{aligned}
$$

It remains to estimate the quantity $\|P_{f,t}T(I-P_{f,t})f\|_{L^1}$. By Lemma 10.14, we obtain

$$\|P_{f,t}T(I-P_{f,t})f\|_{L^1}=\|P_{f,t}T(f-f_t)\|_{L^1}\le c(T)\sum_i\int_{K_i}|f-f_{K_i}|\,.$$

By the construction of the cubes $K_i$, we have (see (9.20) and (9.21))

$$\sum_i\int_{K_i}|f-f_{K_i}|\le c\sum_i t\,|K_i|^\lambda\le ct\left(\left|\{Q_x\}_{x\in\Omega}\right|_\lambda\right),$$

where $\lambda$ is defined in (9.8) with $p=1$. Applying Theorem 9.6, we finally obtain

$$\|P_{f,t}T(I-P_{f,t})f\|_{L^1}\le c(T)L_{1,q}(t^{1-q},Tf;L^1,\dot{W}^k_q).\qquad\square$$

Now, we prove Theorem 10.13.

*Proof.* Since $T$ is bounded on $\dot{C}^{\alpha,k}_1$, we have

$$\|Tf_t\|_{\dot{C}^{\alpha,k}_1}\le c(T)\,\|f_t\|_{\dot{C}^{\alpha,k}_1}\le c(T)t,$$

and we must estimate only the term $\|Tf-Tf_t\|_{L^1}$. Here we choose $u\in\dot{C}^{\alpha,k}_1$ with

$$\|u\|_{\dot{C}^{\alpha,k}_1}\le 2t \text{ and } \|Tf-u\|_{L^1}\le \operatorname{dist}_{L^1}(Tf,B_{\dot{C}^{\alpha,k}_1}(t)).$$

Then, as in the preceding proof and in the proof of Theorem 3.21, we write

$$
\begin{aligned}
\|Tf-Tf_t\|_{L^1}&=\|T(I-P_{f,t})f\|_{L^1}\\
&\le\|P_{f,t}T(I-P_{f,t})f\|_{L^1}+\|(I-P_{f,t})T(I-P_{f,t})f\|_{L^1}\\
&\le\|P_{f,t}T(I-P_{f,t})f\|_{L^1}+\|(I-P_{f,t})(Tf-u)\|_{L^1}\\
&\quad+\|(I-P_{f,t})TP_{f,t}f\|_{L^1}+\|(I-P_{f,t})u\|_{L^1}\,.
\end{aligned}
$$

The operator $I-P_{f,t}$ is bounded on $L^1$ and takes $\dot{C}^{\alpha,k}_1$ to $L^1$ with the norm estimates (10.15) and (10.18). Therefore,

$$\|(I-P_{f,t})(Tf-u)\|_{L^1}\le c\,\|(Tf-u)\|_{L^1}\le c\operatorname{dist}_{L^1}(Tf,B_{\dot{C}^{\alpha,k}_1}(\frac{t}{c}))$$

and

$$\|(I-P_{f,t})u\|_{L^1} \leq \frac{c}{t}\operatorname{dist}_{L^1}(f, B_{\dot{C}_1^{\alpha,k}}(\frac{t}{c}))\,\|u\|_{\dot{C}_1^{\alpha,k}} \leq c\operatorname{dist}_{L^1}(f, B_{\dot{C}_1^{\alpha,k}}(t)).$$

Similarly,

$$\begin{aligned}\|(I-P_{f,t})TP_{f,t}f\|_{L^1} &\leq \frac{c}{t}\operatorname{dist}_{L^1}(f, B_{\dot{C}_1^{\alpha,k}}(\frac{t}{c}))\,\|TP_{f,t}f\|_{\dot{C}_1^{\alpha,k}} \\ &\leq c(T)\frac{c}{t}\operatorname{dist}_{L^1}(f, B_{\dot{C}_1^{\alpha,k}}(\frac{t}{c}))\,\|f_t\|_{\dot{C}_1^{\alpha,k}} \\ &\leq c\operatorname{dist}_{L^1}(f, B_{\dot{C}_1^{\alpha,k}}(\frac{t}{c})).\end{aligned}$$

It remains to estimate the quantity $\|P_{f,t}T(I-P_{f,t})f\|_{L^1} = \|P_{f,t}T(f-f_t)\|_{L^1}$. By Lemma 10.14, we have

$$\|P_{f,t}T(I-P_{f,t})f\|_{L^1} = \|P_{f,t}T(f-f_t)\|_{L^1} \leq c(T)\sum_i \int_{K_i} |f-f_{K_i}|\,.$$

The construction of the cubes $K_i$ and Theorem 9.10 imply ($\lambda = 1+\frac{\alpha}{n}$ in the case in question):

$$\begin{aligned}\sum_i \int_{K_i} |f-f_{K_i}| &\leq ct\sum_i |K_i|^{\lambda} \leq ct\left|\{Q_x\}_{x\in\Omega}\right|_{\lambda} \\ &\leq c\operatorname{dist}_{L^1}(f, B_{\dot{C}_1^{\alpha,k}}(\frac{t}{c}))\end{aligned}$$

and the theorem is proved. □

### 10.3.1 Proof of the main lemma

We act as in the proof of Theorem 3.21 after formula (3.51), with some modifications. The function $f-f_t$ can be represented as the sum of the blocks $b_i = (f-P_{\psi_i}(f))\psi_i$:

$$f - f_t = \sum_{i\in I} b_i,$$

which satisfy

$$\int b_i = 0 \tag{10.22}$$

and $\operatorname{supp} b_i \subset \frac{1+\frac{\delta}{3}}{1+\delta}K_i$. We choose $\delta$ in such a way that $\frac{1+\frac{\delta}{3}}{1+\delta} \leq \frac{1}{2}$ (say, $\delta = 3$), then

$$\operatorname{supp} b_i \subset \frac{1}{2}K_i. \tag{10.23}$$

Now, formally we can write

$$P_{f,t}T(f - f_t) = \sum_{i\in I} P_{\psi_i}\left(T\sum_{j\in I} b_j\right)\psi_i + \left(T\sum_{j\in I} b_j\right)\chi_{\mathbb{R}^n\setminus\cup K_i} \tag{10.24}$$

$$= \sum_{i\in I} P_{\psi_i}\left(\sum_{j\,:\,K_i\cap K_j\neq\varnothing} Tb_j\right)\psi_i + \sum_{i,j\,:\,K_i\cap K_j=\varnothing} P_{\psi_i}(Tb_j)\psi_i + \sum_{j\in I} Tb_j\chi_{\mathbb{R}^n\setminus\cup K_i}.$$

To make this transformation legitimate, we need to understand at least that all terms in the last line make sense, i.e., that all functions to which $P_{\psi_i}$ is applied are integrable over $K_i$. This is done as in the proof of Theorem 3.21. Specifically, this is clear for all terms of the second sum in that line because $T$ is long-range regular. Moreover, the arguments below will show that the sum in question is an absolutely convergent series in $L^1$ (more specifically, the quantity (10.25) is finite). Then, for every $i$, also the function

$$\sum_{j\,:\,K_i\cap K_j\neq\varnothing} Tb_j = T(f - f_t) - \sum_{j\,:\,K_i\cap K_j=\varnothing} P_{\psi_i}(Tb_j)$$

is integrable over $K_i$. (By the way, it is useful to observe that a finite sum occurs here on the left, because the cubes $K_l$ have finite overlap; note also that, separately, the summands on the left may in principle fail to be integrable over $K_i$). But this is precisely the function to which $P_{\psi_i}$ is applied under the sign of the first sum.

Having said this, we simply estimate all summands on the right in (10.24), not dwelling any more on the consistency question. Cf. the proof of Theorem 3.21 for a more methodical treatment of the same matter.

## Estimate of the second and third summands in (10.24)

*Proof.* Clearly,

$$\left\|\sum_{i,j\,:\,K_i\cap K_j=\varnothing} P_{\psi_i}(Tb_j)\psi_i + \sum_{j\in I} Tb_j\chi_{\mathbb{R}^n\setminus\cup K_i}\right\|_{L^1}$$
$$\leq \sum_{i,j\,:\,K_i\cap K_j=\varnothing} \|P_{\psi_i}(Tb_j)\psi_i\|_{L^1} + \sum_{j\in I} \|Tb_j\|_{L^1(\mathbb{R}^n\setminus K_j)}.$$

Since

$$\operatorname{supp} b_j \subset \frac{1}{2}K_j \text{ and } \int b_i = 0,$$

by long-range regularity and (10.11) we obtain

$$\begin{aligned}\sum_{j\in I}\|Tb_j\|_{L^1(\mathbb{R}^n\setminus K_j)} &\le c(T)\sum_{j\in I}\|b_j\|_{L^1} = c(T)\sum_{j\in I}\left\|(f-P_{\psi_j}(f))\psi_j\right\|_{L^1}\\ &= c(T)\sum_{j\in I}\int_{K_j}\left|f-P_{\psi_j}(f)\right| \le ct\sum_{j\in I}\int_{K_j}\left|f-f_{K_j}\right|.\end{aligned}$$

So, it remains to estimate the quantity

$$\sum_{i,j\,:\,K_i\cap K_j=\varnothing}\|P_{\psi_i}(Tb_j)\psi_i\|_{L^1}. \tag{10.25}$$

Now,

$$P_{\psi_i}(Tb_j)\psi_i = \sum_{|m|<k}\left(\frac{\int(Tb_j x^m_{\psi_i})\psi_i}{\int(x^m_{\psi_i})^2\psi_i}\right)x^m_{\psi_i}\psi_i,$$

and Lemma 10.1 shows that

$$\int(x^m_{\psi_i})^2\psi_i \approx |K_i|^{1+\frac{2|m|}{n}},\ \left\|x^m_{\psi}\right\|_{L^\infty(K_i)} \approx |K_i|^{\frac{|m|}{n}}.$$

So, we arrive at

$$\begin{aligned}&\sum_{j\,:\,K_i\cap K_j=\varnothing}\|P_{\psi_i}(Tb_j)\psi_i\|_{L^1}\\ &\le \sum_{j\,:\,K_i\cap K_j=\varnothing}\ \sum_{|m|<k}\int_{K_i}\left|\left(\frac{\int(Tb_j x^m_{\psi_i})\psi_i}{\int(x^m_{\psi_i})^2\psi_i}\right)x^m_{\psi_i}\right|\\ &\le c\sum_{j\,:\,K_i\cap K_j=\varnothing}\ \sum_{|m|<k}\int_{K_i}\left(\frac{\left(\int_{K_i}|Tb_j|\right)|K_i|^{\frac{|m|}{n}}}{|K_i|^{1+\frac{2|m|}{n}}}|K_i|^{\frac{|m|}{n}}\right)\\ &\le c\sum_{j\,:\,K_i\cap K_j=\varnothing}\int_{K_i}|Tb_j|.\end{aligned}$$

Consequently,

$$\begin{aligned}\left\|\sum_{i,j\,:\,K_i\cap K_j=\varnothing}P_{\psi_i}(Tb_j)\psi_i\right\|_{L^1} &\le c\sum_{i,j\,:\,K_i\cap K_j=\varnothing}\int_{K_i}|Tb_j|\\ &\le c\sum_j\left(\sum_{i\,:\,K_i\cap K_j=\varnothing}\int_{K_i}|Tb_j|\right).\end{aligned}$$

Now, the finite multiplicity property implies

$$\sum_{i\,:\,K_i\cap K_j=\varnothing} \int_{K_i} |Tb_j| \le c \int_{\mathbb{R}^n\setminus K_j} |Tb_j| \,.$$

Invoking long-range regularity once again, we obtain

$$\begin{aligned}\int_{\mathbb{R}^n\setminus K_j} |Tb_j| \le c \int_{K_j} |b_j| &= c \int_{K_j} \left|(f - P_{\psi_j}(f))\psi_j\right| \\ &\le c \int_{K_j} \left|(f - P_{\psi_j}(f))\right| \le c \int_{K_j} \left|f - f_{K_j}\right| .\end{aligned}$$

This is the estimate we require. □

## Estimate of the first summand in (10.24)

We formulate the estimate to be proved as a separate lemma.

**Lemma 10.15.** *If an operator $T$ is weakly bounded on $L^1$, then*

$$\left\| \sum_i P_{\psi_i} \left( \sum_{j\,:\,K_i\cap K_j\ne\varnothing} (Tb_j) \right) \psi_i \right\|_{L^1} \le c \sum_i \int_{K_i} |f - f_{K_i}| \,.$$

For the proof, we need an auxiliary statement.

**Proposition 10.16.** *Suppose $P \in P_k$. Let a cube $Q$ and a function $\psi$ satisfy*

$$\operatorname{supp}\psi \subset Q, |\psi| \le 1, \left\|\nabla^l \psi\right\|_{L^\infty} \le c\,|Q|^{-\frac{l}{n}}\,, \quad l = 1, 2, ..., \tag{10.26}$$

*and*

$$\psi(x) \ge c > 0 \;\; \textit{for} \;\; x \in \frac{1}{4}Q. \tag{10.27}$$

*Then* (10.26) *is also satisfied by the function $\varphi = \frac{1}{\|P\psi\|_{L^\infty}} P\psi$ in place of $\psi$.*

*Proof.* Clearly,

$$\operatorname{supp}\varphi \subset Q \text{ and } |\varphi| \le 1.$$

Let $\delta$ denote the operator that maps any function to a fixed partial derivative of it of order $l$. Calculating $\delta(P\psi)$ by the Leibnitz formula, and then applying the Markov inequality (Theorem 0.21) and (10.26) to each term of the resulting expansion, we readily see that

$$\|\delta(\varphi)\|_{L^\infty} \le \frac{c\,\|P\|_{L^\infty(Q)}}{\|P\psi\|_{L^\infty}\,|Q|^{\frac{l}{n}}}.$$

It remains to observe that (10.27), the relation $P \in \mathcal{P}_k$, and Theorem 0.22 on the growth of a polynomial imply $\|P\|_{L^\infty(Q)} \le C\,\|P\psi\|_{L^\infty}$. □

Now, we prove the lemma.

*Proof.* We must estimate the $L^1$-norm of the function

$$\sum_i P_{\psi_i}\left(\sum_{j\,:\,K_i\cap K_j\neq\varnothing} Tb_j\right)\psi_i.$$

We have

$$P_{\psi_i}\left(\sum_{j\,:\,K_i\cap K_j\neq\varnothing} Tb_j\right)\psi_i = \sum_{|m|<k}\left(\frac{\int(\sum_{j\,:\,K_i\cap K_j\neq\varnothing} Tb_j x^m_{\psi_i})\psi_i}{\int(x^m_{\psi_i})^2\psi_i}\right)x^m_{\psi_i}\psi_i,$$

therefore

$$\begin{aligned}\left\|\sum_i P_{\psi_i}\left(\sum_{j\,:\,K_i\cap K_j\neq\varnothing} Tb_j\right)\psi_i\right\|_{L^1} \\ \leq c\sum_{|m|<k}\sum_i\left\|\frac{\int\left(\sum_{j\,:\,K_i\cap K_j\neq\varnothing}(Tb_j x^m_{\psi_i})\right)\psi_i}{\int(x^m_{\psi_i})^2\psi_i}x^m_{\psi_i}\psi_i\right\|_{L^1}.\end{aligned} \tag{10.28}$$

Consequently, it suffices to estimate the sum over $i$ on the right-hand side for fixed $m$. Fixing $i$, we put

$$\varphi = \frac{x^m_{\psi_i}\psi_i}{\left\|x^m_{\psi_i}\right\|_{L^\infty(K_i)}}.$$

Then Proposition 10.16 and condition (10.21) show that we have

$$\begin{aligned}\left|\int_{K_i}\sum_{j\,:\,K_i\cap K_j\neq\varnothing} Tb_j\varphi\right| &\leq c\left\|\sum_{j\,:\,K_i\cap K_j\neq\varnothing} b_j\right\|_{L^1} \\ &\leq c\sum_{j\,:\,K_i\cap K_j\neq\varnothing}\|b_j\|_{L^1} \leq c\sum_{j\,:\,K_i\cap K_j\neq\varnothing}\int_{K_j}\left|f-f_{K_j}\right|.\end{aligned}$$

Therefore, the sum in question over $i$ is dominated by

$$\begin{aligned}\sum_i\left\|\sum_{j\,:\,K_i\cap K_j\neq\varnothing}\left\|x^m_{\psi_i}\right\|_{L^\infty(K_i)}\frac{\int_{K_j}\left|f-f_{K_j}\right|}{\int(x^m_{\psi_i})^2\psi_i}\left|x^m_{\psi_i}\right|\psi_i\right\|_{L^1} \\ \leq \sum_{i,j\,:\,K_i\cap K_j\neq\varnothing}\int_{K_j}\left|f-f_{K_j}\right||K_i|\frac{\left(\|x^m_{\psi_i}\psi_i\|_{L^\infty(K_i)}\right)^2}{\int(x^m_{\psi_i})^2\psi_i}.\end{aligned}$$

Invoking Lemma 10.1 and the fact that only finitely many cubes $K_i$ intersect each $K_j$ (moreover, their number depends only on the dimension), we see that the last quantity does not exceed a constant times the sum

$$\sum_j \int_{K_j} |f - f_{K_j}|,$$

as required. □

## Notes and remarks

The results of this chapter were obtained by the authors jointly and, partly, were published in [KK0] and [KK]. We emphasize that a main application of the stability theorem is the smoothness shift result (Theorem 10.7).

We attract the reader's attention to the restriction $q \geq p$ for couples of the form $(L^p, \dot{W}^{k,q})$ and the restriction $\alpha \in (-\frac{1}{p}, k]$ for couples of the form $(L^p, \dot{C}_p^{\alpha,k})$ in Lemma 10.3. The authors do not know if these restrictions can be weakened.

We want to remark that a (seemingly) natural modification of the proof of Theorem 10.12 leads to the following statement, which is (seemingly) stronger than Theorems 10.4 and 10.5. It is valid for operators satisfying certain "axioms" rather than precisely for singular integrals. Specifically, we modify the long-range regularity and weak $L^1$-boundedness conditions in the following way.

**Definition.** *An operator $T$ is said to be finely long-range regular if there exist positive numbers $\alpha_j$ with $\sum_j \alpha_j < \infty$ such that for every $f$ supported on a cube $Q$ and satisfying $\int f = 0$ we have*

$$|(Tf)(x)| \leq \alpha_j \frac{1}{|2^j Q|} \int_Q |f|\,, x \in 2^{j+1}Q \setminus 2^j Q, \quad j = 1, \ldots.$$

**Definition.** *An operator $T$ is said to be weakly $L^p$-bounded if there exists $k \in \mathbb{N}$ such that for every function $\varphi \in C^\infty$ supported on a cube $Q$ and every $f \in L^1$ such that $Tf$ is integrable over $Q$, we have*

$$\left| \frac{1}{|Q|^{1-\frac{1}{p}}} \int T(f)(s)\varphi(s)ds \right| \leq c(T) \|f\|_{L^p} \left( \|\varphi\|_{L^\infty} + |Q|^{\frac{1}{n}} \sum_{i=1}^n \left\| \frac{\partial}{\partial x_i} \varphi \right\|_{L^\infty} \right).$$

Now, we formulate the result.

**Theorem.** *Suppose $f \in L^p$ and $t > 0$. Let $f_t = P_{t,f}(f)$ be a near-minimizer constructed for the couple $(L^p, \dot{W}_p^k)$, in accordance with the procedures of the present chapter; in particular, we have*

$$\|f - f_t\|_{L^p} + t^k \|f_t\|_{\dot{W}_p^k} \leq cK(t^k, f; L^p, \dot{W}_p^k).$$

*Suppose that $T$ takes $\dot{W}_p^k$ into itself boundedly and is weakly $L^p$-bounded. Suppose also that it is finely long-range regular if $p > 1$ and long-range regular if $p = 1$. If $Tf \in L^p$, then*

$$\|Tf - Tf_t\|_{L^p} + t^k \|Tf_t\|_{\dot{W}_p^k} \le c(T)\left(K(t^k, f; L^p, \dot{W}_p^k) + K(t^k, Tf; L^p, \dot{W}_p^k)\right).$$

For the couples $(L^p, \dot{W}_q^k)$ with $q > p$, a similar result is true for an $L$-functional.

Unfortunately, for $p > 1$ we do not know of any useful example of an operator $T$ that is *not* bounded on $L^p$ but satisfies the above conditions. This was the reason for omission of the statement and the details of the proof in the main part of this chapter.

# Chapter 11

# The omitted case of a limit exponent

In the formulas obtained in Chapter 9, the $L$- and $E$-functionals were expressed in terms of the $\lambda$-capacity of a certain family of cubes. The analysis of these formulas shows that the parameter $\lambda$ may take all real values, except for $\lambda = 0$. Since

$$\lambda = 1 + \frac{k}{n(\frac{1}{p} - \frac{1}{q})}$$

for the couple $\left(L^p, \dot{W}_q^k\right)$, we see that, formally, the case of $\lambda = 0$ corresponds to the limit exponent in the Sobolev embedding theorem:

$$\frac{k}{n} + \frac{1}{p} - \frac{1}{q} = 0. \tag{11.1}$$

Below we shall denote by $q_*$ the parameter $q$ found from (11.1). In this chapter, we suggest an algorithm, which is a result of the limit passage in the algorithm for the couples $\left(L^p, \dot{W}_q^k\right)$ as $q \to q_*$ from the right. It should be noted that, in place of new covering theorems, this algorithm involves the classical statement due to Besicovitch.

The outcome of this algorithm is a piecewise polynomial function $f_t$ possessing a series of interesting properties. Let $N(t)$ denote the number of polynomial pieces that constitute $f_t$. If $f \in \dot{W}_{q_*}^k$, it turns out that

$$\|f - f_t\|_{L^p} \le cN(t)^{-\frac{k}{n}} \|f\|_{\dot{W}_{q*}^k} .$$

In connection with this inequality, we recall the classical result due to Birman and Solomyak [BS] saying that, if $f \in \dot{W}_q^k(Q_0)$ ($Q_0$ is the unit cube in $\mathbb{R}^n$) and $q > q_*$, then for every natural $N$ the cube $Q_0$ can be split into $cN$ dyadic cubes in such a way that for a piecewise polynomial approximant $f_N$ with respect to these cubes we have

$$\|f - f_N\|_{L^p} \le cN^{-\frac{k}{n}} \|f\|_{\dot{W}^k_q} .$$

Thus, the above inequality generalizes the result of Birman and Solomyak to the case of the limit exponent, at the expense of abandoning a dyadic grid. This price must be paid in any case: should we have succeeded in remaining within dyadic cubes, we would have arrived at the compactness of the embedding $\dot{W}^k_{q_*} \subset L^p$, which is not true.

It would be natural to expect that the limiting algorithm provides a near-minimizer for the $L$-functional of the couple $\left(L^p, \dot{W}^k_{q_*}\right)$. It turns out, however, that for a rectangular domain $\Omega$ in $\mathbb{R}^2$ this algorithm yields a near-minimizer for a couple of the form $\left(L^p, \dot{V}^k_{p,p/2}\right)$, where $\dot{V}^k_{p,q}$ is the scale of Brudnyi's generalized variation spaces. This scale joins the quotient $\dot{V}^k_{p,p} = L^p/\mathcal{P}_k$ of $L^p(\Omega)$ over the polynomials of degree smaller than $k$ with the space $\dot{W}^k_{q_*}$ (specifically, $\dot{V}^k_{p,q_*} = \dot{W}^k_{q_*}$). It seems interesting that a new (non-Sobolev) space emerges here. We shall show that in the particular case of $k = 1$ and $p = 2$, the algorithm in question produces a near-minimizer for the couple $\left(L^2, \mathrm{BV}\right)$. This couple is important in applications (see, e.g., [CS]). Here BV is the space of functions of bounded variation on $\Omega$. It should be noted that, in the case of dimension 1, a similar result was obtained by Bergh and Peetre, see [BP].

## 11.1 Description of the algorithm

In what follows, $\Omega$ stands for a rectangular parallelepiped in $\mathbb{R}^n$, and always $0 < p < \infty$. If $\lambda = 0$, it is natural (see (9.9)) to define $\varphi_x(r)$ by the formula

$$\varphi_x(r) = \left(\int_{Q(x,r)\cap\Omega} \left|f(x) - f_{Q(x,r)}(x)\right|^p dx\right)^{\frac{1}{p}} .$$

Here and below, $f_Q$ denotes a polynomial of degree strictly smaller than $k$ that provides the best approximation to $f$ on $Q \cap \Omega$ in the metric of $L^p$, that is,

$$\left(\int_{Q\cap\Omega} |f(x) - f_Q(x)|^p \, dx\right)^{\frac{1}{p}} = \inf_{P\in\mathcal{P}_k} \left(\int_{Q\cap\Omega} |f(x) - P(x)|^p \, dx\right)^{\frac{1}{p}} .$$

Clearly, $\varphi_x(r)$ is a continuous function of $r \in \mathbb{R}_+$ that tends to zero as $r \to 0$ and, for sufficiently large $r$, is equal to

$$t_* = \inf_{P\in\mathcal{P}_k} \left(\int_\Omega |f(x) - P(x)|^p \, dx\right)^{\frac{1}{p}} = \left(\int_\Omega |f(x) - f_\Omega(x)|^p \, dx\right)^{\frac{1}{p}} , \qquad (11.2)$$

where, again, $f_\Omega$ denotes a polynomial of best $L^p$-approximation of the same degree as before, but this time on $\Omega$.

Now we describe the construction of the function $f_t$. If $t \geq t_*$, we put

$$f_t = f_\Omega \ . \tag{11.3}$$

If $t < t_*$, we proceed as follows. The definition of $t_*$ shows that for every $x \in \Omega$ there exists a number $r_x > 0$ (see Figure 11.1) with

$$\varphi_x(r_x) = t \text{ and } \varphi_x(r) < t \text{ for all } r < r_x. \tag{11.4}$$

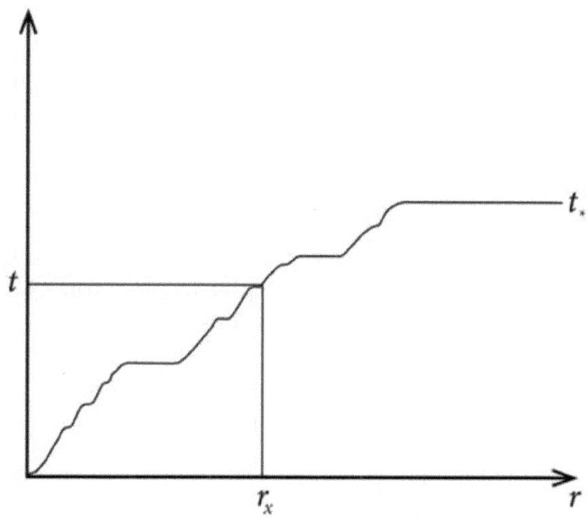


Figure 11.1: The graph of $\varphi_x(r)$.

Consider the family of cubes $\{Q_x\}_{x\in\Omega}$, where

$$Q_x = Q(x, r_x).$$

In what follows, we shall often use the notation

$$N(t) = \left|\{Q_x\}_{x\in\Omega}\right|_0 = \sup \sum_{Q_{x_i}\in\pi} |Q_{x_i}|^0 , \tag{11.5}$$

where the supremum is taken over all packings $\pi$ included in $\{Q_x\}_{x\in\Omega}$. The above family $\{Q_x\}_{x\in\Omega}$ possesses the following property:

$$N(t) \leq \frac{1}{t^p} \|f\|^p_{L^p} \ , \tag{11.6}$$

that is, every subpacking of it contains at most $\frac{1}{t^p} \|f\|^p_{L^p}$ cubes.

Indeed, if $\pi = \{Q_{x_i}\}$ is a packing included in $\{Q_x\}_{x\in\Omega}$, then

$$\sum_{Q_{x_i}\in\pi} |Q_{x_i}|^0 = \sum_{Q_{x_i}\in\pi} \frac{1}{t^p} \int_{Q_{x_i}\cap\Omega} \left|f(x) - f_{Q_{x_i}}(x)\right|^p dx \le \frac{1}{t^p} \|f\|^p_{L^p} .$$

Now, we invoke the Besicovitch covering theorem to construct a covering of $\Omega$ by cubes of the family $\{Q_x\}_{x\in\Omega}$. This covering will be denoted by $\{Q_{x_i}\}_{i\in I}$. Clearly, it depends on $f \in L^p$ and $t > 0$. The number of cubes in this family is finite and admits the estimate

$$|I| \le cN(t) \le \frac{c}{t^p} \|f\|^p_{L^p} \tag{11.7}$$

with a constant $c > 0$ depending on the dimension only. Indeed, the Besicovitch theorem and Proposition 3.5 show that the family $\{Q_{x_i}\}_{i\in I}$ splits in finitely many packings, their number depending only on the dimension $n$, and (11.7) follows from (11.6).

We may assume that the cubes in $\{Q_{x_i}\}_{i\in I}$ are indexed in agreement with the decrease of their volumes, that is, $\left|Q_{x_j}\right| \ge |Q_{x_i}|$ for $j \le i$. If some cube $Q_{x_i}$ is included in the union of the cubes with smaller indices, we delete it from the family $\{Q_{x_i}\}_{i\in I}$. Now we define $f_t$ to be equal to $f_{Q_{x_1}}$ on $Q_{x_1} \cap \Omega$, to $f_{Q_{x_2}}$ on $Q_{x_2} \cap \Omega$ outside $Q_{x_1}$, to $f_{Q_{x_3}}$ on $Q_{x_3} \cap \Omega$ outside $Q_{x_1} \cup Q_{x_2}$, and so on. This process stops after a finite number of steps because the family $\{Q_{x_i}\}_{i\in I}$ is finite, and results in the definition of $f_t$ on the union of the sets $Q_{x_i} \cap \Omega$. This union is equal to $\Omega$. It should be noted that the function constructed above is given by the formula

$$f_t = \sum_{i\in I} f_{Q_{x_i}} \chi_{(Q_{x_i}\cap\Omega)\setminus(\bigcup_{j<i} Q_{x_j})}. \tag{11.8}$$

It turns out that the number of summands in (11.8) that are nonzero on $Q_{x_m} \in \{Q_{x_i}\}_{i\in I}$ depends only on the dimension $n$. This follows from the proposition below.

**Proposition 11.1.** *Let $Q_{x_m}$ be an arbitrary cube in the covering $\{Q_{x_i}\}_{i\in I}$. The number of $i \in I$ such that $\left[(Q_{x_i} \cap \Omega) \setminus (\bigcup_{j<i} Q_{x_j})\right] \cap Q_{x_m} \ne \varnothing$ depends only on the dimension $n$.*

*Proof.* Indeed, if $\left[(Q_{x_i} \cap \Omega) \setminus (\bigcup_{j<i} Q_{x_j})\right] \cap Q_{x_m} \ne \varnothing$, then, clearly, $i \le m$. Therefore, if $Q_{x_i}$ with $i \le m$ meets the cube $Q_{x_m}$, then the inequality $|Q_{x_i}| \ge |Q_{x_m}|$ implies that the former cube contains at least one vertex of the latter. Since the family $\{Q_{x_i}\}_{i\in I}$ is of finite multiplicity, the claim follows. □

We observe yet another interesting property of the family $\{Q_x\}_{x\in\Omega}$: it is Lipschitz with constant 1, i.e.,

$$|r(x) - r(y))| \le \|x - y\| = \max_{i=1-n} |x_i - y_i| . \tag{11.9}$$

Indeed, let $r(Q_x) > r(Q_y)$ for definiteness (if $r(Q_x) = r(Q_y)$, then (11.9) is clear). If a point $y$ is not in the interior of $Q_x$, then $r(Q_x) \le \|x - y\|$, and (11.9) follows. But if $y \in \mathring{Q}_x$ and $r(Q_x) - r(Q_y) > \|x - y\|$, then the cube $Q(x, r)$ centered at $x$ and of radius $r = r(Q_y) + \|x - y\| < r(Q_x)$ includes $Q_y$. Therefore,

$$\varphi_x(r) = \left( \int_{Q(x,r)\cap\Omega} \left| f(x) - f_{Q(x,r)}(x) \right|^p dx \right)^{\frac{1}{p}}$$
$$\ge \left( \int_{Q_y\cap\Omega} \left| f(x) - f_{Q_y}(x) \right|^p dx \right)^{\frac{1}{p}} = t,$$

which contradicts the choice of $Q_x$ (see (11.4)) as the smallest cube centered at $x$ and satisfying

$$\left( \int_{Q(x,r)} \left| f(x) - f_{Q(x,r)}(x) \right|^p dx \right)^{\frac{1}{p}} = t.$$

## 11.2 Principal results, and outlines of the proofs

### 11.2.1 Statement of the main results

In what follows, our basic domain will be a fixed rectangular parallelepiped $\Omega$ in $\mathbb{R}^n$, which will often be omitted in the notation of function spaces. In particular, now $L^p = L^p(\Omega)$, $0 < p < \infty$. The Brudnyi generalized variation space (see [Br]) $\dot{V}^k_{p,q}$, on $\Omega$ is determined by the (semi)norm

$$\|f\|_{\dot{V}^k_{p,q}} = \sup_{\pi = \{Q_i\}} \left( \sum_{Q_i\in\pi} \left( \int_{Q_i\cap\Omega} |f - f_{Q_i}|^p \right)^{\frac{q}{p}} \right)^{\frac{1}{q}}, \tag{11.10}$$

where the supremum is taken over all packings $\pi = \{Q_i\}$ composed of cubes with centers in $\Omega$, and $f_Q \in \mathcal{P}_k$ denotes a polynomial (of degree strictly smaller than $k$) that provides the best approximation to $f$ in $L^p$ on the set $Q \cap \Omega$.

*Remark* 11.2. It is clear that $\dot{V}^k_{p,p} = L^p/\mathcal{P}_k$. Moreover, the Sobolev–Poincaré inequality (9.2) for the limit exponent $q_*$ readily implies the inclusion $\dot{V}^k_{p,q_*} \supset \dot{W}^k_{q_*}$. In fact, Brudnyi showed that $\dot{V}^k_{p,q_*} = \dot{W}^k_{q_*}$ (see [Br]). This will not be used below.

The following theorems are the main results of this chapter.

**Theorem 11.3.** *Suppose* $f \in \dot{W}^k_{q_*}$. *Then*

$$\|f - f_t\|_{L^p} \le cN(t)^{-\frac{k}{n}} \|f\|_{\dot{W}^k_{q_*}} \tag{11.11}$$

*with a constant* $c > 0$ *independent of* $f \in L^p$ *and* $t > 0$.

**Theorem 11.4.** *Let $\Omega$ be a rectangular domain in $\mathbb{R}^2$, and let $f \in L^p$, $0 < p < \infty$. Then for $q = \frac{p}{2}$ the function $f_t$ constructed above satisfies*

$$\|f - f_t\|_{L^p}^p + t^{p-q}\,\|f_t\|_{\dot{V}_{p,q}^k}^q \le cL_{p,q}(t^{p-q}, f; L^p, \dot{V}_{p,q}^k).$$

*Thus, $f_t$ is a near-minimizer for the $L_{p,q}$-functional of $f$ for the couple $\left(L^p, \dot{V}_{p,q}^k\right)$ at the point $t^{p-q}$ with $q = \frac{p}{2}$. Here $c > 0$ is a constant independent of $f \in L^p$ and $t > 0$.*

**Theorem 11.5.** *Let $\Omega$ be a rectangular domain in $\mathbb{R}^2$, and let $f \in L^p$, $p < \infty$. Put $q = \frac{p}{2}$. Then for $t \ge t_* = \left(\int_\Omega |f(x) - f_\Omega(x)|^p\,dx\right)^{\frac{1}{p}}$ we have the following equivalence:*

$$L_{p,q}(t^{p-q}, f; L^p, \dot{V}_{p,q}^k) \approx \int_\Omega |f(x) - f_\Omega(x)|^p\,dx; \tag{11.12}$$

*next, for $t < t_*$ we have*

$$L_{p,q}(t^{p-q}, f; L^p, \dot{V}_{p,q}^k) \approx t^{p-q} \sup \sum_{Q\in\pi} \left(\int_{Q\cap\Omega} |f(x) - f_Q(x)|^p\,dx\right)^{\frac{q}{p}}. \tag{11.13}$$

*In (11.13), the supremum is taken over all packings $\pi$ composed of at most $N(t)$ cubes with centers in $\Omega$.*

**Problem 11.6.** *Find analogs of Theorems 11.4 and 11.5 in the dimension $n$. For example, do they remain true if we put $q = \frac{p(n-1)}{n}$?*

### Outline of the proofs

The arguments being fairly lengthy, we first outline them. The starting point is the verification of (11.12), which shows that $f_t = f_\Omega$ is a near-minimizer for $t \ge t_*$. This part of the proof will be presented in the subsection "The case of $t \ge t_*$".

Now, we explain the case of $t < t_*$. Note that it suffices to establish the inequalities

$$\begin{aligned}\|f - f_t\|_{L^p}^p + t^{p-q}\,\|f_t\|_{\dot{V}_{p,q}^k}^q &\le ct^{p-q} \sup_{|\pi|\le N(t)} \sum_{Q\in\pi} \left(\int_{Q\cap\Omega} |f(x) - f_Q(x)|^p\,dx\right)^{\frac{q}{p}} \\ &\le cL_{p,q}(t^{p-q}, f; L^p, \dot{V}_{p,q}^k).\end{aligned}$$

In the subsection "Lemmas valid in the multidimensional case and Theorem 11.3" we prove Theorem 11.3 and show that

$$\begin{aligned}\|f - f_t\|_{L^p}^p &\le ct^{p-q} \sup_{|\pi|\le N(t)} \sum_{Q\in\pi} \left(\int_{Q\cap\Omega} |f(x) - f_Q(x)|^p\,dx\right)^{\frac{q}{p}} \\ &\le cL_{p,q}(t^{p-q}, f; L^p, \dot{V}_{p,q}^k).\end{aligned} \tag{11.14}$$

This part of the proof is valid in an arbitrary dimension for $q < p$ and is based on the estimate

$$N(t) \leq c\frac{1}{t^p} L_{p,q}(t^{p-q}, f; L^p, \dot{V}^k_{p,q}). \tag{11.15}$$

This is even better than (11.6), which follows from the next obvious inequality:

$$\|f\|^p_{L^p} \geq \inf_{g \in \dot{V}^k_{p,q}} (\|f-g\|^p_{L^p} + t^{p-q}(\|g\|_{\dot{V}^k_{p,q}})^q) = L_{p,q}(t^{p-q}, f; L^p, \dot{V}^k_{p,q}).$$

Once we have the estimates (11.14), the proof is finished if we show that

$$t^{p-q} \|f_t\|^q_{\dot{V}^k_{p,q}} \leq c t^{p-q} \sup_{|\pi| \leq N(t)} \sum_{Q \in \pi} \left( \int_{Q \cap \Omega} |f(x) - f_Q(x)|^p \, dx \right)^{\frac{q}{p}}.$$

The verification of this inequality involves the geometry of the plane in a crucial way. Since $q = \frac{p}{2}$ and the seminorm in $\dot{V}^k_{p,q}$ is defined by (11.10), we must show that for an *arbitrary* packing $\tilde{\pi}$ made of cubes with centers in $\Omega$ (in fact, of squares, because we work in dimension 2) we have

$$\sum_{Q \in \tilde{\pi}} \left( \int_{Q \cap \Omega} |f_t - (f_t)_Q|^p \right)^{\frac{1}{2}} \leq c \sup_{|\pi| \leq N(t)} \sum_{Q \in \pi} \left( \int_{Q \cap \Omega} |f(x) - f_Q(x)|^p \, dx \right)^{\frac{1}{2}}. \tag{11.16}$$

This part of the proof is presented in the subsection "Geometric lemmas".

## 11.3 Proofs

### 11.3.1 The case of $t \geq t_*$

In this subsection we show that, for $t \geq t_*$, the role of a near-minimizer can be played by the polynomial $f_\Omega$ and therefore

$$L_{p,q}(t^{p-q}, f; L^p, \dot{V}^k_{p,q}) \approx \|f - f_\Omega\|^p_{L^p} + t^{p-q} \|f_\Omega\|^q_{\dot{V}^k_{p,q}} = \|f - f_\Omega\|^p_{L^p}.$$

This is established in the following lemma.

**Lemma 11.7.** *Suppose $q < p$ and $t \geq t_*$. Then the function $f_t = f_\Omega \chi_\Omega$ is a near-minimizer for the L-functional of the couple $(L^p, \dot{V}^k_{p,q})$ at the point $t^{p-q}$:*

$$\|f - f_t\|^p_{L^p} + t^{p-q} \|f_t\|^q_{\dot{V}^k_{p,q}} \leq c L_{p,q}(t^{p-q}, f; L^p, \dot{V}^k_{p,q})$$

*with a constant $c > 0$ independent of $t$ and $f$.*

*Proof.* Since

$$\|f - f_\Omega\|^p_{L^p} + t^{p-q} \|f_\Omega\|^q_{\dot{V}^k_{p,q}} = \|f - f_\Omega\|^p_{L^p} = \int_\Omega |f(x) - f_\Omega(x)|^p \, dx, \tag{11.17}$$

it suffices to show that, for an arbitrary function $g \in \dot{V}^k_{p,q}$ and $t \geq t_*$, we have

$$\|f-g\|^p_{L^p} + t^{p-q}\,\|g\|^q_{\dot{V}^k_{p,q}} \geq \varepsilon \int_\Omega |f(x) - f_\Omega(x)|^p\,dx \tag{11.18}$$

with a positive constant $\varepsilon$, whose value will be clear from the proof. Since $q < p$, we have $p - q > 0$, and (11.18) is clearly fulfilled if

$$t_*^{p-q}\,\|g\|^q_{\dot{V}^k_{p,q}} \geq \varepsilon \int_\Omega |f(x) - f_\Omega(x)|^p\,dx.$$

So, there is no loss of generality in assuming that

$$t_*^{p-q}\,\|g\|^q_{\dot{V}^k_{p,q}} \leq \varepsilon \int_\Omega |f(x) - f_\Omega(x)|^p\,dx.$$

Since $t_* = \left(\int_\Omega |f(x) - f_\Omega(x)|^p\,dx\right)^{\frac{1}{p}}$, the last inequality can be rewritten in the form

$$\|g\|_{\dot{V}^k_{p,q}} \leq \varepsilon^{\frac{1}{q}} \left(\int_\Omega |f(x) - f_\Omega(x)|^p\,dx\right)^{\frac{1}{p}}.$$

By the definition of $\dot{V}^k_{p,q}$, we have

$$\left(\int_\Omega |g(x) - g_\Omega(x)|^p\,dx\right)^{\frac{1}{p}} \leq \|g\|_{\dot{V}^k_{p,q}}.$$

Using this and the triangle inequality, we obtain

$$\begin{aligned}
\|f-g\|_{L^p} &= \left(\int_\Omega |f(x) - g(x)|^p\,dx\right)^{\frac{1}{p}} \\
&\geq c_p^{-1}\left(\int_\Omega |f(x) - g_\Omega(x)|^p\,dx\right)^{\frac{1}{p}} - \left(\int_\Omega |g(x) - g_\Omega(x)|^p\,dx\right)^{\frac{1}{p}} \\
&\geq c_p^{-1}\left(\int_\Omega |f(x) - f_\Omega(x)|^p\,dx\right)^{\frac{1}{p}} - \|g\|_{\dot{V}^k_{p,q}} \\
&\geq (c_p^{-1} - \varepsilon^{\frac{1}{q}})\left(\int_\Omega |f(x) - f_\Omega(x)|^p\,dx\right)^{\frac{1}{p}}.
\end{aligned}$$

Thus, for the validity of (11.18), it suffices that

$$(c_p^{-1} - \varepsilon^{\frac{1}{q}})^p \geq \varepsilon.$$

But this is really so if $\varepsilon$ is a sufficiently small positive number, and consequently, the function

$$f_t = f_\Omega \chi_\Omega$$

is a near-minimizer. □

### 11.3.2 Lemmas valid in the multidimensional case and Theorem 11.3

First, we estimate the quantity $N(t)$ (by definition, see (11.5), this is the maximal possible cardinality of a packing composed of cubes of the family $\{Q_x\}_{x\in\Omega}$) in terms of the $L$-functional. By the Besicovitch theorem and Proposition 3.5, the covering $\{Q_{x_i}\}_{i\in I}$ splits into finitely many packings, and their number depends only on the dimension. Therefore,

$$|I| \leq cN(t) \tag{11.19}$$

with a constant $c > 0$ determined by the dimension.

**Lemma 11.8.** *Let $q < p$. For a sufficiently small positive $\varepsilon$ depending only on the dimension $n$, we have*

$$L_{p,q}(t^{p-q}, f; L^p, \dot{V}^k_{p,q}) \geq \varepsilon t^p N(t). \tag{11.20}$$

*Proof.* Choose $g \in \dot{V}^k_{p,q}$ in such a way that

$$\|f-g\|^p_{L^p} + t^{p-q}\,\|g\|^q_{\dot{V}^k_{p,q}} \leq 2L_{p,q}(t^{p-q}, f; L^p, \dot{V}^k_{p,q}).$$

We fix a positive $\varepsilon$, whose precise value will be specified later. If $\|g\|^q_{\dot{V}^k_{p,q}} \geq 2\varepsilon t^q N(t)$, then (11.20) is obvious. So, it suffices to treat the case where

$$\|g\|^q_{\dot{V}^k_{p,q}} < 2\varepsilon t^q N(t). \tag{11.21}$$

Since the quantity $N(t)$ is finite (see (11.6)), there is a packing $\pi$ composed of cubes of the family $\{Q_x\}_{x\in\Omega}$ and containing precisely $N(t)$ cubes. The definition of the space $\dot{V}^k_{p,q}$ shows that

$$\sum_{Q_{x_j}\in\pi}\left(\int_{Q_{x_j}\cap\Omega}\left|g(x) - g_{Q_{x_j}}(x)\right|^p dx\right)^{\frac{q}{p}} \leq \|g\|^q_{\dot{V}^k_{p,q}} < 2\varepsilon t^q N(t). \tag{11.22}$$

We consider the subpacking $\tilde{\pi}$ of $\pi$ that consists of the cubes satisfying

$$\int_{Q_{x_j}\cap\Omega}\left|g(x) - g_{Q_{x_j}}(x)\right|^p dx \geq (4\varepsilon)^{\frac{p}{q}}\, t^p.$$

By (11.22), we have

$$4\varepsilon t^q \sum_{Q_{x_j}\in\tilde{\pi}} \left|Q_{x_j}\right|^0 < 2\varepsilon t^q N(t)$$

and, consequently,

$$\sum_{Q_{x_j}\in\tilde{\pi}} \left|Q_{x_j}\right|^0 < \frac{1}{2}N(t).$$

So, for the cubes in $\pi \setminus \tilde{\pi}$ (there are at least $\frac{1}{2}N(t)$ of them) we have

$$\int_{Q_{x_j}\cap\Omega} \left|g(x) - g_{Q_{x_j}}(x)\right|^p dx \le (4\varepsilon)^{\frac{p}{q}}\, t^p.$$

The triangle inequality implies

$$\begin{aligned}
\|f-g\|_{L^p} &\ge \left(\sum_{Q_{x_j}\in\pi\setminus\tilde{\pi}} \int_{Q_{x_j}\cap\Omega} |f(x)-g(x)|^p\,dx\right)^{\frac{1}{p}} \\
&\ge c_p^{-1}\left(\sum_{Q_{x_j}\in\pi\setminus\tilde{\pi}} \int_{Q_{x_j}\cap\Omega} \left|f(x)-g_{Q_{x_j}}(x)\right|^p dx\right)^{\frac{1}{p}} \\
&\quad - \left(\sum_{Q_{x_j}\in\pi\setminus\tilde{\pi}} \int_{Q_{x_j}\cap\Omega} \left|g(x)-g_{Q_{x_j}}(x)\right|^p dx\right)^{\frac{1}{p}} \\
&\ge c_p^{-1}\left(\sum_{Q_{x_j}\in\pi\setminus\tilde{\pi}} \int_{Q_{x_j}\cap\Omega} \left|f(x)-f_{Q_{x_j}}(x)\right|^p dx\right)^{\frac{1}{p}} \\
&\quad - \left(\sum_{Q_{x_j}\in\pi\setminus\tilde{\pi}} \int_{Q_{x_j}\cap\Omega} \left|g(x)-g_{Q_{x_j}}(x)\right|^p dx\right)^{\frac{1}{p}} \\
&\ge tc_p^{-1}\left(\sum_{Q_{x_i}\in\pi\setminus\tilde{\pi}} |Q_{x_i}|^0\right)^{\frac{1}{p}} - (4\varepsilon)^{\frac{1}{q}}\, t\left(\sum_{Q_{x_i}\in\pi\setminus\tilde{\pi}} |Q_{x_i}|^0\right)^{\frac{1}{p}} \\
&\ge (c_p^{-1} - (4\varepsilon)^{\frac{1}{q}})(\frac{1}{2})^{\frac{1}{p}} tN(t)^{\frac{1}{p}}.
\end{aligned}$$

Consequently, if $\varepsilon > 0$ is chosen in such a way that

$$\left((c_p^{-1} - (4\varepsilon)^{\frac{1}{q}})(\frac{1}{2})^{\frac{1}{p}}\right)^p \ge 2\varepsilon,$$

then

$$\|f-g\|_{L^p}^p \ge 2\varepsilon t^p N(t).$$

Therefore (11.20) is fulfilled. It remains to observe that any sufficiently small positive $\varepsilon$ fits. □

*Remark* 11.9 (Proof of Theorem 11.3). Suppose $f \in \dot{V}^k_{p,q}$, then

$$L_{p,q}(t^{p-q}, f; L^p, \dot{V}^k_{p,q}) \le t^{p-q} \|f\|_{\dot{V}^k_{p,q}} .$$

So from the lemma we have $N(t) \le ct^{-q} \|f\|^q_{\dot{V}^k_{p,q}}$. Since the Sobolev–Poincaré inequality (9.2) for the limit exponent implies the embedding $\dot{V}^k_{p,q_*} \supset \dot{W}^k_{q_*}$, we see that

$$N(t) \le ct^{-q_*} \|f\|^{q_*}_{\dot{W}^k_{q_*}}$$

and, consequently,

$$t \le cN(t)^{-\frac{1}{q_*}} \|f\|_{\dot{W}^k_{q_*}} .$$

Since $\frac{k}{n} + \frac{1}{p} - \frac{1}{q_*} = 0$, we have

$$\begin{aligned}
\|f - f_t\|_{L^p} &\le \left( \sum_{i \in I} \int_{Q_{x_i} \cap \Omega} |f - f_{Q_{x_i}}|^p \right)^{\frac{1}{p}} \le (t^p |I|)^{\frac{1}{p}} \le ctN(t)^{\frac{1}{p}} \\
&\le cN(t)^{\frac{1}{p} - \frac{1}{q_*}} \|f\|_{\dot{W}^k_{q_*}} = cN(t)^{-\frac{k}{n}} \|f\|_{\dot{W}^k_{q_*}} .
\end{aligned}$$

The next lemma implies an upper estimate for the right-hand side of the display in Theorem 11.5.

**Lemma 11.10.** *Let $q < p$, and let $\pi = \{Q\}$ be an arbitrary family of mutually disjoint cubes with centers in $\Omega$ and such that*

$$|\pi| = \sum_{Q \in \pi} |Q|^0 \le N(t) . \tag{11.23}$$

*Then*

$$\sum_{Q \in \pi} \left( \int_{Q \cap \Omega} |f(x) - f_Q(x)|^p \, dx \right)^{\frac{q}{p}} \le c \frac{1}{t^{p-q}} L_{p,q}(t^{p-q}, f; L^p, \dot{V}^k_{p,q}) ,$$

*where $c > 0$ is a constant depending only on the dimension $n$.*

*Proof.* Choose $g \in \dot{V}^k_{p,q}$ in such a way that

$$\|f - g\|^p_{L^p} + t^{p-q} \|g\|^q_{\dot{V}^k_{p,q}} \le 2L_{p,q}(t^{p-q}, f; L^p, \dot{V}^k_{p,q}). \tag{11.24}$$

Since $q < p$, the Hölder inequality shows that

$$\sum_{i=1}^N a_i \le \left( \sum_{i=1}^N a_i^{\frac{p}{q}} \right)^{\frac{q}{p}} N^{1-\frac{q}{p}}, \quad a_i \ge 0 . \tag{11.25}$$

Next, the definition of $\dot{V}^k_{p,q}$ implies (at the last step, we use (11.25) with $a_i = \left(\int_{Q\cap\Omega} |f(x)-g(x)|^p dx\right)^{q/p}$):

$$
\begin{aligned}
&\sum_{Q\in\pi}\left(\int_{Q\cap\Omega} |f(x)-f_Q(x)|^p\,dx\right)^{\frac{q}{p}}\\
&\le \sum_{Q\in\pi}\left(\int_{Q\cap\Omega} |f(x)-g_Q(x)|^p\,dx\right)^{\frac{q}{p}}\\
&\le c\sum_{Q\in\pi}\left(\int_{Q\cap\Omega} |f(x)-g(x)|^p\,dx\right)^{\frac{q}{p}} + c\sum_{Q\in\pi}\left(\int_{Q\cap\Omega} |g(x)-g_Q(x)|^p\,dx\right)^{\frac{q}{p}}\\
&\le c\left(\sum_{Q\in\pi}\int_{Q\cap\Omega} |f(x)-g(x)|^p\,dx\right)^{\frac{q}{p}}\left(\sum_{Q\in\pi}|Q|^0\right)^{1-\frac{q}{p}} + c\,\|g\|^q_{\dot{V}^k_{p,q}}.
\end{aligned}
$$

From this, (11.24), (11.23), and Lemma 11.8, we deduce that

$$
\begin{aligned}
&\sum_{Q\in\pi}\left(\int_{Q\cap\Omega} |f(x)-f_Q(x)|^p\,dx\right)^{\frac{q}{p}}\\
&\le cL_{p,q}(t^{p-q},f;L^p,\dot{V}^k_{p,q})^{\frac{q}{p}}\left(\frac{L_{p,q}(t^{p-q},f;L^p,\dot{V}^k_{p,q})}{t^p}\right)^{1-\frac{q}{p}}\\
&\quad + c\frac{1}{t^{p-q}}L(t^{p-q},f;L^p,\dot{V}^k_{p,q})\\
&\le c\frac{1}{t^{p-q}}L(t^{p-q},f;L^p,\dot{V}^k_{p,q}).
\end{aligned}
$$

□

Lemma (11.10) gives the second inequality in (11.14). The first inequality in (11.14) is a consequence of the following statement.

**Proposition 11.11.** *Let $q<p$. Then*

$$
\|f-f_t\|^p_{L^p} \le ct^{p-q}\sup_{|\pi|\le N(t)}\sum_{Q\in\pi}\left(\int_{Q\cap\Omega} |f(x)-f_Q(x)|^p\,dx\right)^{\frac{q}{p}},
$$

*where $c>0$ depends only on the dimension $n$.*

*Proof.* Since $\int_{Q_{x_i}\cap\Omega} \left|f(x)-f_{Q_{x_i}}(x)\right|^p dx = t^p$ and the family $\{Q_{x_i}\}_{i\in I}$ splits into finitely many packings (the number of which depends only on the dimension), we have

$$\|f - f_t\|_{L^p}^p \leq \sum_{Q_{x_i} \in \{Q_{x_i}\}_{i \in I}} \int_{Q_{x_i} \cap \Omega} |f(x) - f_{Q_{x_i}}(x)|^p \, dx$$

$$\leq \sum_{Q_{x_i} \in \{Q_{x_i}\}_{i \in I}} \left( \int_{Q_{x_i} \cap \Omega} |f(x) - f_{Q_{x_i}}(x)|^p \, dx \right)^{\frac{q}{p}} (t^p)^{1-\frac{q}{p}}$$

$$\leq ct^{p-q} \sup_{|\pi| \leq N(t)} \sum_{Q \in \pi} \left( \int_{Q \cap \Omega} |f(x) - f_Q(x)|^p \, dx \right)^{\frac{q}{p}}. \qquad \square$$

### 11.3.3 Geometric lemmas and the proofs of Theorems 11.4 and 11.5

So (see the subsection "Outline of the proofs"), it remains to verify inequality (11.16). We must show that, for every packing $\tilde{\pi}$ composed of cubes (in fact, squares) with centers in $\Omega$, we have

$$\sum_{Q \in \tilde{\pi}} \left( \int_{Q \cap \Omega} |f_t - (f_t)_Q|^p \right)^{\frac{1}{2}} \leq c \sup_{|\pi| \leq N(t)} \sum_{Q \in \pi} \left( \int_{Q \cap \Omega} |f(x) - f_Q(x)|^p \, dx \right)^{\frac{1}{2}}. \tag{11.26}$$

(Here and below, $c > 0$ denotes various constants depending only on $p$ and the ratio of the side lengths of the rectangle $\Omega$.)

Precisely this will be done in this subsection. First, we split the packing $\tilde{\pi}$ in three parts. To the first part $\pi_1$, we attribute all cubes in $\tilde{\pi}$ that contain at least one vertex of the covering $\{Q_{x_i}\}_{i \in I}$. To the second part $\pi_2$, we attribute all cubes $Q$ in $\tilde{\pi} \setminus \pi_1$ for which there exists a cube $Q_{x_i}$ in the covering such that $Q \cap \Omega \subset Q_{x_i}$. The remaining cubes in $\tilde{\pi}$ are attributed to $\pi_3$.

We start with establishing (11.26) for $\pi_1$. The number of cubes in $\pi_1$ is at most the number of vertices of the cubes in the family $\{Q_{x_i}\}_{i \in I}$. Consequently (see (11.19)), we have

$$|\pi_1| \leq cN(t).$$

So, (11.26) for $\pi_1$ is a consequence of the following lemma, which is valid also in the $n$-dimensional case.

**Lemma 11.12.** *Let $q < p$, and let $\tilde{\pi}$ be an arbitrary family of mutually disjoint cubes with centers in $\Omega$ and such that*

$$|\tilde{\pi}| \leq N(t).$$

*Then*

$$\sum_{Q \in \tilde{\pi}} \left( \int_{Q \cap \Omega} |f_t - (f_t)_Q|^p \right)^{\frac{q}{p}} \leq c \sup_{|\pi| \leq N(t)} \sum_{Q \in \pi} \left( \int_{Q \cap \Omega} |f(x) - f_Q(x)|^p \, dx \right)^{\frac{q}{p}}.$$

*Proof.* The Hölder inequality (11.25) shows that

$$\sum_{Q\in\tilde\pi}\left(\int_{Q\cap\Omega}|f_t(x)-(f_t)_Q(x)|^p\,dx\right)^{\frac{q}{p}} \le \sum_{Q\in\tilde\pi}\left(\int_{Q\cap\Omega}|f_t(x)-f_Q(x)|^p\,dx\right)^{\frac{q}{p}}$$
$$\le c\sum_{Q\in\tilde\pi}\left(\int_{Q\cap\Omega}|f_t(x)-f(x)|^p\,dx\right)^{\frac{q}{p}} + c\sum_{Q\in\tilde\pi}\left(\int_{Q\cap\Omega}|f(x)-f_Q(x)|^p\right)^{\frac{q}{p}}$$
$$\le c\left(\sum_{Q\in\tilde\pi}\int_{Q\cap\Omega}|f(x)-f_t(x)|^p\,dx\right)^{\frac{q}{p}} N(t)^{1-\frac{q}{p}} + c\sum_{Q\in\tilde\pi}\left(\int_{Q\cap\Omega}|f(x)-f_Q(x)|^p\right)^{\frac{q}{p}}$$
$$\le \|f-f_t\|_{L^p}^q\cdot N(t)^{1-\frac{q}{p}} + c\sum_{Q\in\tilde\pi}\left(\int_{Q\cap\Omega}|f(x)-f_Q(x)|^p\right)^{\frac{q}{p}}.$$

Thus, it suffices to estimate the first term on the right. Let $\bar\pi$ be a packing formed by precisely $N(t)$ cubes of the family $\{Q_x\}_{x\in\Omega}$. Then

$$N(t) = \sum_{Q\in\bar\pi}\left(\frac{1}{t^p}\int_{Q\cap\Omega}|f(x)-f_Q(x)|^p\right)^{\frac{q}{p}} \tag{11.27}$$
$$\le \frac{1}{t^q}\sup_{|\pi|\le N(t)}\sum_{Q\in\pi}\left(\int_{Q\cap\Omega}|f(x)-f_Q(x)|^p\,dx\right)^{\frac{q}{p}}.$$

Therefore, Proposition 11.11 implies

$$\|f-f_t\|_{L^p}^q\,N(t)^{1-\frac{q}{p}}$$
$$\le c\left(t^{p-q}\sup_{|\pi|\le N(t)}\sum_{Q\in\pi}\left(\int_{Q\cap\Omega}|f(x)-f_Q(x)|^p\,dx\right)^{\frac{q}{p}}\right)^{\frac{q}{p}}$$
$$\times\left(\frac{1}{t^q}\sup_{|\pi|\le N(t)}\sum_{Q\in\pi}\left(\int_{Q\cap\Omega}|f(x)-f_Q(x)|^p\,dx\right)^{\frac{q}{p}}\right)^{1-\frac{q}{p}}$$
$$\le c\sup_{|\pi|\le N(t)}\sum_{Q\in\pi}\left(\int_{Q\cap\Omega}|f(x)-f_Q(x)|^p\,dx\right)^{\frac{q}{p}}.$$

□

*Remark* 11.13. For $q=\frac{p}{2}$, inequality (11.27) yields

$$N(t)\le\frac{1}{t^{\frac{p}{2}}}\sup_{|\pi|\le N(t)}\sum_{Q\in\pi}\left(\int_{Q\cap\Omega}|f(x)-f_Q(x)|^p\,dx\right)^{\frac{1}{2}}. \tag{11.28}$$

To estimate the summands in (11.26) that correspond to the cubes in $\bar\pi\setminus\pi_1$, we need auxiliary constructions. We remind the reader that the case of $t\ge t_*$ has been treated earlier. Here we only consider the case where $t < t_*$. Therefore, none of the cubes $Q_{x_i}$ contains $\Omega$. Also, it will be used repeatedly that, if $Q$ is a cube centered at a point of $\Omega$, then the ratio of the greater side of the rectangle $Q \cap \Omega$ to its smaller side is dominated by the ratio of the side lengths of $\Omega$ and, therefore, is bounded by a constant independent of $Q$. Consequently, on $Q \cap \Omega$ the $L^p$-norms are equivalent on the space of polynomials of degree strictly smaller than $k$, the equivalence constants depending only on $k$, the ratio of the side lengths of $\Omega$, and the exponents "$p$" involved. In particular, these constants do not depend on $Q$ if $\Omega$ is fixed.

### Auxiliary constructions

Below we only consider the two-dimensional case. Suppose $Q_{x_i} \cap Q_{x_j} \neq \varnothing$, $i \neq j$. Consider the rectangle $Q_{x_i} \cap Q_{x_j}$ and define $Q_{i,j}$ to be the cube centered at the center of this rectangle and of radius equal to $\frac{1}{4}$ the length of the greater side of the rectangle. Next, let $\mu_{i,j}$ be the length of the boundary of the rectangle $Q_{x_i} \cap Q_{x_j}$. Suppose that $j < i$, then $|Q_{x_j}| \geq |Q_{x_i}|$. It is clear from the construction that the cube $Q_{i,j}$ can be split into two congruent rectangles $Q^+_{i,j}$ and $Q^-_{i,j}$ in such a way that

$$Q^-_{i,j} \subset Q_{x_i} \text{ and } Q^+_{i,j} \subset Q_{x_j} \tag{11.29}$$

(the rectangle with the upper index + is included in the greater cube; see Figure 11.2).

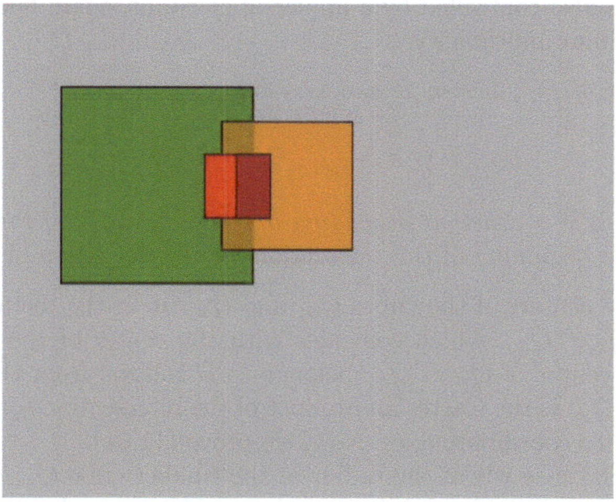

Figure 11.2: The cubes $Q_{x_j}$ and $Q_{x_i}$, and the rectangles $Q^+_{i,j}$ and $Q^-_{i,j}$.

We establish some properties of the family $\{Q_{i,j}\}$. Since $Q_{i,j} = Q_{j,i}$, in the proofs we shall assume that $j < i$.

(a) *The number of cubes in the family* $\{Q_{i,j}\}$ *obeys the estimate*

$$|\{Q_{i,j}\}| \leq cN(t) \,, \tag{11.30}$$

*where* $c$ *is a universal constant (largely, it must depend on the dimension, which however is equal to* 2 *here).*

*Proof.* Indeed, fix a cube $Q_{x_i}$ and consider all cubes $Q_{x_j}$ with $j < i$ that intersect $Q_{x_i}$. Since $j < i$, we have $\left|Q_{x_j}\right| \geq |Q_{x_i}|$, and therefore the cube $Q_{x_j}$ contains at least one vertex of $Q_{x_i}$. Since the family $\{Q_{x_i}\}_{i\in I}$ is of finite multiplicity, it follows that the number of such cubes $Q_{x_j}$ depends only on the dimension, which is 2. Therefore, $|\{Q_{i,j}\}| \leq c\,|I|$, where $c > 0$ is a universal constant. It remains to refer to (11.19). □

(b) *The family* $\{Q_{i,j}\}$ *is of finite multiplicity with a universal multiplicity constant.*

*Proof.* Taking an arbitrary point $x$, we estimate the number of cubes $Q_{i,j}$ with $j < i$ such that $x \in Q_{i,j}$. Since $Q_{i,j} \subset Q_{x_i} \cup Q_{x_j}$, we have $x \in Q_{x_i} \cup Q_{x_j}$. If $x \in Q_{x_i}$, then, much as in (a), for every $i$ we prove that the number of possible $j$'s is controlled by a universal constant. By the finite multiplicity property, the set of $i$'s is finite. But if $x \in Q_{x_j} \setminus Q_{x_i}$, the construction of the cube $Q_{i,j}$ shows that at least one of the four projections of $x$ to the sides of $Q_{x_j}$ is in $Q_{x_i}$. Since the family $\{Q_{x_i}\}_{i\in I}$ is of finite multiplicity, it follows that the number of possible $i$'s is controlled by a universal constant. It remains to observe that the number of possible $j$'s is also controlled by a universal constant, again because the family $\{Q_{x_i}\}_{i\in I}$ is of finite multiplicity. □

(c) *The center of* $Q_{i,j}$ *lies in* $\Omega$ *and we have*

$$\left|Q^{+}_{i,j} \cap \Omega\right|^{\frac{1}{2}} \geq c\mu_{i,j}, \ \left|Q^{-}_{i,j} \cap \Omega\right|^{\frac{1}{2}} \geq c\mu_{i,j} \,, \tag{11.31}$$

*where* $c > 0$ *is a constant depending only on the ratio of the side lengths of* $\Omega$. (See the beginning of this subsection for the definition of $\mu_{i,j}$.)

*Proof.* Since the centers of the cubes $Q_{x_i}$ and $Q_{x_j}$ lie in the rectangle $\Omega$, so does the center of $Q_{x_i} \cap Q_{x_j}$, which coincides with the center of the cube $Q_{i,j}$. The claim that the center of $Q_{x_i} \cap Q_{x_j}$ belongs to $\Omega$ follows from the fact that the rectangle $Q_{x_i} \cap Q_{x_j}$ is the Cartesian product of the intersections of the projections of $Q_{x_i}$ and $Q_{x_j}$ to coordinate axes. Now, we prove (11.31).

Consider the case where the radius of the smaller cube $Q_{x_i}$ does not exceed the length of the smaller side of $\Omega$. Geometric considerations show that the volumes $\left|Q^{+}_{i,j} \cap \Omega\right|$ and $\left|Q^{-}_{i,j} \cap \Omega\right|$ are at least $\frac{1}{4}$ times the volume of $Q_{i,j}$. Since the radius of $Q_{i,j}$ is $\frac{1}{4}$ times the length of the greater side of the rectangle $Q_{x_i} \cap Q_{x_j}$, we arrive at (11.31). If the radii of $Q_{x_i}$ and $Q_{x_j}$ are greater than the length of the smaller side of $\Omega$, we observe that, for $t < t_*$ (we only consider this case), the cubes $Q_{x_i}$ and $Q_{x_j}$ do not contain $\Omega$, and all the quantities in (11.31) are proportional to the length of the smaller side of $\Omega$. The corresponding ratios depend only on the ratio of the side lengths of $\Omega$. $\square$

Below, we shall need the following statement.

**Lemma 11.14.** *We have*

$$\sum_{i,j\in I} \left\| f_{Q_{x_j}} - f_{Q_{x_i}} \right\|_{L^\infty(Q_{x_j}\cap Q_{x_i})}^{\frac{p}{2}} \cdot \mu_{i,j} \tag{11.32}$$
$$\leq c \sup_{|\pi|\leq N(t)} \sum_{Q\in\pi} \left( \int_{Q\cap\Omega} |f(x) - f_Q(x)|^p \, dx \right)^{\frac{1}{2}},$$

*where $c > 0$ is a universal constant.*

*Proof.* If $Q_{x_j} \cap Q_{x_i} \neq \varnothing$, then $Q_{i,j} \neq \varnothing$ and therefore

$$\sum_{i,j\in I} \left\| f_{Q_{x_j}} - f_{Q_{x_i}} \right\|_{L^\infty(Q_{x_j}\cap Q_{x_i})}^{\frac{p}{2}} \cdot \mu_{i,j} = \sum_{Q_{i,j}\in\{Q_{i,j}\}} \left\| f_{Q_{x_j}} - f_{Q_{x_i}} \right\|_{L^\infty(Q_{x_j}\cap Q_{x_i})}^{\frac{p}{2}} \cdot \mu_{i,j}.$$

Since the multiplicity of the family $Q_{i,j}$ is controlled by a universal constant, this family splits into packings whose number is also controlled by a universal constant. Therefore, it suffices to prove (11.32) for each packing. Let $\bar{\pi} = \{Q_{i,j}\}$ be any of them. Consider the auxiliary function

$$g = \sum_{Q_{i,j}\in\bar{\pi}} (f_{Q_{x_j}} \chi_{Q_{i,j}^+} + f_{Q_{x_i}} \chi_{Q_{i,j}^-}).$$

To prove (11.32), it suffices to show that

$$c \sum_{Q_{i,j}\in\bar{\pi}} \left\| f_{Q_{x_j}} - f_{Q_{x_i}} \right\|_{L^\infty(Q_{x_j}\cap Q_{x_i})}^{\frac{p}{2}} \cdot \mu_{i,j} \leq \sum_{Q_{i,j}\in\bar{\pi}} \left( \int_{Q_{i,j}\cap\Omega} \left| g(x) - g_{Q_{i,j}}(x) \right|^p dx \right)^{\frac{1}{2}} \tag{11.33}$$

and

$$\sum_{Q_{i,j}\in\bar{\pi}} \left( \int_{Q_{i,j}\cap\Omega} \left| g(x) - g_{Q_{i,j}}(x) \right|^p dx \right)^{\frac{1}{2}}$$
$$\leq c \sup_{|\pi|\leq N(t)} \sum_{Q\in\pi} \left( \int_{Q\cap\Omega} |f(x) - f_Q(x)|^p \, dx \right)^{\frac{1}{2}}.$$

We start with the last inequality. We have

$$\begin{aligned}
&\sum_{Q_{i,j}\in\bar\pi}\left(\int_{Q_{i,j}\cap\Omega}|g(x)-g_{Q_{i,j}}(x)|^p\,dx\right)^{\frac12}\\
&\le\sum_{Q_{i,j}\in\bar\pi}\left(\int_{Q_{i,j}\cap\Omega}|g(x)-f_{Q_{i,j}}(x)|^p\,dx\right)^{\frac12}\\
&\le c\sum_{Q_{i,j}\in\bar\pi}\left(\int_{Q_{i,j}\cap\Omega}|g(x)-f(x)|^p\,dx\right)^{\frac12}\\
&\quad+c\sum_{Q_{i,j}\in\bar\pi}\left(\int_{Q_{i,j}\cap\Omega}|f(x)-f_{Q_{i,j}}(x)|^p\,dx\right)^{\frac12}.
\end{aligned}$$

Since by (a) the cardinality of $\bar\pi$ does not exceed $cN(t)$, the second summand is estimated in an obvious way. The first summand is controlled with the help of (11.29), (11.28), and the following inequalities:

$$\begin{aligned}
&\sum_{Q_{i,j}\in\bar\pi}\left(\int_{Q_{i,j}\cap\Omega}|g(x)-f(x)|^p\,dx\right)^{\frac12} \qquad (11.34)\\
&\le\sum_{Q_{i,j}\in\bar\pi}\left(\int_{Q^+_{i,j}\cap\Omega}\left|f_{Q_{x_j}}(x)-f(x)\right|^p\,dx\right)^{\frac12}\\
&\quad+\sum_{Q_{i,j}\in\bar\pi}\left(\int_{Q^-_{i,j}\cap\Omega}\left|f_{Q_{x_i}}(x)-f(x)\right|^p\,dx\right)^{\frac12}\\
&\le 2t^{\frac p2}\,|\bar\pi|\le ct^{\frac p2}N(t)\le c\sup_{|\pi|\le N(t)}\sum_{Q\in\pi}\left(\int_{Q\cap\Omega}|f(x)-f_Q(x)|^p\,dx\right)^{\frac12}.
\end{aligned}$$

In order to prove (11.33), it suffices to show that

$$\left(\int_{Q_{i,j}\cap\Omega}|g(x)-g_{Q_{i,j}}(x)|^p\,dx\right)^{\frac12}\ge c\left\|f_{Q_{x_j}}-f_{Q_{x_i}}\right\|^{\frac p2}_{L^\infty(Q_{x_j}\cap Q_{x_i})}\cdot\mu_{i,j}.$$

Since $g_{Q_{i,j}}(x)$, $f_{Q_j}(x)$, and $f_{Q_i}(x)$ are polynomials of degree strictly smaller than $k$, we use the equivalence of various norms on the space of polynomials on rectangles with controlled ratio of side lengths, (11.31), and Theorem 0.22 on the growth of polynomials to conclude that

$$\begin{aligned}
&\int_{Q_{i,j}\cap\Omega} \left|g(x) - g_{Q_{i,j}}(x)\right|^p dx \\
&= \int_{Q^+_{i,j}\cap\Omega} \left|f_{Q_{x_j}}(x) - g_{Q_{i,j}}(x)\right|^p dx + \int_{Q^-_{i,j}\cap\Omega} \left|f_{Q_{x_i}}(x) - g_{Q_{i,j}}(x)\right|^p dx \\
&\geq c\left\|f_{Q_{x_j}} - g_{Q_{i,j}}\right\|^p_{L^\infty(Q^+_{i,j}\cap\Omega)} \cdot \mu^2_{i,j} + c\left\|f_{Q_{x_i}} - g_{Q_{i,j}}\right\|^p_{L^\infty(Q^-_{i,j}\cap\Omega)} \cdot \mu^2_{i,j} \\
&\geq c\left\|f_{Q_{x_j}} - g_{Q_{i,j}}\right\|^p_{L^\infty(Q_{i,j}\cap\Omega)} \cdot \mu^2_{i,j} + c\left\|f_{Q_{x_i}} - g_{Q_{i,j}}\right\|^p_{L^\infty(Q_{i,j}\cap\Omega)} \cdot \mu^2_{i,j} \\
&\geq c\left\|f_{Q_{x_j}} - f_{Q_{x_i}}\right\|^p_{L^\infty(Q_{i,j}\cap\Omega)} \cdot \mu^2_{i,j}.
\end{aligned}$$

It remains to observe that the intersection of the set $Q_{i,j} \cap \Omega$ with $Q_{x_i} \cap Q_{x_j}$ has measure of at least $c\left|Q_{x_i} \cap Q_{x_j}\right|$ with a constant $c > 0$ depending only on the ratio of the side lengths of $\Omega$. Therefore, the theorem on the growth of polynomials implies $\left\|f_{Q_{x_j}} - f_{Q_{x_i}}\right\|^p_{L^\infty(Q_{i,j}\cap\Omega)} \geq c\left\|f_{Q_{x_j}} - f_{Q_{x_i}}\right\|^p_{L^\infty(Q_i\cap Q_j)}$. □

*Remark* 11.15. In the course of the proof (see (11.34)), we have established the inequality

$$\begin{aligned}
\sum_{Q_{x_i}\cap Q_{x_j}\neq\varnothing} &\left[\left(\int_{Q^+_{i,j}\cap\Omega} \left|f_{Q_{x_j}}(x) - f_{Q_{i,j}}(x)\right|^p dx\right)^{\frac{1}{2}}\right. \qquad (11.35)\\
&\left.+\left(\int_{Q^-_{i,j}\cap\Omega} \left|f_{Q_{x_i}}(x) - f_{Q_{i,j}}(x)\right|^p dx\right)^{\frac{1}{2}}\right] \\
&\leq c \sup_{|\pi|\leq N(t)} \sum_{Q\in\pi} \left(\int_{Q\cap\Omega} |f(x) - f_Q(x)|^p dx\right)^{\frac{1}{2}}.
\end{aligned}$$

**Proof of** (11.26) **for** $\pi_2$

Suppose $Q \in \pi_2$, and let $i \in I$ be the smallest index with $Q \cap \Omega \subset Q_{x_i}$. The set of cubes $Q \in \pi_2$ corresponding to one and the same $i$ will be denoted by $\pi_{2,i}$. If a cube $Q \in \pi_{2,i}$ does not intersect any of the $Q_{x_j}$ with $j < i$, the function $f_t$ is equal to $f_{Q_{x_i}}$ on $Q \cap \Omega$, and therefore

$$\int_{Q\cap\Omega} |f_t(x) - (f_t)_Q(x)|^p dx = 0.$$

Now, we consider the case where $Q \in \pi_{2,i}$ intersects some cubes $Q_{x_j}$ with $j < i$. Since $Q \cap \Omega \subset Q_{x_i}$, in this case we have

$$\begin{aligned}\int_{Q\cap\Omega} |f_t(x) - (f_t)_Q(x)|^p \, dx &\le \int_{Q\cap\Omega} \left|f_t(x) - f_{Q_{x_i}}(x)\right|^p dx \\ &\le \sum_{j<i} \int_{Q_{x_j}\cap Q\cap\Omega} \left|f_{Q_{x_j}}(x) - f_{Q_{x_i}}(x)\right|^p dx \\ &\le \sum_{j<i, Q_{x_j}\cap Q\neq\varnothing} \left\|f_{Q_{x_j}} - f_{Q_{x_i}}\right\|^p_{L^\infty(Q_{x_j}\cap Q_{x_i})} |Q| .\end{aligned}$$

Therefore, for the summands in (11.26) that correspond to cubes $Q \in \pi_{2,i}$, we have

$$\sum_{Q\in\pi_{2,i}} \left(\int_{Q\cap\Omega} |f_t(x) - (f_t)_Q(x)|^p \, dx\right)^{\frac{1}{2}} \tag{11.36}$$

$$\le \sum_{Q\in\pi_{2,i}} \left( \sum_{j<i, Q_{x_j}\cap Q\neq\varnothing} \left\|f_{Q_{x_j}} - f_{Q_{x_i}}\right\|^{\frac{p}{2}}_{L^\infty(Q_{x_j}\cap Q_{x_i})} |Q|^{\frac{1}{2}} \right)$$

$$\le \sum_{j<i} \left( \left\|f_{Q_{x_j}} - f_{Q_{x_i}}\right\|^{\frac{p}{2}}_{L^\infty(Q_{x_j}\cap Q_{x_i})} \cdot \sum_{Q\in\pi_{2,i}; Q_{x_j}\cap Q\neq\varnothing} |Q|^{\frac{1}{2}} \right) .$$

**Lemma 11.16.** *For fixed $j < i$, we have*

$$\sum_{Q\in\pi_{2,i}; Q_{x_j}\cap Q\neq\varnothing} |Q|^{\frac{1}{2}} \le 2\mu_{i,j} . \tag{11.37}$$

*Proof.* The argument below is illustrated by Figure 11.3. Recall that the cube $Q \in \pi_{2,i}$ contains no vertices of $Q_{x_j}$. Moreover, the rectangle $Q \cap \Omega$ is included in $Q_{x_i}$ and is not included in $Q_{x_j}$. Consequently, $Q$ intersects $Q_{x_j}$ by a side, and the length of this intersection is equal to $|Q|^{\frac{1}{2}}$. Since $Q_{x_i}$ includes $Q \cap \Omega$, the length of the intersection of $Q$ with the boundary of the rectangle $Q_{x_i} \cap Q_{x_j}$ is at least $\frac{1}{2}|Q|^{\frac{1}{2}}$. Since the cubes in $\pi_{2,i}$ are disjoint, we obtain

$$\sum_{Q\in\pi_{2,i}; Q_{x_j}\cap Q\neq\varnothing} \frac{1}{2} |Q|^{\frac{1}{2}} \le \mu_{i,j},$$

which is (11.37). □

Now, we estimate the right-hand side of (11.36) with the help of this lemma, and then sum over all $i \in I$, obtaining

$$\sum_{Q\in\pi_2} \left(\int_{Q\cap\Omega} |f_t(x) - (f_t)(x)|^p \, dx\right)^{\frac{1}{2}} \le 2\sum_{j<i} \left\|f_{Q_{x_j}} - f_{Q_{x_i}}\right\|^{\frac{p}{2}}_{L^\infty(Q_{x_j}\cap Q_{x_i})} \cdot \mu_{i,j}. \tag{11.38}$$

So, Lemma 11.14 implies (11.26) for the packing $\pi_2$.

To complete the proof, it remains to verify (11.26) for $\pi_3$.

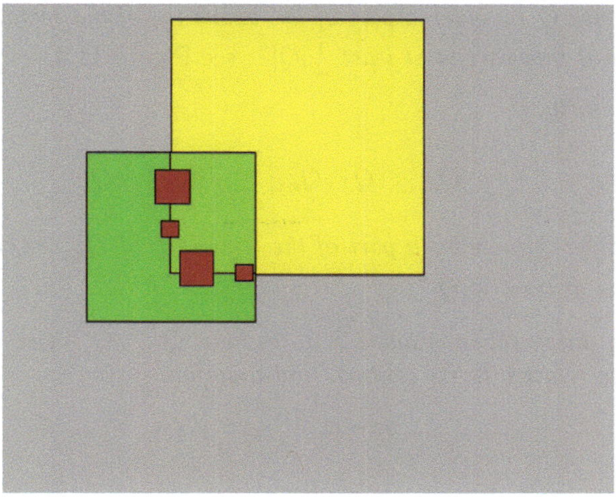

Figure 11.3: Small cubes are in $\pi_2$.

**Proof of inequality** (11.26) **for the packing** $\pi_3$

First, we claim that, whenever $Q \in \pi_3$, there exist two cubes $Q_{x_i}$ and $Q_{x_j}$ in the covering $\{Q_{x_i}\}_{i\in I}$ such that

$$Q \subset Q_{x_i} \cup Q_{x_j} \ . \tag{11.39}$$

Surely, the indices $i = i(Q)$ and $j = j(Q)$ depend on $Q$. To prove the claim, consider all cubes $Q_{x_m} \in \{Q_{x_i}\}_{i\in I}$ satisfying $Q_{x_m} \cap Q \neq \varnothing$. Since $Q$ does not belong to $\pi_1$ or $\pi_2$, it is easily seen that $Q_{x_m}$ contains precisely one side of $Q$. We split all $Q_{x_m}$ into four groups, depending on which particular side of $Q$ is included in $Q_{x_m}$. In each group, we take a cube having the largest intersection with $Q$. It is clear from the geometry that these four cubes cover $Q$. Moreover, $Q$ is covered by those two of these cubes which have nonempty intersection and include opposite sides of $Q$. These two cubes are the required cubes $Q_{x_i}$ and $Q_{x_j}$. There is no loss of generality in assuming that $j < i$. Since the intersection of these cubes is nonempty, the cube $Q_{i,j}$ is well defined.

We need three properties of the cubes $Q_{x_i}$ and $Q_{x_j}$ constructed above; these properties follow from geometric considerations.

(a) *The cube* $Q$ *includes a part of the boundary of the rectangle* $Q_{x_i} \cap Q_{x_j}$*; the length of this part is at least* $|Q|^{\frac{1}{2}}$.

(b) *Let* $m$ *be different from* $i$ *and* $j$*, and let* $Q_{x_m} \cap Q \neq \varnothing$*; if*

$$|Q_{x_m} \cap Q \cap Q_{x_i}| \geq \frac{1}{2}\,|Q_{x_m} \cap Q|\,,$$

*then the cube* $Q$ *includes a part of the boundary of the rectangle* $Q_{x_i} \cap Q_{x_m}$; *the length of this part is at least* $\frac{1}{2}|Q|^{\frac{1}{2}}$; see Figure 11.4.

*Similarly, if*

$$\left|Q_{x_m} \cap Q \cap Q_{x_j}\right| \geq \frac{1}{2}\left|Q_{x_m} \cap Q\right|,$$

*then the cube* $Q$ *includes a part of the boundary of* $Q_{x_j} \cap Q_{x_m}$; *the length of this part is at least* $\frac{1}{2}|Q|^{\frac{1}{2}}$.

(c) *For some universal constant* $c > 0$, *we have* $Q \subset cQ^{-}_{i,j}$ *(we mean dilation of a rectangle relative to its center), and therefore*

$$Q \cap \Omega \subset (cQ^{-}_{i,j} \cap \Omega).$$

Figure 11.4: The smallest cube is $Q$. The other are $Q_{x_i}$, $Q_{x_j}$, and $Q_{x_m}$.

We pass to the proof of the required estimate. Below, $Q \in \pi_3$ and $i$, $j$ are chosen in such a way that (11.39) is fulfilled. We denote by $I_Q$ the set of $m$ with $[(Q_{x_m} \cap \Omega) \setminus (\cup_{l<m} Q_{x_l})] \cap Q \neq \varnothing$. This condition means that, on a part of $Q$, the function $f_t$ (see (11.8)) is equal to the polynomial $f_{Q_{x_m}}$. By (11.39) and Proposition 11.1, we see that

$$|I_Q| \leq c, \tag{11.40}$$

where $c > 0$ is a universal constant. The construction of $f_t$ (see (11.8)) shows that

$$\left(\int_{Q\cap\Omega} |f_t(x) - (f_t)_Q(x)|^p \, dx\right)^{\frac{1}{2}} \le \left(\int_{Q\cap\Omega} \left|f_t(x) - f_{Q_{i,j}}(x)\right|^p dx\right)^{\frac{1}{2}} \tag{11.41}$$
$$\le \sum_{m\in I_Q} \left(\int_{(Q\cap\Omega)\cap Q_{x_m}} \left|f_{Q_{x_m}}(x) - f_{Q_{i,j}}(x)\right|^p dx\right)^{\frac{1}{2}}.$$

Clearly, at least one of the cubes $Q_{x_i}$ and $Q_{x_j}$ includes at least half of the area of $(Q \cap \Omega) \cap Q_{x_m}$. We put $\varepsilon^Q_{i,m}$ to be equal to 1 if $Q_{x_i}$ includes at least half of the area of $(Q \cap \Omega) \cap Q_{x_m}$ and to 0 otherwise ($\varepsilon^Q_{j,m}$ is defined similarly). Therefore, the one-dimensional version of Theorem 0.22 (about the growth of polynomials) shows that

$$\begin{aligned}
&\left(\int_{(Q\cap\Omega)\cap Q_{x_m}} \left|f_{Q_{x_m}}(x) - f_{Q_{i,j}}(x)\right|^p dx\right)^{\frac{1}{2}} \\
&\le c\varepsilon^Q_{i,m} \left(\int_{(Q\cap\Omega)\cap Q_{x_m}\cap Q_{x_i}} \left|f_{Q_{x_m}}(x) - f_{Q_{i,j}}(x)\right|^p dx\right)^{\frac{1}{2}} \\
&\quad + c\varepsilon^Q_{j,m} \left(\int_{(Q\cap\Omega)\cap Q_{x_m}\cap Q_{x_j}} \left|f_{Q_{x_m}}(x) - f_{Q_{i,j}}(x)\right|^p dx\right)^{\frac{1}{2}} \\
&\le c\varepsilon^Q_{i,m} \left(\int_{(Q\cap\Omega)\cap Q_{x_m}\cap Q_{x_i}} \left|f_{Q_{x_m}}(x) - f_{Q_{x_i}}(x)\right|^p dx\right)^{\frac{1}{2}} \qquad (11.42)\\
&\quad + c\varepsilon^Q_{i,m} \left(\int_{(Q\cap\Omega)\cap Q_{x_m}\cap Q_{x_i}} \left|f_{Q_{x_i}}(x) - f_{Q_{i,j}}(x)\right|^p dx\right)^{\frac{1}{2}} \\
&\quad + c\varepsilon^Q_{j,m} \left(\int_{(Q\cap\Omega)\cap Q_{x_m}\cap Q_{x_j}} \left|f_{Q_{x_m}}(x) - f_{Q_{x_j}}(x)\right|^p dx\right)^{\frac{1}{2}} \\
&\quad + c\varepsilon^Q_{j,m} \left(\int_{(Q\cap\Omega)\cap Q_{x_m}\cap Q_{x_j}} \left|f_{Q_{x_j}}(x) - f_{Q_{i,j}}(x)\right|^p dx\right)^{\frac{1}{2}}.
\end{aligned}$$

We estimate the terms on the right. Let $\mu_{i,m}(Q)$ denote the length of the boundary of the rectangle $Q_{x_m} \cap Q_{x_i}$, which is included in $Q$. By (b) and the definition of the numbers $\varepsilon^Q_{i,m}$, we have

$$\begin{aligned}
&\varepsilon^Q_{i,m} \left(\int_{(Q\cap\Omega)\cap Q_{x_m}\cap Q_{x_i}} \left|f_{Q_{x_m}}(x) - f_{Q_{x_i}}(x)\right|^p dx\right)^{\frac{1}{2}} \qquad (11.43)\\
&\le c\varepsilon^Q_{i,m} \left\|f_{Q_{x_m}} - f_{Q_{x_i}}\right\|^{\frac{p}{2}}_{L^\infty(Q_{x_m}\cap Q_{x_i})} |Q|^{\frac{1}{2}} \\
&\le c\left\|f_{Q_{x_m}} - f_{Q_{x_i}}\right\|^{\frac{p}{2}}_{L^\infty(Q_{x_m}\cap Q_{x_i})} \cdot \mu_{i,m}(Q)
\end{aligned}$$

and similarly

$$\begin{aligned}
&\varepsilon^Q_{j,m}\left(\int_{(Q\cap\Omega)\cap Q_{x_m}\cap Q_{x_j}}\left|f_{Q_{x_m}}(x)-f_{Q_{x_j}}(x)\right|^p dx\right)^{\frac{1}{2}} \qquad (11.44)\\
&\le c\varepsilon^Q_{j,m}\left\|f_{Q_{x_m}}-f_{Q_{x_j}}\right\|^{\frac{p}{2}}_{L^\infty(Q_{x_m}\cap Q_{xj})}|Q|^{\frac{1}{2}}\\
&\le c\left\|f_{Q_{x_m}}-f_{Q_{x_j}}\right\|^{\frac{p}{2}}_{L^\infty(Q_{x_m}\cap Q_{x_i})}\cdot\mu_{j,m}(Q).
\end{aligned}$$

When estimating the second and fourth summands, we employ property (c), the equivalence of norms on the space of polynomials of degree strictly smaller than $k$ on rectangles with controlled ratio of the side lengths, and the theorem on the growth of polynomials:

$$\begin{aligned}
&\left(\int_{Q\cap\Omega}|f_{Q_{x_i}}(x)-f_{Q_{i,j}}(x)|^p dx\right)^{\frac{1}{2}}\le c\left\|f_{Q_{x_i}}-f_{Q_{i,j}}\right\|^{\frac{p}{2}}_{L^\infty(Q\cap\Omega)}|Q|^{\frac{1}{2}} \qquad (11.45)\\
&\le c\left\|f_{Q_{x_i}}-f_{Q_{i,j}}\right\|^{\frac{p}{2}}_{L^\infty(Q^-_{i,j}\cap\Omega)}|Q|^{\frac{1}{2}}\\
&\le c\frac{|Q|^{\frac{1}{2}}}{|Q_{i,j}\cap\Omega|^{\frac{1}{2}}}\left(\int_{Q^-_{i,j}\cap\Omega}|f_{Q_{x_i}}(x)-f_{Q_{i,j}}(x)|^p dx\right)^{\frac{1}{2}}\\
&\le c\frac{|Q|^{\frac{1}{2}}}{|Q_{i,j}|^{\frac{1}{2}}}\left(\int_{Q^-_{i,j}\cap\Omega}|f_{Q_{x_i}}(x)-f_{Q_{i,j}}(x)|^p dx\right)^{\frac{1}{2}}.
\end{aligned}$$

In the last inequality, we have used the estimate $|Q_{i,j}|\le c\,|Q_{i,j}\cap\Omega|$ with a constant independent of the ratio of side lengths of $\Omega$.

Similarly,

$$\begin{aligned}
&\left(\int_{Q\cap\Omega}\left|f_{Q_{x_j}}(x)-f_{Q_{i,j}}(x)\right|^p dx\right)^{\frac{1}{2}}\le c\left\|f_{Q_{x_j}}-f_{Q_{i,j}}\right\|^{\frac{p}{2}}_{L^\infty(Q\cap\Omega)}|Q|^{\frac{1}{2}} \qquad (11.46)\\
&\le c\left\|f_{Q_{x_j}}-f_{Q_{i,j}}\right\|^{\frac{p}{2}}_{L^\infty(Q^+_{i,j}\cap\Omega)}|Q|^{\frac{1}{2}}\\
&\le c\frac{|Q|^{\frac{1}{2}}}{|Q_{i,j}|^{\frac{1}{2}}}\left(\int_{Q^+_{i,j}\cap\Omega}\left|f_{Q_{x_j}}(x)-f_{Q_{i,j}}(x)\right|^p dx\right)^{\frac{1}{2}}.
\end{aligned}$$

Thus, taking (11.40) into account, we obtain

$$\sum_{m\in I_Q}\left(\int_{(Q\cap\Omega)\cap Q_{x_m}}|f_{Q_{x_m}}(x)-f_{Q_{i,j}}(x)|^p dx\right)^{\frac{1}{2}} \qquad (11.47)$$

$$\leq c \sum_{m\in I_Q} \left\|f_{Q_{x_m}} - f_{Q_{x_i}}\right\|^{\frac{p}{2}}_{L^\infty(Q_{x_m}\cap Q_{x_i})} \cdot \mu_{i,m}(Q)$$
$$+ c \sum_{m\in I_Q} \left\|f_{Q_{x_m}} - f_{Q_{x_j}}\right\|^{\frac{p}{2}}_{L^\infty(Q_{x_m}\cap Q_{x_j})} \cdot \mu_{j,m}(Q)$$
$$+ c\frac{|Q|^{\frac{1}{2}}}{|Q_{i,j}|^{\frac{1}{2}}}\left(\int_{Q^-_{i,j}\cap\Omega} \left|f_{Q_{x_i}}(x) - f_{Q_{i,j}}(x)\right|^p dx\right)^{\frac{1}{2}}$$
$$+ c\frac{|Q|^{\frac{1}{2}}}{|Q_{i,j}|^{\frac{1}{2}}}\left(\int_{Q^+_{i,j}\cap\Omega} \left|f_{Q_{x_j}}(x) - f_{Q_{i,j}}(x)\right|^p dx\right)^{\frac{1}{2}}.$$

We denote by $\pi_{i,j}$ the collection of cubes $Q \in \pi_3$ for which $Q \subset Q_{x_i} \cup Q_{x_j}$. Property (a) shows that

$$\sum_{Q\in\pi_{i,j}} |Q|^{\frac{1}{2}} \leq \mu_{i,j} \leq c\,|Q_{i,j}|^{\frac{1}{2}} .$$

Thus, by (11.41) and (11.47), we have

$$\sum_{Q\in\pi_{i,j}} \left(\int_{Q\cap\Omega} |f_t(x) - (f_t)_Q(x)|^p\,dx\right)^{\frac{1}{2}}$$
$$\leq c \sum_{Q\in\pi_{i,j}} \sum_{m\in I} \left\|f_{Q_{x_m}} - f_{Q_{x_i}}\right\|^{\frac{p}{2}}_{L^\infty(Q_{x_m}\cap Q_{x_i})} \cdot \mu_{i,m}(Q)$$
$$+ c \sum_{Q\in\pi_{i,j}} \sum_{m\in I} \left\|f_{Q_{x_m}} - f_{Q_{x_j}}\right\|^{\frac{p}{2}}_{L^\infty(Q_{x_m}\cap Q_{x_j})} \cdot \mu_{j,m}(Q)$$
$$+ c\left(\int_{Q^-_{i,j}\cap\Omega} \left|f_{Q_{x_i}}(x) - f_{Q_{i,j}}(x)\right|^p dx\right)^{\frac{1}{2}}$$
$$+ c\left(\int_{Q^+_{i,j}\cap\Omega} \left|f_{Q_{x_j}}(x) - f_{Q_{i,j}}(x)\right|^p dx\right)^{\frac{1}{2}}.$$

Summing over all packings $\pi_{i,j}$ and observing that

$$\sum_{Q\in\pi_3} \mu_{l,m}(Q) \leq \mu_{l,m},$$

we arrive at

$$\sum_{Q\in\pi_3} \left(\int_{Q\cap\Omega} |f_t(x) - (f_t)_Q(x)|^p\,dx\right)^{\frac{1}{2}}$$

$$\leq c \sum_{l,m \in I} \left\| f_{Q_{x_m}} - f_{Q_{x_l}} \right\|^{\frac{p}{2}}_{L^\infty(Q_{x_m} \cap Q_{x_l})} \cdot \mu_{m,l}$$

$$+ c \sum_{l<m} \left( \int_{Q^-_{m,l} \cap \Omega} \left| f_{Q_{x_m}}(x) - f_{Q_{m,l}}(x) \right|^p dx \right)^{\frac{1}{2}}$$

$$+ c \sum_{l<m} \left( \int_{Q^+_{m,l} \cap \Omega} \left| f_{Q_{x_l}}(x) - f_{Q_{m,l}}(x) \right|^p dx \right)^{\frac{1}{2}}.$$

The first summand on the right was estimated in Lemma 11.14, and the second and third summands were estimated in (11.35).

### The case of $k = 1$ and spaces of functions of bounded $p$-variation

As before, in this subsection we assume that $\Omega = [a, b] \times [c, d]$ is a rectangular domain on the plane. Consider the spaces $L^p = L^p(\Omega)$ and $\mathrm{BV}_q$. The latter is the space of all functions on $\Omega$ for which the following seminorm is finite:

$$\|f\|_{\mathrm{BV}_q} = \left( \int_a^b \mathrm{Var}_{q,y} f(x, y) dx + \int_c^d \mathrm{Var}_{q,x} f(x, y) dx \right)^{\frac{1}{q}},$$

here

$$\mathrm{Var}_{q,y} f(x, y) = \sup_{c \leq y_0 < y_1 < \ldots < y_m \leq d} \sum_{i=1}^{m} |f(x, y_i) - f(x, y_{i-1})|^q$$

and $\mathrm{Var}_{q,x} f(x, y)$ is defined similarly. For $q = 1$, the space $\mathrm{BV}_q$ is none other than the well-known space BV of functions of bounded variation.[1]

Since for smooth functions we have

$$\mathrm{Var}_{1,y} f(x, y) \leq \int_0^1 \left| \frac{\partial}{\partial y} f(x, y) \right| dy \text{ and } \mathrm{Var}_{1,x} f(x, y) \leq \int_0^1 \left| \frac{\partial}{\partial x} f(x, y) \right| dx,$$

it follows that $\|f\|_{\mathrm{BV}} \leq \|f\|_{\dot{W}^1_1}$ and, consequently, $\dot{W}^1_1 \subset \mathrm{BV}$.

In many issues, BV is a good substitute for $\dot{W}^1_1$. Below, $f_Q$ denotes the constant that provides the best approximation to $f$ on $Q \cap \Omega$ in the metric of $L^p$.

**Proposition 11.17.** *Let $f \in \mathrm{BV}_{p/2}$. Then for every square $Q$ with center in $\Omega$ we have*

$$\left( \int_{Q \cap \Omega} |f - f_Q|^p \right)^{\frac{1}{p}} \leq c \|f\|_{\mathrm{BV}_{p/2}(Q \cap \Omega)} \tag{11.48}$$

*with a constant $c > 0$ independent of $f$ and $Q$. It should be noted that $c$ depends on the ratio of the side lengths of $\Omega$.*

[1] In the case of dimension 1, the spaces of functions of bounded $q$-variation with $1 < q < \infty$ have some peculiar properties as Banach spaces; see, e.g., [Kis0].

*Proof.* Let the rectangle $Q \cap \Omega$ have the form $[A, B] \times [C, D]$. It is easy to choose points $x_0 \in [A, B]$ and $y_0 \in [C, D]$ in such a way that the quantities $\operatorname{Var}_{p/2,y} f(x_0, y)$ and $\operatorname{Var}_{p/2,x} f(x, y_0)$ are well defined, and

$$\operatorname{Var}_{p/2,y} f(x_0, y) \le \frac{1}{B-A} \int_A^B \operatorname{Var}_{p/2,y} f(x, y)dx,$$

$$\operatorname{Var}_{p/2,x} f(x, y_0) \le \frac{1}{D-C} \int_C^D \operatorname{Var}_{p/2,x} f(x, y)dy.$$

Then

$$\begin{aligned}
&\int_{Q\cap\Omega} |f(x,y) - f_Q|^p \, dxdy \le \int_{Q\cap\Omega} \left(|f(x,y) - f(x_0,y_0)|^{\frac{p}{2}}\right)^2 dxdy \\
&\le c \int\int_{Q\cap\Omega} (\operatorname{Var}_{p/2,x} f(x,y_0) + \operatorname{Var}_{p/2,y} f(x,y)) \cdot (\operatorname{Var}_{p/2,y} f(x_0,y) \\
&\qquad + \operatorname{Var}_{p/2,x} f(x,y))dxdy \\
&\le c \int\int_{Q\cap\Omega} \operatorname{Var}_{p/2,x} f(x,y_0) \operatorname{Var}_{p/2,y} f(x_0,y)dxdy \\
&\quad + c \int\int_{Q\cap\Omega} (\operatorname{Var}_{p/2,x} f(x,y_0) \operatorname{Var}_{p/2,x} f(x,y) \\
&\qquad + \operatorname{Var}_{p/2,y} f(x,y) \operatorname{Var}_{p/2,y} f(x_0,y))dxdy \\
&\quad + c \int\int_{Q\cap\Omega} \operatorname{Var}_{p/2,y} f(x,y) \operatorname{Var}_{p/2,x} f(x,y)dxdy \; .
\end{aligned}$$

Clearly,

$$\begin{aligned}
&\int\int_{Q\cap\Omega} \operatorname{Var}_{p/2,x} f(x,y_0) \operatorname{Var}_{p/2,y} f(x_0,y)dxdy \\
&\le \int_A^B \operatorname{Var}_{p/2,y} f(x,y)dx \cdot \int_C^D \operatorname{Var}_{p/2,x} f(x,y)dy \le c\,\|f\|^p_{\mathrm{BV}_{p/2}(Q\cap\Omega)}
\end{aligned}$$

and

$$\begin{aligned}
&\int\int_{Q\cap\Omega} \operatorname{Var}_{p/2,y} f(x,y) \operatorname{Var}_{p/2,x} f(x,y))dxdy \\
&\le \int_A^B \operatorname{Var}_{p/2,y} f(x,y)dx \cdot \int_C^D \operatorname{Var}_{p/2,x} f(x,y)dy \\
&\le c\,\|f\|^p_{\mathrm{BV}_{p/2}(Q\cap\Omega)} \, .
\end{aligned}$$

Since $\Omega$ is a rectangular domain and $[A, B] \times [C, D] = Q \cap \Omega$, where $Q$ is a square with center in $\Omega$, we see that

$$\max\left(\frac{B-A}{D-C}, \frac{D-C}{B-A}\right) \le \gamma \; ,$$

where $\gamma > 0$ is a constant depending only on the ratio of the side lengths of $\Omega$. Therefore,

$$\int\int_{Q\cap\Omega} (\mathrm{Var}_{p/2,x} f(x,y_0)\,\mathrm{Var}_{p/2,x} f(x,y) + \mathrm{Var}_{p/2,y} f(x,y)\,\mathrm{Var}_{p/2,y} f(x_0,y))dxdy$$
$$\leq \frac{B-A}{D-C}\int_C^D \mathrm{Var}_{p/2,x} f(x,y)dy \int_C^D \mathrm{Var}_{p/2,x} f(x,y)dy$$
$$+ \frac{D-C}{B-A}\int_A^B \mathrm{Var}_{p/2,y} f(x,y)dx \cdot \int_A^B \mathrm{Var}_{p/2,y} f(x,y)dx$$
$$\leq c\,\|f\|^p_{\mathrm{BV}_{p/2}(Q\cap\Omega)}\,.$$

Consequently,

$$\left(\int_{Q\cap\Omega} |f - f_Q|^p\right)^{\frac{1}{p}} \leq c\,\|f\|_{\mathrm{BV}_{p/2}(Q\cap\Omega)}.$$ □

From (11.48) we readily deduce an analog of inequality (9.2).

**Lemma 11.18.** *Suppose $f \in \mathrm{BV}_{p/2}$, and let $\pi$ be a family of mutually nonintersecting cubes with centers in $\Omega$. Then*

$$\sum_{Q\in\pi}\left(\int_{Q\cap\Omega} |f - f_Q|^p\right)^{\frac{1}{2}} \leq c\left(\|f\|_{\mathrm{BV}_{p/2}}\right)^{\frac{p}{2}} \tag{11.49}$$

*with a constant $c > 0$ that depends on the ratio of the side lengths of $\Omega$.*

*Proof.* The definition of the seminorm in $\mathrm{BV}_{p/2}$ shows that

$$\left(\|f\|_{\mathrm{BV}_{p/2}}\right)^{\frac{p}{2}} \geq \sum_{Q\in\pi}\left(\|f\|_{\mathrm{BV}_{p/2}(Q\cap\Omega).}\right)^{\frac{p}{2}}.$$

It remains to apply inequality (11.48) for every $Q \in \pi$. □

Inequality (11.49) implies that

$$\mathrm{BV}_{p/2} \subset \dot{V}^1_{p,p/2}$$

and therefore

$$L_{p,p/2}(t^{\frac{p}{2}}, f; L^p, \mathrm{BV}_{p/2}) \geq c L_{p,p/2}(t^{\frac{p}{2}}, f; L^p, \dot{V}^1_{p,p/2}) \tag{11.50}$$

with a constant $c > 0$ independent of $f \in L^p$ and $t > 0$.

In fact, (11.50) can be improved.

**Theorem 11.19.** *We have the following equivalence with constants independent of* $f \in L^p$ *and* $t > 0$:

$$L_{p,p/2}(t^{\frac{p}{2}}, f; L^p, \mathrm{BV}_{p/2}) \approx L_{p,p/2}(t^{\frac{p}{2}}, f; L^p, \dot{V}^1_{p,p/2}).$$

*Moreover, the piecewise-constant function* $f_t$ (see (11.8)) *constructed for* $k = 1$ *is a near-minimizer for the couple* $(L^p, \mathrm{BV}_{p/2})$*, that is,*

$$\|f - f_t\|^p_{L^p} + t^{\frac{p}{2}} \left(\|f_t\|_{\mathrm{BV}_{p/2}}\right)^{\frac{p}{2}} \leq cL_{p,p/2}(t^{\frac{p}{2}}, f; L^p, \mathrm{BV}_{p/2}).$$

*Proof.* Theorem 11.4 and formula (11.50) show that

$$\|f - f_t\|^p_{L^p} \leq cL_{p,p/2}(t^{\frac{p}{2}}, f; L^p, \dot{V}^1_{p,p/2}) \leq cL_{p,p/2}(t^{\frac{p}{2}}, f; L^p, \mathrm{BV}_{p/2}).$$

Thus, it suffices to verify the inequality

$$\left(\|f_t\|_{\mathrm{BV}_{p/2}}\right)^{\frac{p}{2}} \leq \frac{c}{t^{\frac{p}{2}}} L_{p,p/2}(t^{\frac{p}{2}}, f; L^p, \dot{V}^1_{p,p/2}).$$

For this, we trace the change of the variation in the course of building $f_t$, i.e., we estimate the difference

$$\left(\|f_t\|_{\mathrm{BV}_{p/2}(\bigcup_{j\leq i} Q_{x_j})}\right)^{\frac{p}{2}} - \left(\|f_t\|_{\mathrm{BV}_{p/2}(\bigcup_{j\leq i-1} Q_{x_j})}\right)^{\frac{p}{2}}.$$

Since $f_t$ does not change on the set $\bigcup_{j\leq i-1} Q_{x_j}$, and on the set $Q_{x_i} \setminus \bigcup_{j\leq i-1} Q_{x_j}$ it is made equal to $f_{Q_{x_i}}$, variation increases only at the expense of jumps at the boundary of $\bigcup_{j\leq i-1} Q_{x_j}$. Since the boundary of $\bigcup_{j\leq i-1} Q_{x_j}$ is included in the union of the boundaries of the cubes $Q_{x_j}$, variation increases at most by

$$\sum_{j\leq i-1} \mu_{j,i} \left|f_{Q_{x_j}} - f_{Q_{x_i}}\right|^{\frac{p}{2}},$$

where $\mu_{j,i}$ is the length of the boundary of $Q_{x_j} \cap Q_{x_i}$. Continuing in the same manner, we obtain the estimate

$$\left(\|f_t\|_{\mathrm{BV}_{p/2}}\right)^{\frac{p}{2}} \leq \sum_{j,i} \mu_{j,i} \left|f_{Q_{x_j}} - f_{Q_{x_i}}\right|^{\frac{p}{2}}.$$

It remains to observe that the quantity on the right was estimated in Lemma 11.14, and we arrive at

$$\left(\|f_t\|_{\mathrm{BV}_{p/2}}\right)^{\frac{p}{2}} \leq \frac{c}{t^{\frac{p}{2}}} L_{p,p/2}(t^{\frac{p}{2}}, f; L^p, \dot{V}^1_{p,p/2}).$$

So, (11.50) yields

$$L_{p,p/2}(t^{\frac{p}{2}}, f; L^p, \mathrm{BV}_{p/2}) \leq cL_{p,p/2}(t^{\frac{p}{2}}, f; L^p, V^1_{p,p/2}).$$ □

*Remark* 11.20. For $p = 2$ the theorem gives the $L$-functional and a near-minimizer for the couple $(L^2, \mathrm{BV})$.

## Notes and remarks

The results of this chapter are due to Kruglyak. Theorem 11.19 is a generalization of the result in [AK], where the case of $p = 2$ was considered.

It is surprising that spaces of functions of bounded $p$-variation emerge here in place of Sobolev spaces. We repeat that a generalization of these results to the dimension $n > 2$ is an open problem.

# Chapter A

# Appendix. Near-minimizers for Brudnyi and Triebel–Lizorkin spaces

In Chapter 9, near-minimizers were constructed for couples of the form $(L^p, \dot{W}^k_q)$ and $(L^p, \dot{C}^{\alpha,k}_p)$. It is natural to ask whether it is possible to replace $L^p$ in these results by a more general function space $X$. Our purpose in this chapter is to supplement the previous constructions by a generalization of the algorithm of Chapter 9. We shall test this general algorithm on couples of the form $(X, Y)$, where $X$ is a Triebel–Lizorkin space (or Brudnyi's generalized variation spaces), and $Y$ is a Morrey space constructed on the basis of $X$. It will turn out that the algorithm in question really provides near-minimizers for such couples.

An outline of the algorithm in an "abstract" setting will be given in §1. The last two sections (§§2 and 3) are devoted to the specific couples of spaces mentioned above. Fairly bulky formulas occurring in those sections seem to be unavoidable.

## A.1 Description of the general algorithm

Let $X$ be a (Banach or quasi-Banach) function space on $\mathbb{R}^n$. Suppose that for every cube $Q$ some space $X(Q)$ is defined (heuristically, it is obtained by localization of the norm of $X$ to $Q$). In particular, if $f \in X$, then $f\chi_Q \in X(Q)$ and

$$\|f\chi_Q\|_{X(Q)} \leq \|f\|_X .$$

For example, if $X = L^p$, then $X(Q) = L^p(Q)$; if $X = \dot{W}^k_p$, then $X(Q)$ is determined by the seminorm

$$\|f\|_{X(Q)} = \sup_{|m|=k} \|D^m f\|_{L^p(Q)} .$$

In what follows, we shall present more examples related to the Brudnyi and Triebel–Lizorkin spaces.

Suppose we are given a function $f \in X$ and a number $t > 0$. The algorithm in question depends on two parameters $\lambda \in \mathbb{R} \setminus \{0\}$ and $k \in \mathbb{Z}_+$ (the symbol $\mathbb{Z}_+$ stands for the set of nonnegative integers). As before, we denote by $\mathcal{P}_k$ the space of algebraic polynomials of degree strictly smaller than $k$, in particular, $\mathcal{P}_0 = \{0\}$.

For every $x \in \mathbb{R}^n$, we consider the following function of $r > 0$:

$$\varphi_x(r) = \inf_{P \in \mathcal{P}_k} \frac{1}{|Q(x,r)|^\lambda} \|f - P\|_{X(Q(x,r))} \tag{A.1}$$

and put

$$\Omega = \left\{ x \in \mathbb{R}^n : \sup_{r>0} \varphi_x(r) > t \right\}. \tag{A.2}$$

First, we describe the algorithm in the case where $\lambda > 0$. In this case, since $\|f\chi_Q\|_{X(Q)} \le \|f\|_X < \infty$, we see that $\varphi_x(r) \to 0$ as $r \to \infty$. Therefore, for every $x \in \Omega$, there exists a positive number $r_x$ with

$$\sup_{r \ge r_x} \varphi_x(r) \le t \tag{A.3}$$

and

$$\sup_{r \ge \frac{1}{2} r_x} \varphi_x(r) > t. \tag{A.4}$$

We put

$$Q_x = Q(x, r_x).$$

Thus, we have constructed a family $\{Q_x\}_{x \in \Omega}$ of cubes.

Now, conditions (A.3)—(A.4) imply the existence of $r \in \left[\frac{1}{2} r_x, r_x\right]$ such that

$$t \le \frac{1}{|Q(x,r)|^\lambda} \inf_{P \in \mathcal{P}_k} \|f - P\|_{X(Q(x,r))} \le \frac{1}{|Q(x,r)|^\lambda} \|f\|_X \,.$$

Consequently,

$$|Q_x| \le 2^n |Q(x,r)| \le 2^n \left(\frac{1}{t} \|f\|_X\right)^{\frac{1}{\lambda}}.$$

So, the cubes $\{Q_x\}_{x \in \Omega}$ possess the Besicovitch property

$$\sup_{x \in \Omega} |Q_x| < \infty,$$

which makes it possible to apply the controlled extension theorem. As a result, we obtain a family $\{K_i\}_{i \in I}$ of cubes that is a WB-covering. We take a smooth partition $\{\psi_i\}_{i \in I}$ of unity adjusted to this WB-covering. Also, let $f_{K_i}$, $i \in I$, be polynomials of degree strictly smaller than $k$ yielding the best approximation on the cubes $K_i, i \in I$, that is

$$\|f - f_{K_i}\|_{X(K_i)} = \inf_{P \in \mathcal{P}_k} \|f - P\|_{X(K_i)} \,.$$

Such polynomials exist because the space $\mathcal{P}_k$ is finite-dimensional.

We define a function $f_t$ by the formula

$$f_t = \sum_{i \in I} f_{K_i} \psi_i + f\, \chi_{\mathbb{R}^n \setminus \cup K_i}. \tag{A.5}$$

It will be shown below that in many specific cases formula (A.5) defines a near-minimizer for a certain couple $(X, Y)$. Finding the space $Y$ is a nontrivial problem.

Now we describe the construction of $f_t$ for $\lambda < 0$. In this case, the function $\varphi_x(r)$ given by (A.1) tends to zero as $r \to 0$. We shall also assume that the function $f \in X$ is such that $\varphi_x(r)$ is continuous in the variable $r > 0$ (if $X = L^p$, all functions have this property). Therefore, if $x \in \Omega$, where $\Omega$ is defined by (A.2), then there exists $r_x > 0$ such that

$$\sup_{r \le r_x} \varphi_x(r) \le t \text{ and } \varphi_x(r_x) = \frac{1}{|Q(x,r)|^\lambda} \inf_{P \in \mathcal{P}_k} \|f \chi_{Q(x,r)} - P\|_{X(Q(x,r))} \ge \frac{1}{2} t. \tag{A.6}$$

We put $Q_x = Q(x, r_x)$, obtaining a family $\{Q_x\}_{x \in \Omega}$ of cubes.

From (A.6) and the relation

$$\inf_{P \in \mathcal{P}_k} \|f - P\|_{X(Q)} \le \|f\|_X$$

we deduce that

$$|Q_x| \ge \left(\frac{t}{2\,\|f\|_X}\right)^{-\frac{1}{\lambda}}$$

for every $x \in \Omega$ and, consequently, the assumption of the controlled contraction theorem is fulfilled:

$$\inf_{x \in \Omega} |Q_x| > 0.$$

Application of that theorem yields a family $\{K_i\}_{i \in I}$ of cubes that forms a WB-covering of $\mathbb{R}^n$.

The function $f_t$ is defined by the formula

$$f_t = \sum_{i \in I} f_{K_i} \psi_i, \tag{A.7}$$

where the $f_{K_i}$, $i \in I$, are polynomials of degree strictly smaller than $k$ that provide the best approximations on the cubes $K_i$, $i \in I$. The absence of the term $f\, \chi_{\mathbb{R}^n \setminus \cup K_i}$ in the formula for $f_t$ is related to the fact that $\cup K_i = \mathbb{R}^n$ in the controlled contraction theorem.

Below we consider two examples in which the algorithm described above does lead to near-minimizers. The proofs are quite similar in both cases. First, we consider the case where $X$ is Brudnyi's generalized variation space, see [Br1] and the references therein, because then the proof is slightly less involved than for the more well-known Triebel–Lizorkin spaces. For instance, the Fefferman–Stein maximal theorem can be avoided for Brudnyi spaces.

## A.2 Near-minimizers for Morrey spaces built on the basis of Brudnyi spaces

Putting

$$E_k(f,Q) = \inf_{P\in\mathcal{P}_k} \int_Q |f-P|,$$

we define the generalized variation space $\dot{V}_p^{s,k}$ (see [Br0]) to be the space of functions on $\mathbb{R}^n$ for which the following seminorm is finite:

$$\|f\|_{\dot{V}_p^{s,k}} = \sup_{\pi=\{Q_i\}} \left(\sum_i \left(\frac{1}{|Q_i|^{1+\frac{s}{n}}} E_k(f,Q_i)\right)^p |Q_i|\right)^{\frac{1}{p}}. \tag{A.8}$$

Here the supremum is taken over all packings $\pi = \{Q_i\}$ that consist of cubes.[1] In what follows, we shall assume that

$$k \geq s \geq 0,\ p \in (0,\infty) \text{ and } 1 + \frac{s}{n} - \frac{1}{p} > 0. \tag{A.9}$$

Comparison with the seminorm in the Triebel–Lizorkin spaces $\dot{F}_{p,q}^{s,k}$ to be treated below shows that the parameter $s$ plays the role of "smoothness", and the quantity $|Q_i|$ in each summand in (A.8) replaces integration over $Q_i$.

The localization of the space $\dot{V}_p^{s,k}$ (denoted by $\dot{V}_p^{s,k}(Q)$) is defined by formula (A.8) with the additional restriction that the supremum is taken over packings $\pi = \{Q_i\}$ that consist only of cubes included in $Q$.

The Morrey space built on the basis of the Brudnyi space $\dot{V}_p^{s,k}$ is defined with the help of the seminorm

$$\|f\|_{\dot{C}^{\lambda,k}(\dot{V}_p^{s,k})} = \sup_Q \left(\frac{1}{|Q|^\lambda} \inf_{P\in\mathcal{P}_k} \|f-P\|_{\dot{V}_p^{s,k}(Q)}\right), \tag{A.10}$$

where $\lambda > 0$ is some positive number.

*Remark* A.1. The results (with slightly more complicated proofs) can be extended to the case of more general Morrey–Campanato spaces determined by the seminorms

$$\|f\|_{\dot{C}^{\lambda,k_1}(\dot{V}_p^{s,k})} = \sup_Q \left(\frac{1}{|Q|^\lambda} \inf_{P\in\mathcal{P}_{k_1}} \|f-P\|_{\dot{V}_p^{s,k}(Q)}\right),$$

where $k_1 \geq k$. We have omitted this improvement because it involves even bulkier formulas.

[1] In Chapter 11 we encountered another generalized variation space, also depending on three indices, which were arranged differently. The point is that, in fact, there are *four* indices in the most general definition of these spaces, see [T]. Two cases in question involve different triples out of the quad.

*Remark* A.2. If $k = s = 0$, it is known that $\dot{V}_p^{s,k} = L^p$. Next, taking $\lambda = \frac{1}{p} - \frac{\sigma}{n}$, $\sigma \in \left[0, \frac{n}{p}\right)$, we obtain the Morrey spaces (see (9.26)) built on the basis of $L^p$. This justifies the name given above to the spaces $\dot{C}^{\lambda,k}(\dot{V}_p^{s,k})$. If $k = s = 1$, the space $\dot{V}_p^{s,k}$ coincides with the Sobolev space $\dot{W}_p^1$ for $p > 1$ and with BV for $p = 1$. Thus, in this case the seminorm (A.10) generates what can naturally be called "the Morrey space built on the basis of a Sobolev space or BV".

In the sequel, a polynomial $P$ of degree strictly smaller than $k$ at which the infimum in (A.10) is attained will be denoted by $f_Q$. So,

$$\|f\|_{\dot{C}^{\lambda,k}(\dot{V}_p^{s,k})} = \sup_Q \left( \frac{1}{|Q|^\lambda} \|f - f_Q\|_{\dot{V}_p^{s,k}(Q)} \right).$$

We take $\dot{V}_p^{s,k}$ for the role of $X$ and apply the general algorithm of constructing a near-minimizer with parameters $\lambda$ and $k$ to a function $f \in \dot{V}_p^{s,k}$ and a number $t > 0$. That is, for arbitrary $x \in \mathbb{R}^n$, we consider the function

$$\varphi_x(r) = \frac{1}{|Q(x,r)|^\lambda} \left\|f - f_{Q(x,r)}\right\|_{\dot{V}_p^{s,k}(Q(x,r))}.$$

Since $\lambda > 0$, this function tends to zero as $r \to \infty$. Next, we consider the set

$$\Omega = \left\{ x \in \mathbb{R}^n : \sup_{r>0} \varphi_x(r) > t \right\}$$

and for $x \in \Omega$ choose $r_x > 0$ such that

$$\sup_{r \geq r_x} \varphi_x(r) \leq t \text{ and } \sup_{r \geq \frac{1}{2} r_x} \varphi_x(r) > t. \tag{A.11}$$

Putting

$$Q_x = Q(x, r_x),$$

we obtain a family $\{Q_x\}_{x \in \Omega}$ of cubes.

It is important that this family has finite $\alpha$-capacity for $\alpha = \lambda p$. Indeed, let $\pi = \{Q_{x_i}\}$ be a packing consisting of some cubes among $\{Q_x\}_{x \in \Omega}$. By (A.11), for some $r \in \left[\frac{1}{2} r_{x_i}, r_{x_i}\right]$ we have

$$\begin{aligned} \|f\|_{\dot{V}_p^{s,k}(Q_{x_i})}^p &\geq \|f\|_{\dot{V}_p^{s,k}(Q(x_i,r))}^p \geq \left\|f - f_{Q(x_i,r)}\right\|_{\dot{V}_p^{s,k}(Q(x_i,r))}^p \geq t^p \left|Q(x_i,r)\right|^{\lambda p} \\ &\geq c t^p \left|Q_{x_i}\right|^{\lambda p}. \end{aligned}$$

Since the cubes $Q_{x_i}$ are mutually disjoint, the definition of the seminorm in $\dot{V}_p^{s,k}$ implies

$$\|f\|_{\dot{V}_p^{s,k}}^p \geq \sum_{Q_{x_i} \in \pi} \|f\|_{\dot{V}_p^{s,k}(Q_{x_i})}^p \geq c \sum_{Q_{x_i} \in \pi} t^p \left|Q_{x_i}\right|^{\lambda p}$$

and, consequently,

$$\left|\{Q_x\}_{x\in\Omega}\right|_{\lambda p} \le c\frac{\|f\|^p_{\dot{V}^{s,k}_p}}{t^p} < \infty.$$

Since $\lambda > 0$, we see that

$$\sup_{x\in\Omega} |Q_x| < \infty,$$

and the controlled extension theorem can be applied. This results in a WB-covering $\{K_i\}_{i\in I}$. We define a function $f_t$ by the formula

$$f_t = \sum_{i\in I} f_{K_i}\psi_i + f\,\chi_{\mathbb{R}^n\setminus\cup K_i}, \tag{A.12}$$

where $\{\psi_i\}_{i\in I}$ is a smooth partition of unity adjusted to the WB-covering mentioned above, and the $f_{K_i} \in \mathcal{P}_k$ are polynomials of best approximation of $f$ on the cubes $K_i$, $i \in I$.

To state the result, we introduce yet another family of cubes:

$$\Omega_{f,t} = \left\{Q : \frac{1}{|Q|^\lambda} \inf_{P\in\mathcal{P}_k} \|f-P\|_{\dot{V}^{s,k}_p(Q)} \ge t\right\}. \tag{A.13}$$

**Theorem A.3.** *Suppose we are given a function $f \in \dot{V}^{s,k}_p$ and a number $t > 0$. If* (A.9) *is fulfilled and $\lambda \le \frac{1}{p} + \frac{k-s}{n}$, then*

$$\mathrm{dist}_{\dot{V}^{s,k}_p}(f, B_{\dot{C}^{\lambda,k}(\dot{V}^{s,k}_p)}(t)) \approx t\left(|\Omega_{f,t}|_{\lambda p}\right)^{\frac{1}{p}}, \tag{A.14}$$

*and the function $f_t$* (see (A.12)) *is a near-minimizer for the distance functional of the couple $(\dot{V}^{s,k}_p, \dot{C}^{\lambda,k}(\dot{V}^{s,k}_p))$.*

*Remark* A.4. For $k = s = 0$, the restriction $\lambda \le \frac{1}{p} + \frac{k-s}{n}$ turns into $\lambda \le \frac{1}{p}$, which coincides with the restriction imposed in the case of classical Morrey spaces. We remind the reader that the equivalence (A.14) means the existence of constants $c_1$, $c_2 > 0$ independent of $f \in \dot{V}^{s,k}_p$ and $t > 0$ such that

$$c_1\,\mathrm{dist}_{\dot{V}^{s,k}_p}(f, B_{\dot{C}^{\lambda,k}(\dot{V}^{s,k}_p)}\left(\frac{t}{c_1}\right)) \le t\left(|\Omega_{f,t}|_{\lambda p}\right)^{\frac{1}{p}} \le c_2\,\mathrm{dist}_{\dot{V}^{s,k}_p}(f, B_{\dot{C}^{\lambda,k}(\dot{V}^{s,k}_p)}\left(\frac{t}{c_2}\right)).$$

We also remind the reader that the statement that $f_t$ is a near-minimizer for the above distance functional means that, for some $c > 0$ independent of $f \in \dot{V}^{s,k}_p$ and $t > 0$, we have

$$\|f_t\|_{\dot{C}^{\lambda,k}(\dot{V}^{s,k}_p)} \le ct \text{ and } \|f - f_t\|_{\dot{V}^{s,k}_p} \le c\,\mathrm{dist}_{\dot{V}^{s,k}_p}(f, B_{\dot{C}^{\lambda,k}(\dot{V}^{s,k}_p)}\left(\frac{t}{c}\right)).$$

### A.2.1 Auxiliary lemmas

In the proof, we shall need three lemmas. The first two will express the following observation. Suppose we are given an arbitrary WB-covering $\{K_i\}_{i\in I}$, and let $\{\psi_i\}_{i\in I}$ be a smooth partition of unity adjusted to this covering. For a function $h \in \dot{V}_p^{s,k}$, consider the decomposition

$$h = h_{\{K_i\}} + (h - h_{\{K_i\}}),$$

where

$$h_{\{K_i\}} = \sum_{i\in I} h_{K_i}\psi_i + h\chi_{\mathbb{R}^n/\cup K_i} \tag{A.15}$$

and the $h_{K_i}$ are polynomials of degree strictly smaller than $k$ that provide the best approximation to $h$ on the cubes $K_i$ in the metric of the space $\dot{V}_p^{s,k}(K_i)$, $i \in I$.

It turns out that the function $h_{\{K_i\}}$ admits a good estimate on cubes included in some cube of the family $\{K_i\}_{i\in I}$, whereas $h - h_{\{K_i\}}$ admits a good estimate on cubes that are not included in any cube of this family. We give precise definitions.

**Definition A.5.** A cube $Q$ is said to be *small* relative to the family $\{K_i\}_{i\in I}$ if there exists a cube $K_i$ such that $Q \subset K_i$. If a cube $Q$ is not small, it is said to be *big*.

In the sequel, packings $\pi = \{Q_i\}$ composed of small cubes will be denoted by $\pi_s$, and those composed of big cubes will be denoted by $\pi_b$.

The next lemma shows that the function $h_{\{K_i\}}$ admits a good estimate on packings $\pi_s$.

**Lemma A.6.** *For every packing $\pi_s$ composed of small cubes we have*

$$\left(\sum_{Q\in\pi_s}\left(\frac{1}{|Q|^{1+\frac{s}{n}}}\int_Q \left|h_{\{K_i\}} - \left(h_{\{K_i\}}\right)_Q\right|\right)^p |Q|\right)^{\frac{1}{p}}$$
$$\leq c\left(\sum_{i\in I}\left(\frac{1}{|K_i|^{1+\frac{s}{n}}}\int_{K_i} |h - h_{K_i}|\right)^p |K_i|\right)^{\frac{1}{p}}.$$

*Proof.* Consider a packing $\pi_s = \{Q_j\}$ as described above. The function $h_{\{K_i\}}$ defined by (A.15) is infinitely differentiable on $\cup K_i$. Therefore, if for every $Q_j$ we replace the best approximation polynomial $(h_{\{K_i\}})_{Q_j}$ by the Taylor polynomial of order $k-1$ (relative to the center of $Q_j$) for $h_{\{K_i\}}$, we obtain (see Theorem 0.20)

$$\int_{Q_j} |h_{\{K_i\}} - (h_{\{K_i\}})_{Q_j}| \leq c|Q_j|^{1+\frac{k}{n}} \max_{|m|=k} \left\|D^m(h_{\{K_i\}})\right\|_{L^\infty(Q_j)}.$$

Consequently,

$$\left(\sum_{Q_j\in\pi_s}\left(\frac{1}{|Q_j|^{1+\frac{s}{n}}}\int_{Q_j} \left|h_{\{K_i\}} - \left(h_{\{K_i\}}\right)_{Q_j}\right|\right)^p |Q_j|\right)^{\frac{1}{p}}$$

$$\leq \left( \sum_{Q_j \in \pi_s} \left( \frac{1}{|Q_j|^{1+\frac{s}{n}}} |Q_j|^{1+\frac{k}{n}} \max_{|m|=k} \left\| D^m(h_{\{K_i\}}) \right\|_{L^\infty(Q_j)} \right)^p |Q_j| \right)^{\frac{1}{p}}$$
$$= \left( \sum_{Q_j \in \pi_s} \left( |Q_j|^{\frac{1}{p}+\frac{k}{n}-\frac{s}{n}} \max_{|m|=k} \left\| D^m(h_{\{K_i\}}) \right\|_{L^\infty(Q_j)} \right)^p \right)^{\frac{1}{p}}.$$

Since $k \geq s$ (see (A.9)), we have $(\frac{1}{p} + \frac{k}{n} - \frac{s}{n})p \geq 1$, and for the cubes $Q_j$ included in $K_i$ we have

$$\sum_{Q_j \subset K_i} \left( |Q_j|^{\frac{1}{p}+\frac{k}{n}-\frac{s}{n}} \right)^p \leq \left( |K_i|^{\frac{1}{p}+\frac{k}{n}-\frac{s}{n}} \right)^p.$$

Now, all cubes $Q_j$ are small, so each of them is included in some $K_i$. Therefore,

$$\left( \sum_{Q_j \in \pi_s} \left( |Q_j|^{\frac{1}{p}+\frac{k}{n}-\frac{s}{n}} \max_{|m|=k} \left\| D^m(h_{\{K_i\}}) \right\|_{L^\infty(Q_j)} \right)^p \right)^{\frac{1}{p}}$$
$$\leq \left( \sum_{i \in I} \left( |K_i|^{\frac{1}{p}+\frac{k}{n}-\frac{s}{n}} \max_{|m|=k} \left\| D^m(h_{\{K_i\}}) \right\|_{L^\infty(K_i)} \right)^p \right)^{\frac{1}{p}}.$$

Since $|m| = k$, we have $D^m h_{K_i} = 0$ and, consequently, on $K_{i_0}$ we have

$$D^m(h_{\{K_i\}}) = \sum_{|l|<k} \sum_{i: K_i \cap K_{i_0} \neq \varnothing} c_l D^l(h_{K_i}) D^{m-l} \psi_i,$$
$$D^m h_{K_{i_0}} = D^m(\sum_i h_{K_{i_0}} \psi_i) = \sum_{|l|<k} \sum_{i: K_i \cap K_{i_0} \neq \varnothing} c_l D^l(h_{K_{i_0}}) D^{m-l} \psi_i = 0.$$

Hence,

$$D^m(h_{\{K_i\}}) = \sum_{|l|<k} \sum_{i: K_i \cap K_{i_0} \neq \varnothing} c_l \left( (D^l(h_{K_i}) - (D^l(h_{K_{i_0}})\right) D^{m-l} \psi_i$$

on $K_{i_0}$. Applying the Markov inequality (Theorem 0.21) to the right-hand side and using the finite multiplicity and engagement properties for the family $\{K_i\}_{i \in I}$, as well as Theorem 0.22 and estimates for derivatives of the $\psi_i$, we obtain

$$\left\| D^m(h_{\{K_i\}}) \right\|_{L^\infty(K_{i_0})} \tag{A.16}$$
$$\leq c \sum_{|l|<k} \sum_{i: K_i \cap K_{i_0} \neq \varnothing} \frac{1}{|K_{i_0}|^{\frac{|l|}{n}}} \left\| h_{K_i} - h_{K_{i_0}} \right\|_{L^\infty(K_{i_0})} \frac{1}{|K_{i_0}|^{\frac{k-|l|}{n}}}$$
$$\leq c \frac{1}{|K_{i_0}|^{\frac{k}{n}}} \cdot \frac{1}{|K_{i_0}|} \sum_{i: K_i \cap K_{i_0} \neq \varnothing} \int_{K_i} |h - h_{K_i}| \leq c \sum_{i: K_i \cap K_{i_0} \neq \varnothing} \frac{1}{|K_i|^{1+\frac{k}{n}}} \int_{K_i} |h - h_{K_i}|.$$

Since we are dealing with a WB-covering, the number of cubes $K_i$ that intersect $K_{i_0}$ is finite and is controlled by a constant depending only on the dimension $n$. So, finally we arrive at

$$\left(\sum_{Q_j\in\pi_s}\left(\frac{1}{|Q_j|^{1+\frac{s}{n}}}\int_Q\left|h_{\{K_i\}}-\left(h_{\{K_i\}}\right)_{Q_j}\right|\right)^p|Q_j|\right)^{\frac{1}{p}}$$
$$\le c\left(\sum_{i\in I}\left(|K_i|^{\frac{1}{p}+\frac{k}{n}-\frac{s}{n}}\frac{1}{|K_i|^{1+\frac{k}{n}}}\int_{K_i}|h-h_{K_i}|\right)^p\right)^{\frac{1}{p}}$$
$$\le c\left(\sum_{i\in I}\left(\frac{1}{|K_i|^{1+\frac{s}{n}}}\int_{K_i}|h-h_{K_i}|\right)^p|K_i|\right)^{\frac{1}{p}}. \qquad \square$$

To prove the next lemma, we need some geometric properties of big cubes.

**Proposition A.7.** a) *If a cube $Q$ is big, then*

$$\bigcup_{i:K_i\cap Q\neq\varnothing} K_i\subset\gamma Q. \tag{A.17}$$

*with some constant $\gamma>0$ depending only on the dimension $n$.* b) *Let a packing $\pi_b=\{Q_j\}$ consist of big cubes. Then the number of cubes in $\pi_b$ that intersect $K_i$ is finite and is controlled in terms of the dimension $n$.*

*Proof.* a) Suppose $K_i\cap Q\neq\varnothing$, and take an arbitrary point $x\in K_i\cap Q$. By the Whitney property, there exists a cube $K_j$ such that $x\in\frac{1}{1+\delta}K_j$. Since $Q$ is not included in $K_j$, it follows that $2r(Q)\ge\frac{\delta}{1+\delta}r(K_j)$. By the strong engagement property, we have $r(K_i)\le cr(K_j)$ with a constant $c>0$ depending only on the dimension $n$. Consequently,

$$r(K_i)\le cr(Q), \tag{A.18}$$

and, therefore, (A.17) is fulfilled.

b) Consider the cube $2K_i$. If a cube $Q_j\in\pi_b$ intersects $K_i$, formula (A.18) shows that the volume of the intersection of $Q_j$ with $2K_i$ is at least $c|K_i|$, where $c>0$ is a constant depending only on the dimension $n$. Since the cubes in $\pi_b$ are mutually disjoint, the claim follows. $\square$

The next lemma shows that the function $h-h_{\{K_i\}}$ admits a good estimate on packings $\pi_b$ that consist of big cubes.

**Lemma A.8.** *If* (A.9) *is fulfilled, then for every packing $\pi_b=\{Q_j\}$ that consists of big cubes we have*

$$\left(\sum_{Q_j\in\pi}\left(\frac{1}{|Q_j|^{1+\frac{s}{n}}}\int_{Q_j}|h-h_{\{K_i\}}|\right)^p|Q_j|\right)^{\frac{1}{p}}$$

$$\leq c\left(\sum_{i:K_i\cap(\bigcup Q_j)\neq\varnothing}\left(\frac{1}{|K_i|^{1+\frac{s}{n}}}\int_{K_i}|h-h_{K_i}|\right)^p|K_i|\right)^{\frac{1}{p}}$$

*with a constant $c>0$ depending only on the dimension $n$.*

*Proof.* The arguments are different in the cases where $p\geq 1$ and $p<1$. Suppose first that $p\geq 1$. If $Q$ is an arbitrarily big cube, we have

$$\int_Q|h-h_{\{K_i\}}|\leq\sum_{j:Q\cap K_j\neq\varnothing}\int_{K_j}|h-h_{K_j}| \tag{A.19}$$

$$\leq\sum_{j:Q\cap K_j\neq\varnothing}\left(\frac{|K_j|^{\frac{1}{p}}}{|K_j|^{1+\frac{s}{n}}}\int_{K_j}|h-h_{K_j}|\right)|K_j|^{1+\frac{s}{n}-\frac{1}{p}}$$

$$\leq\left(\sum_{j:Q\cap K_j\neq\varnothing}\left(\frac{|K_j|^{\frac{1}{p}}}{|K_j|^{1+\frac{s}{n}}}\int_{K_j}|h-h_{K_j}|\right)^p\right)^{\frac{1}{p}}\Big(\sum_{j:Q\cap K_j\neq\varnothing}|K_j|^{(1+\frac{s}{n}-\frac{1}{p})p'}\Big)^{\frac{1}{p'}}.$$

Now, $(1+\frac{s}{n}-\frac{1}{p})p'\geq 1$ because $s\geq 0$. Next, Proposition A.7 and the finite multiplicity property of the family $\{K_i\}_{i\in I}$ show that

$$\sum_{j:Q\cap K_j\neq\varnothing}|K_j|\leq c\,|Q|\,,$$

implying

$$\sum_{j:Q\cap K_j\neq\varnothing}|K_j|^{(1+\frac{s}{n}-\frac{1}{p})p'}\leq c\,|Q|^{(1+\frac{s}{n}-\frac{1}{p})p'}.$$

Therefore, from (A.19) we obtain

$$\frac{|Q|^{\frac{1}{p}}}{|Q|^{1+\frac{s}{n}}}\int_Q|h-h_{\{K_i\}}|\leq c\left(\sum_{j:Q\cap K_j\neq\varnothing}\left(\frac{|K_j|^{\frac{1}{p}}}{|K_j|^{1+\frac{s}{n}}}\int_{K_j}|h-h_{K_j}|\right)^p\right)^{\frac{1}{p}}$$

and, consequently,

$$\left(\frac{|Q|^{\frac{1}{p}}}{|Q|^{1+\frac{s}{n}}}\int_Q|h-h_{\{K_i\}}|\right)^p\leq c\sum_{j:Q\cap K_j\neq\varnothing}\left(\frac{|K_j|^{\frac{1}{p}}}{|K_j|^{1+\frac{s}{n}}}\int_{K_j}|h-h_{K_j}|\right)^p. \tag{A.20}$$

Let $\pi_b$ be a packing that consists of big cubes. By statement b) of Proposition A.7, the cube $K_j$ may intersect only a controlled finite number of cubes in $\pi_b$. Therefore, summation of (A.20) over cubes in $\pi_b$ implies the required estimate.

Now, we treat the case of $p < 1$. Since $1 + \frac{s}{n} - \frac{1}{p} > 0$ and $p < 1$, statement a) of Proposition A.7 implies

$$\begin{aligned}
&\left(\sum_{Q_j\in\pi_b}\left(\frac{1}{|Q_j|^{1+\frac{s}{n}}}\int_{Q_j}\left|h-h_{\{K_i\}}\right|\right)^p|Q_j|\right)^{\frac{1}{p}}\\
&\le c\left(\sum_{Q_j\in\pi_b}\left(\frac{1}{|Q_j|^{1+\frac{s}{n}-\frac{1}{p}}}\sum_{i:Q_j\cap K_i\neq\varnothing}\int_{K_i}|h-h_{K_i}|\right)^p\right)^{\frac{1}{p}}\\
&\le c\left(\sum_{Q_j\in\pi_b}\left(\sum_{i:Q_j\cap K_i\neq\varnothing}\frac{1}{|K_i|^{1+\frac{s}{n}-\frac{1}{p}}}\int_{K_i}|h-h_{K_i}|\right)^p\right)^{\frac{1}{p}}\\
&\le c\left(\sum_{Q_j\in\pi_b}\sum_{i:Q_j\cap K_i\neq\varnothing}\left(\frac{1}{|K_i|^{1+\frac{s}{n}-\frac{1}{p}}}\int_{K_i}|h-h_{K_i}|\right)^p\right)^{\frac{1}{p}}.
\end{aligned}$$

It remains to observe that, by statement b) of Proposition A.7, the number of occurrences of every cube $K_i$ on the right in this inequality is finite and is controlled in terms of the dimension $n$ only. □

Lemmas A.6 and A.8 above imply the following statement.

**Proposition A.9.** *If $h \in \dot{V}_p^{s,k}$ and $1 + \frac{s}{n} - \frac{1}{p} > 0$, then*

$$\max(\|h_{\{K_i\}}\|_{\dot{V}_p^{s,k}}, \|h - h_{\{K_i\}}\|_{\dot{V}_p^{s,k}}) \le c\,\|h\|_{\dot{V}_p^{s,k}},$$

*where $c > 0$ depends only on the parameters $p, s, k$.*

*Proof.* It suffices to show that

$$\|h_{\{K_i\}}\|_{\dot{V}_p^{s,k}} \le c\,\|h\|_{\dot{V}_p^{s,k}}.$$

Let $\pi_s$ be a packing that consists of small cubes. Then the claim follows from Lemma A.6 and the finite multiplicity property of the family $\{K_i\}_{i\in I}$. For a packing $\pi_b$ of big cubes, we write $h_{\{K_i\}}$ as the difference of $h$ and $h - h_{\{K_i\}}$, and use the triangle inequality and Lemma A.8. □

We need yet another lemma.

**Lemma A.10.** *There is $c > 0$ independent of $f \in \dot{V}_p^{s,k}$ and $t > 0$ such that*

$$t\left(|\Omega_{f,t}|_{\lambda p}\right)^{\frac{1}{p}} \le c\,\mathrm{dist}_{\dot{V}_p^{s,k}}(f, B_{\dot{C}^{\lambda,k}(\dot{V}_p^{s,k})}\left(\frac{t}{c}\right)).$$

*Proof.* Suppose $\{Q_j\}$ is a packing that consists of cubes in $\Omega_{f,t}$. The definition of $\Omega_{f,t}$ (see (A.13)) shows that

$$\frac{\|f - f_{Q_j}\|_{\dot V_p^{s,k}(Q_j)}}{|Q_j|^\lambda} \geq t.$$

Now, for every $j$ the formula for the seminorm in $\dot V_p^{s,k}(Q_j)$ implies the existence of a packing $\pi_j$ composed of subcubes of $Q_j$ such that

$$\left( \sum_{Q \in \pi_j} \left( \frac{1}{|Q|^{1+\frac{s}{n}}} E_k(f - f_{Q_j}, Q) \right)^p |Q| \right)^{\frac{1}{p}} \geq ct\, |Q_j|^\lambda ,$$

or, equivalently,

$$\sum_{Q\in\pi_j} \frac{1}{|Q|^{(1+\frac{s}{n})p-1}} E_k(f - f_{Q_j}, Q)^p \geq ct^p\, |Q_j|^{\lambda p} .$$

Since the cubes $Q_j$ are mutually disjoint, the union $\pi = \bigcup \pi_j$ is a packing. Moreover,

$$\sum_{Q\in\pi} \frac{1}{|Q|^{(1+\frac{s}{n})p-1}} E_k(f - f_{Q_j}, Q)^p \geq ct^p \sum_j |Q_j|^{\lambda p} .$$

Now, let $g \in \dot C^{\lambda,k}(\dot V_p^{s,k})$ satisfy $\|g\|_{\dot C^{\lambda,k}(\dot V_p^{s,k})} \leq \varepsilon t$, where $\varepsilon > 0$ will be chosen later. For every cube $Q_j$, there is a polynomial $g_{Q_j}$ of degree strictly smaller than $k$ and such that

$$\sum_{Q\in\pi_j} \frac{1}{|Q|^{(1+\frac{s}{n})p-1}} E_k(g - g_{Q_j}, Q)^p \leq \varepsilon^p t^p\, |Q_j|^{\lambda p} ,$$

whence

$$\sum_j \sum_{Q\in\pi_j} \frac{1}{|Q|^{(1+\frac{s}{n})p-1}} E_k(g - g_{Q_j}, Q)^p \leq \varepsilon^p t^p \sum_j |Q_j|^{\lambda p} .$$

The quasinorm

$$\|h\| = \left( \sum_j \sum_{Q\in\pi_j} \frac{1}{|Q|^{(1+\frac{s}{n})p-1}} E_k(h, Q)^p \right)^{\frac{1}{p}}$$

obeys the triangle inequality with a constant $c_p > 0$. Therefore,

$$\|f - g\|_{\dot V_p^{s,k}} \geq \left( \sum_{Q\in\pi} \frac{1}{|Q|^{(1+\frac{s}{n})p-1}} E_k(f - g, Q)^p \right)^{\frac{1}{p}}$$

$$
\begin{aligned}
&\geq c_p^{-1}\left(\sum_j \sum_{Q\in\pi_j} \frac{1}{|Q|^{(1+\frac{s}{n})p-1}} E_k(f-g_{Q_j},Q)^p\right)^{\frac{1}{p}} \\
&\quad -\left(\sum_j \sum_{Q\in\pi_j} \frac{1}{|Q|^{(1+\frac{s}{n})p-1}} E_k(g-g_{Q_j},Q)^p\right)^{\frac{1}{p}} \\
&\geq c_p^{-1}\left(\sum_j \sum_{Q\in\pi_j} \frac{1}{|Q|^{(1+\frac{s}{n})p-1}} E_k(f-f_{Q_j},Q)^p\right)^{\frac{1}{p}} \\
&\quad -\left(\sum_j \sum_{Q\in\pi_j} \frac{1}{|Q|^{(1+\frac{s}{n})p-1}} E_k(g-g_{Q_j},Q)^p\right)^{\frac{1}{p}} \\
&\geq c_p^{-1}\left(ct^p \sum_j |Q_j|^{\lambda p}\right)^{\frac{1}{p}} - \left(\varepsilon^p t^p \sum_j |Q_j|^{\lambda p}\right)^{\frac{1}{p}} \\
&= (c_p^{-1}c^{\frac{1}{p}} - \varepsilon)t\left(\sum_j |Q_j|^{\lambda p}\right)^{\frac{1}{p}}.
\end{aligned}
$$

Since $g \in B_{\dot{C}^{\lambda,k}(\dot{V}_p^{s,k})}(\varepsilon t)$ and the packing $\{Q_j\}$ are arbitrary, for sufficiently small $\varepsilon > 0$ we obtain

$$
\operatorname{dist}_{\dot{V}_p^{s,k}}(f, B_{\dot{C}^{\lambda,k}(\dot{V}_p^{s,k})}(\varepsilon t)) \geq \varepsilon t \left(|\Omega_{f,t}|_{\lambda p}\right)^{\frac{1}{p}}.
$$

Putting $c = \frac{1}{\varepsilon}$, we arrive at the required statement. □

### A.2.2 Proof of the main result (Theorem A.3)

*Proof.* First, we show that

$$
\left|\{Q_x\}_{x\in\Omega}\right|_{\lambda p} \leq c\,|\Omega_{f,t}|_{\lambda p}. \tag{A.21}
$$

Let $\pi = \{Q_{x_j}\}$ be a packing formed by cubes of the family $\{Q_x\}_{x\in\Omega}$. By (A.11), for every cube $Q_{x_j}$ there exists $\lambda_j \in [\frac{1}{2}, 1]$ such that $\lambda_j Q_{x_j} \in \Omega_{f,t}$. Since $\lambda_j \leq 1$, the cubes $\{\lambda_j Q_{x_j}\}$ are mutually disjoint. Therefore,

$$
\sum_j |Q_{x_j}|^{\lambda p} \leq 2^{n\lambda p} \sum_j |\lambda_j Q_{x_j}|^{\lambda p} \leq 2^{n\lambda p}\, |\Omega_{f,t}|_{\lambda p}.
$$

This implies (A.21).

Looking at (A.21) and Lemma A.10, we see that the theorem will be proved if we verify the following inequalities:

$$\|f_t\|_{\dot{C}^{\lambda,k}(\dot{V}_p^{s,k})} \leq ct \text{ and } \|f - f_t\|_{\dot{V}_p^{s,k}} \leq ct\left(\left|\{Q_x\}_{x\in\Omega}\right|_{\lambda p}\right)^{\frac{1}{p}}. \tag{A.22}$$

We start with the second inequality. Let $\pi$ be a packing; we split it into the families $\pi_b$ and $\pi_s$ of big (respectively, small) cubes. It suffices to show that

$$\left(\sum_{Q_j\in\pi_b}\left(\frac{1}{|Q_j|^{1+\frac{s}{n}}}E_k(f-f_t,Q_j)\right)^p|Q_j|\right)^{\frac{1}{p}} \leq ct\left(\left|\{Q_x\}_{x\in\Omega}\right|_{\lambda p}\right)^{\frac{1}{p}}, \tag{A.23}$$

$$\left(\sum_{Q_j\in\pi_s}\left(\frac{1}{|Q_j|^{1+\frac{s}{n}}}E_k(f-f_t,Q_j)\right)^p|Q_j|\right)^{\frac{1}{p}} \leq ct\left(\left|\{Q_x\}_{x\in\Omega}\right|_{\lambda p}\right)^{\frac{1}{p}}. \tag{A.24}$$

We prove (A.23). Recall that $f_t = f_{\{K_i\}}$. In Lemma A.8, we put $h = f$, leaving a packing $\pi_b$ of big cubes arbitrary. This yields

$$\begin{aligned}&\left(\sum_{Q_j\in\pi_b}\left(\frac{1}{|Q_j|^{1+\frac{s}{n}}}E_k(f-f_t,Q_j)\right)^p|Q_j|\right)^{\frac{1}{p}}\\ &\leq\left(\sum_{Q_j\in\pi_b}\left(\frac{1}{|Q_j|^{1+\frac{s}{n}}}\int_{Q_j}|f-f_t|\right)^p|Q_j|\right)^{\frac{1}{p}}\\ &\leq c\left(\sum_{i\in I}\left(\frac{1}{|K_i|^{1+\frac{s}{n}}}\int_{K_i}|f-f_{K_i}|\right)^p|K_i|\right)^{\frac{1}{p}}.\end{aligned}$$

To estimate the last quantity, we use the fact that the WB-covering $\{K_i\}_{i\in I}$ is controlled by some subfamily $\{Q_{x_j}\}_{j\in J}$ of $\{Q_x\}_{x\in\Omega}$, and this subfamily is of finite multiplicity, the corresponding constant being controlled in terms of the dimension.

So, for every cube $K_i$ there is a cube $Q_{x_j}$ in the controlling family such that

$$K_i \subset \gamma Q_{x_j}$$

with a numerical constant $\gamma \geq 1$.

The definition of the quasinorm in the space $\dot{V}_p^{s,k}(\gamma Q_{x_j})$ implies

$$\sum_{K_i\subset\gamma Q_j}\left(\frac{1}{|K_i|^{1+\frac{s}{n}}}\int_{K_i}|f-f_{K_i}|\right)^p|K_i| \leq \left\|f - f_{\gamma Q_{x_j}}\right\|^p_{\dot{V}_p^{s,k}(\gamma Q_{x_j})}.$$

Furthermore, the construction of the family $\{Q_x\}_{x\in\Omega}$ (see (A.11)) shows that

$$\frac{\left\|f - f_{\gamma Q_{x_j}}\right\|_{\dot{V}_p^{s,k}(\gamma Q_{x_j})}}{\left|\gamma Q_{x_j}\right|^{\lambda}} \leq t. \tag{A.25}$$

Therefore, the finite multiplicity property of the controlling family $\{Q_{x_j}\}_{j\in J}$ implies

$$\left(\sum_{i\in I}\left(\frac{1}{|K_i|^{1+\frac{s}{n}}}\int_{K_i}|f-f_{K_i}|\right)^p|K_i|\right)^{\frac{1}{p}} \leq ct\left(\sum_{j\in J}|Q_{x_j}|^{\lambda p}\right)^{\frac{1}{p}}$$
$$\leq ct\left(\left|\{Q_x\}_{x\in\Omega}\right|_{\lambda p}\right)^{\frac{1}{p}}. \qquad \text{(A.26)}$$

This proves (A.23). We pass to (A.24). By the triangle inequality, we have

$$\left(\sum_{Q_j\in\pi_s}\left(\frac{1}{|Q_j|^{1+\frac{s}{n}}}E_k(f-f_t,Q_j)\right)^p|Q_j|\right)^{\frac{1}{p}}$$
$$\leq c\left(\sum_{Q_j\in\pi_s}\left(\frac{1}{|Q_j|^{1+\frac{s}{n}}}E_k(f,Q_j)\right)^p|Q_j|\right)^{\frac{1}{p}}$$
$$+c\left(\sum_{Q_j\in\pi_s}\left(\frac{1}{|Q_j|^{1+\frac{s}{n}}}E_k(f_t,Q_j)\right)^p|Q_j|\right)^{\frac{1}{p}}$$
$$=c\left(\sum_{Q_j\in\pi_s}\left(\frac{1}{|Q_j|^{1+\frac{s}{n}}}\int_{Q_j}|f-f_{Q_j}|\right)^p|Q_j|\right)^{\frac{1}{p}}$$
$$+c\left(\sum_{Q_j\in\pi_s}\left(\frac{1}{|Q_j|^{1+\frac{s}{n}}}\int_{Q_j}\left|f_t-(f_t)_{Q_j}\right|\right)^p|Q_j|\right)^{\frac{1}{p}}.$$

Since $f_t = f_{\{K_i\}}$ (cf. (A.12) and (A.15)), Lemma A.6 shows that the second summand in the last expression is dominated by

$$\left(\sum_{i\in I}\left(\frac{1}{|K_i|^{1+\frac{s}{n}}}\int_{K_i}|f-f_{K_i}|\right)^p|K_i|\right)^{\frac{1}{p}},$$

which has already been estimated (see (A.26)). So, it remains to handle the first summand, that is, the quantity

$$\left(\sum_{Q_i\in\pi_s}\left(\frac{1}{|Q_i|^{1+\frac{s}{n}}}\int_{Q_i}|f-f_{Q_i}|\right)^p|Q_i|\right)^{\frac{1}{p}}.$$

Since the cubes $Q_i$ are small, for each of them there exists a cube $Q_{x_j}$ of the controlling family $\{Q_{x_j}\}_{j\in J}$ with $Q_i\subset\gamma Q_{x_j}$. The definition of the quasinorm in

$\dot{V}_p^{s,k}(\gamma Q_{x_j})$ yields

$$\sum_{Q_i \subset \gamma Q_{x_j}} \left( \frac{1}{|Q_i|^{1+\frac{s}{n}}} \int_{Q_i} |f - f_{Q_i}| \right)^p |Q_i| \leq \left\| f - f_{\gamma Q_{x_j}} \right\|^p_{\dot{V}_p^{s,k}(\gamma Q_{x_j})} .$$

So, as above, we obtain

$$\left( \sum_{Q_i \in \pi_s} \left( \frac{1}{|Q_i|^{1+\frac{s}{n}}} \int_{Q_i} |f - f_{Q_i}| \right)^p |Q_i| \right)^{\frac{1}{p}} \leq ct \left( \left| \{Q_x\}_{x \in \Omega} \right|_{\lambda p} \right)^{\frac{1}{p}},$$

proving the inequality

$$\|f - f_t\|_{\dot{V}_p^{s,k}} \leq ct \left( \left| \{Q_x\}_{x \in \Omega} \right|_{\lambda p} \right)^{\frac{1}{p}} .$$

Next, we verify the inequality

$$\|f_t\|_{\dot{C}^{\lambda,k}(\dot{V}_p^{s,k})} \leq ct,$$

which means that

$$\frac{\|f_t - (f_t)_Q\|_{\dot{V}_q^{s,k}(Q)}}{|Q|^\lambda} \leq ct \tag{A.27}$$

for every cube $Q$.

Now, $Q$ is either big or small. If it is small, it is included in some $K_{i_0}$. Then for every packing $\pi$ composed of subcubes of $Q$, we have

$$\left( \sum_{Q_i \in \pi} \left( \frac{1}{|Q_i|^{1+\frac{s}{n}}} E_k(f_t - (f_t)_Q, Q_i) \right)^p |Q_i| \right)^{\frac{1}{p}} \tag{A.28}$$

$$\leq \left( \sum_{Q_i \in \pi} \left( \frac{1}{|Q_i|^{1+\frac{s}{n}}} \int_{Q_i} |f_t - (f_t)_{Q_i}| \right)^p |Q_i| \right)^{\frac{1}{p}}$$

$$\leq c \left( \sum_{Q_i \in \pi} \left( \frac{1}{|Q_i|^{1+\frac{s}{n}}} |Q_i|^{1+\frac{k}{n}} \max_{|m|=k} \|D^m(f_t)\|_{L^\infty(Q_i)} \right)^p |Q_i| \right)^{\frac{1}{p}}$$

$$\leq c \left( \sum_{Q_i \in \pi} |Q_i|^{1+\frac{k-s}{n}p} \left( \max_{|m|=k} \|D^m(f_t)\|_{L^\infty(K_{i_0})} \right)^p \right)^{\frac{1}{p}}$$

$$\leq c \left( \sum_{Q_i \in \pi} |Q_i|^{1+\frac{k-s}{n}p} \right)^{\frac{1}{p}} \max_{|m|=k} \|D^m(f_t)\|_{L^\infty(K_{i_0})}$$

$$\leq c\,|Q|^{\frac{1}{p}+\frac{k-s}{n}} \max_{|m|=k} \|D^m(f_t)\|_{L^\infty(K_{i_0})}.$$

To estimate the last quantity, we observe that the formula for $f_t$, the strong engagement property of the family $\{K_i\}_{i\in I}$, and an estimate of the remainder term in the Taylor formula imply

$$\max_{|m|=k} \|D^m(f_t)\|_{L^\infty(K_{i_0})} \leq c \sum_{K_i\cap K_{i_0}\neq\varnothing} \frac{1}{|K_i|^{1+\frac{k}{n}}} \int_{K_i} |f - f_{K_i}|.$$

Furthermore, the definition of the quasinorm $\|f - f_{K_i}\|_{\dot{V}_p^{s,k}(K_i)}$ shows that

$$\frac{1}{|K_i|^{1+\frac{s}{n}-\frac{1}{p}}} \int_{K_i} |f - f_{K_i}| \leq \|f - f_{K_i}\|_{\dot{V}_p^{s,k}(K_i)},$$

and

$$\|f - f_{K_i}\|_{\dot{V}_p^{s,k}(K_i)} \leq t\,|K_i|^\lambda$$

by the controlled extension theorem. Thus, the last quantity in (A.28) is dominated by

$$c\,|Q|^{\frac{1}{p}+\frac{k-s}{n}} \max_{|m|=k} \|D^m(f_t)\|_{L^\infty(K_{i_0})} \leq c\,|Q|^{\frac{1}{p}+\frac{k-s}{n}} \sum_{K_i\cap K_{i_0}\neq\varnothing} \frac{1}{|K_i|^{1+\frac{k}{n}}} |K_i|^{1+\frac{s}{n}-\frac{1}{p}+\lambda}\, t.$$

By assumption, $\frac{1}{p} + \frac{k-s}{n} \geq \lambda$. Since the family $\{K_i\}_{i\in I}$ is strongly engaged and has finite multiplicity, we arrive at the required inequality

$$c\,|Q|^{\frac{1}{p}+\frac{k-s}{n}} \sum_{K_i\cap K_{i_0}\neq\varnothing} \frac{1}{|K_i|^{1+\frac{k}{n}}} |K_i|^{1+\frac{s}{n}-\frac{1}{p}+\lambda}\, t \leq c \frac{|Q|^{\frac{1}{p}+\frac{k-s}{n}-\lambda}}{|K_{i_0}|^{\frac{1}{p}+\frac{k-s}{n}-\lambda}} |Q|^\lambda\, t \leq c\,|Q|^\lambda\, t.$$

It remains to prove (A.27) in the case where $Q$ is big. Let $\pi = \{Q_j\}$ be a packing composed of subcubes of $Q$. Splitting it into two subpackings $\pi_b$ and $\pi_s$ that consist, respectively, of big and small cubes, we see that it suffices to establish the following inequalities:

$$\left(\sum_{Q_j\in\pi_s} \left(\frac{1}{|Q_j|^{1+\frac{s}{n}}} \int_{Q_j} |f_t - (f_t)_{Q_j}|\right)^p |Q_j|\right)^{\frac{1}{p}} \leq c\,|Q|^\lambda\, t, \tag{A.29}$$

$$\left(\sum_{Q_j\in\pi_b} \left(\frac{1}{|Q_j|^{1+\frac{s}{n}}} \int_{Q_j} |f_t - (f_t)_{Q_j}|\right)^p |Q_j|\right)^{\frac{1}{p}} \leq c\,|Q|^\lambda\, t. \tag{A.30}$$

We verify (A.29). Lemma A.6 and Proposition A.7 show that

$$\left(\sum_{Q_j\in\pi_s} \left(\frac{1}{|Q_j|^{1+\frac{s}{n}}} \int_{Q_j} |f_t - (f_t)_{Q_j}|\right)^p |Q_j|\right)^{\frac{1}{p}}$$

$$\leq c\left(\sum_{K_i\cap Q\neq\varnothing}\left(\frac{1}{|K_i|^{1+\frac{s}{n}}}\int_{K_i}|f-f_{K_i}|\right)^p|K_i|\right)^{\frac{1}{p}}\leq c\,\|f-f_{\gamma Q}\|_{\dot V_p^{s,k}(\gamma Q)}\leq c\,|Q|^{\lambda}\,t.$$

The last inequality is a consequence of the controlled extension theorem. To prove (A.30), we write

$$\begin{aligned}
&\left(\sum_{Q_j\in\pi_b}\left(\frac{1}{|Q_j|^{1+\frac{s}{n}}}\int_{Q_j}|f_t-(f_t)_{Q_j}|\right)^p|Q_j|\right)^{\frac{1}{p}}\\
&\leq\left(\sum_{Q_j\in\pi_b}\left(\frac{1}{|Q_j|^{1+\frac{s}{n}}}\int_{Q_j}|f_t-f_{Q_j}|\right)^p|Q_j|\right)^{\frac{1}{p}}\\
&\leq c_p\left(\sum_{Q_j\in\pi_b}\left(\frac{1}{|Q_j|^{1+\frac{s}{n}}}\int_{Q_j}|f-f_t|\right)^p|Q_j|\right)^{\frac{1}{p}}\\
&\quad+c_p\left(\sum_{Q_j\in\pi_b}\left(\frac{1}{|Q_j|^{1+\frac{s}{n}}}\int_{Q_j}|f-f_{Q_j}|\right)^p|Q_j|\right)^{\frac{1}{p}}.
\end{aligned}$$

By Lemma A.8, the first summand in the last expression is controlled by

$$\begin{aligned}
&\left(\sum_{i:K_i\cap(\bigcup Q_j)\neq\varnothing}\left(\frac{1}{|K_i|^{1+\frac{s}{n}}}\int_{K_i}|f-f_{K_i}|\right)^p|K_i|\right)^{\frac{1}{p}}\\
&\leq\left(\sum_{i:K_i\cap Q\neq\varnothing}\left(\frac{1}{|K_i|^{1+\frac{s}{n}}}\int_{K_i}|f-f_{K_i}|\right)^p|K_i|\right)^{\frac{1}{p}}.
\end{aligned}$$

Now, Proposition A.7 shows that if $K_i\cap Q\neq\varnothing$, then $K_i\subset\gamma Q$. Thus, since the cube $\gamma Q$ is big, we obtain

$$\begin{aligned}
&\left(\sum_{i:K_i\cap Q\neq\varnothing}\left(\frac{1}{|K_i|^{1+\frac{s}{n}}}\int_{K_i}|f-f_{K_i}|\right)^p|K_i|\right)^{\frac{1}{p}}\\
&\leq\left(\sum_{i:K_i\subset\gamma Q}\left(\frac{1}{|K_i|^{1+\frac{s}{n}}}\int_{K_i}|f-f_{K_i}|\right)^p|K_i|\right)^{\frac{1}{p}}\leq\|f-f_{\gamma Q}\|_{\dot V_p^{s,k}(\gamma Q)}\leq ct\,|Q|^{\lambda}.
\end{aligned}$$

To estimate the second summand, we use the assumption that $Q$ is big. Therefore,

$$\left(\sum_{Q_j\in\pi_b}\left(\frac{1}{|Q_j|^{1+\frac{s}{n}}}\int_{Q_j}|f-f_{Q_j}|\right)^p|Q_j|\right)^{\frac{1}{p}}\leq\|f-f_Q\|_{\dot V_p^{s,k}(Q)}\leq ct\,|Q|^{\lambda}.\qquad\square$$

## A.3 Near-minimizers for Morrey spaces built on the basis of Triebel–Lizorkin spaces

The seminorm in the Triebel–Lizorkin space $\dot{F}^{s,k}_{p,q}$ (see [T])[2] is given by the formula

$$\|f\|_{\dot{F}^{s,k}_{p,q}} = \left\| \left( \sum_{i\in\mathbb{Z}} \left( \frac{1}{|Q(x,2^i)|^{1+\frac{s}{n}}} E_k(f, Q(x,2^i)) \right)^q \right)^{\frac{1}{q}} \right\|_{L_p}; \tag{A.31}$$

$$k > s > 0, \quad 0 < p < \infty, \quad 1 + \frac{s}{n} - \frac{1}{p} > 0; \tag{A.32}$$

the localization (denoted by $\dot{F}^{s,k}_{p,q}(Q)$) of this space to a cube $Q$ is defined by formula (A.31) with an additional restriction; namely, the cubes $Q(x,2^i)$ must be included in $Q$.

We introduce a Morrey-type seminorm on the basis of $\dot{F}^{s,k}_{p,q}$ by

$$\|f\|_{\dot{C}^{\lambda,k}(\dot{F}^{s,k}_{p,q})} = \sup_Q \frac{1}{|Q|^\lambda} \inf_{P\in\mathcal{P}_k} \|f - P\|_{\dot{F}^{s,k}_{p,q}(Q)}, \lambda > 0. \tag{A.33}$$

We are interested in construction of a near-minimizer for the couple

$$(\dot{F}^{s,k}_{p,q}, \ \dot{C}^{\lambda,k}(\dot{F}^{s,k}_{p,q})).$$

It will turn out that the general algorithm described at the beginning of this chapter can be used for that. In the case in question, this algorithm looks like this.

We take $\dot{F}^{s,k}_{p,q}$ for the role of $X$. Suppose we are given a function $f \in \dot{F}^{s,k}_{p,q}$ and a number $t > 0$. For arbitrary $x \in \mathbb{R}^n$, we consider the function

$$\varphi_x(r) = \frac{1}{|Q(x,r)|^\lambda} \left\| f - f_{Q(x,r)} \right\|_{\dot{F}^{s,k}_{p,q}(Q(x,r))},$$

which tends to zero as $r \to \infty$, because $\lambda > 0$. Also, we consider the set

$$\Omega = \left\{ x \in \mathbb{R}^n : \sup_{r>0} \varphi_x(r) > t \right\}.$$

For $x \in \Omega$, we choose $r_x > 0$ in such a way that

$$\sup_{r\geq r_x} \varphi_x(r) \leq t \quad \text{and} \quad \sup_{r\geq \frac{1}{2}r_x} \varphi_x(r) > t. \tag{A.34}$$

Putting

$$Q_x = Q(x, r_x),$$

[2] This seminorm is not quite standard, see Notes and remarks at the end of this appendix.

we obtain a family $\{Q_x\}_{x\in\Omega}$.

Now, by (A.34) we obtain

$$\|f\|_{\dot F^{s,k}_{p,q}} \geq \|f\|_{\dot F^{s,k}_{p,q}(Q_x)} \geq ct\,|Q_x|^{\lambda} \tag{A.35}$$

and, since $\lambda > 0$, we see that

$$\sup_{x\in\Omega} |Q_x| < \infty.$$

So, the controlled extension theorem is applicable, yielding a WB-covering $\{K_i\}_{i\in I}$. We define a function $f_t$ by the formula

$$f_t = \sum_{i\in I} f_{K_i}\psi_i + f\,\chi_{\mathbb{R}^n\setminus\cup K_i}, \tag{A.36}$$

where $\{\psi_i\}_{i\in I}$ is a smooth partition of unity that is adjusted to the WB-covering $\{K_i\}_{i\in I}$ , and the $f_{K_i} \in \mathcal{P}_k$ are polynomials of the best approximation of $f$ on the cubes $K_i$, $i \in I$.

**Proposition A.11.** *We have*

$$\left|\{Q_x\}_{x\in\Omega}\right|_{\lambda p} \leq c\frac{\|f\|^p_{\dot F^{s,k}_{p,q}}}{t^p} < \infty. \tag{A.37}$$

*Proof.* Taking into account the fact that the cubes $Q_{x_j}$ are mutually disjoint, for every $x \in \mathbb{R}^n$ we obtain:

$$\begin{aligned}&\left(\sum_{i\in\mathbb{Z}}\left(\frac{1}{|Q(x,2^i)|^{1+\frac{s}{n}}}E_k(f,Q(x,2^i))\right)^q\right)^{\frac{p}{q}}\\ &\geq \sum_{j\in J}\left(\sum_{i:Q(x,2^i)\subset Q_{x_j}}\left(\frac{1}{|Q(x,2^i)|^{1+\frac{s}{n}}}E_k(f,Q(x,2^i))\right)^q\right)^{\frac{p}{q}}.\end{aligned}$$

The definition of the seminorm in $\dot F^{s,k}_{p,q}$ and (A.35) show that

$$\|f\|^p_{\dot F^{s,k}_{p,q}} \geq \sum_{Q_{x_j}\in\pi} \|f\|^p_{\dot F^{s,k}_{p,q}(Q_{x_j})} \geq c \sum_{Q_{x_j}\in\pi} t^p\,|Q_{x_j}|^{\lambda p},$$

and this is (A.37). □

In order to state the result, we need the following family of cubes:

$$\Omega_{f,t} = \left\{\frac{1}{|Q|^{\lambda}}\inf_{P\in\mathcal{P}_k}\|f-P\|_{\dot F^{s,k}_{p,q}(Q)} \geq t\right\}.$$

**Theorem A.12.** *Suppose we are given a function $f \in \dot F^{s,k}_{p,q}$ and a number $t > 0$. If $1 \le q$, $s > 0$, $\frac{1}{p} + \frac{k-s}{n} \ge \lambda > 0$, and* (A.32) *is fulfilled, then*

$$\operatorname{dist}_{\dot F^{s,k}_{p,q}}(f, B_{\dot C^{\lambda,k}(\dot F^{s,k}_{p,q})}(t)) \approx t\left(|\Omega_{f,t}|_{\lambda p}\right)^{\frac{1}{p}}$$

*and the function $f_t$* (see (A.36)) *is a near-minimizer for the distance functional of the couple $(\dot F^{s,k}_{p,q}, \dot C^{\lambda,k}(\dot F^{s,k}_{p,q}))$.*

## A.3.1 Auxiliary lemmas

Below, we denote by $f_Q$ the polynomial of degree strictly smaller then $k$ and such that

$$E_k(f,Q) = \|f - f_Q\|_{\dot F^{s,k}_{p,q}(Q)} = \inf_{P\in\mathcal{P}_k} \|f-P\|_{\dot F^{s,k}_{p,q}(Q)}\,.$$

We prove the following analog of Lemma A.10.

**Lemma A.13.** *For some $c > 0$ independent of $f \in \dot F^{s,k}_{p,q}$ and $t > 0$, we have*

$$t\left(|\Omega_{f,t}|_{\lambda p}\right)^{\frac{1}{p}} \le c \operatorname{dist}_{\dot F^{s,k}_{p,q}}(f, B_{\dot C^{\lambda,k}(\dot F^{s,k}_{p,q})}\left(\frac{t}{c}\right)).$$

*Proof.* Let $\pi = \{Q_j\}$ be a packing consisting of cubes of the family $\Omega_{f,t}$. The definition of this family shows that

$$\frac{\|f - f_{Q_j}\|_{\dot F^{s,k}_{p,q}(Q_j)}}{|Q_j|^{\lambda}} \ge t.$$

Therefore, the formula for the seminorm in the space $\dot F^{s,k}_{p,q}(Q_j)$ implies

$$\begin{aligned}&\|f - f_{Q_j}\|^p_{\dot F^{s,k}_{p,q}(Q_j)}\\ &= \int_{\mathbb{R}^n}\left(\sum_{i:Q(x,2^i)\subset Q_j}\left(\frac{1}{|Q(x,2^i)|^{1+\frac{s}{n}}}E_k(f-f_{Q_j},Q(x,2^i))\right)^q\right)^{\frac{p}{q}} \ge t^p\,|Q_j|^{\lambda p}.\end{aligned}$$

Now, let $g \in \dot C^{\lambda,k}(\dot F^{s,k}_{p,q})$, and let

$$\|g\|_{\dot C^{\lambda,k}(\dot F^{s,k}_{p,q})} \le \varepsilon t,$$

where $\varepsilon > 0$ will be chosen later. For every cube $Q_j$ there exists a polynomial $g_{Q_j}$ of degree strictly smaller than $k$ such that

$$\begin{aligned}\|g - g_{Q_j}\|^p_{\dot F^{s,k}_{p,q}(Q_j)} &= \int_{\mathbb{R}^n}\left(\sum_{i:Q(x,2^i)\subset Q_j}\left(\frac{1}{|Q(x,2^i)|^{1+\frac{s}{n}}}E_k(g-g_{Q_j},Q(x,2^i))\right)^q\right)^{\frac{p}{q}}\\ &\le \varepsilon^p t^p\,|Q_j|^{\lambda p},\end{aligned}$$

whence

$$\sum_{Q_j\in\pi}\int_{\mathbb{R}^n}\left(\sum_{i:Q(x,2^i)\subset Q_j}\left(\frac{1}{|Q(x,2^i)|^{1+\frac{s}{n}}}E_k(g-g_{Q_j},Q(x,2^i))\right)^q\right)^{\frac{p}{q}}$$
$$\leq \varepsilon^p t^p \sum_j |Q_j|^{\lambda p}.$$

Next, since $f_{Q_j}$ is a best approximation polynomial for $f$ on $Q_j$, we have

$$\|f-f_{Q_j}\|^p_{\dot{F}^{s,k}_{p,q}(Q_j)} \leq \|f-g_{Q_j}\|^p_{\dot{F}^{s,k}_{p,q}(Q_j)}.$$

Therefore,

$$\int_{\mathbb{R}^n}\left(\sum_{i:Q(x,2^i)\subset Q_j}\left(\frac{1}{|Q(x,2^i)|^{1+\frac{s}{n}}}E_k(f-g_{Q_j},Q(x,2^i))\right)^q\right)^{\frac{p}{q}} \geq t\,|^pQ_j|^{\lambda p}.$$

Since the cubes of the family $\{Q_j\}$ are mutually disjoint, we obtain

$$\sum_{Q_j\in\pi}\int_{\mathbb{R}^n}\left(\sum_{i:Q(x,2^i)\subset Q_j}\left(\frac{1}{|Q(x,2^i)|^{1+\frac{s}{n}}}E_k(f-g_{Q_j},Q(x,2^i))\right)^q\right)^{\frac{p}{q}} \geq t^p\sum_j|Q_j|^{\lambda p}.$$

Next,

$$\|f-g\|_{\dot{F}^{s,k}_{p,q}} \geq \left(\int_{\mathbb{R}^n}\left(\sum_{i:Q(x,2^i)\subset\bigcup Q_j}\left(\frac{1}{|Q(x,2^i)|^{1+\frac{s}{n}}}E_k(f-g,Q(x,2^i))\right)^q\right)^{\frac{p}{q}}\right)^{\frac{1}{p}} \tag{A.38}$$
$$= \left(\sum_{j\in J}\int_{\mathbb{R}^n}\left(\sum_{i:Q(x,2^i)\subset Q_j}\left(\frac{1}{|Q(x,2^i)|^{1+\frac{s}{n}}}E_k(f-g,Q(x,2^i))\right)^q\right)^{\frac{p}{q}}\right)^{\frac{1}{p}}.$$

We estimate the last quantity from below with the help of the triangle inequality for the seminorm

$$\|h\| = \left\|\sum_{j\in J}\int_{\mathbb{R}^n}\left(\sum_{i:Q(x,2^i)\subset Q_j}\left(\frac{1}{|Q(x,2^i)|^{1+\frac{s}{n}}}E_k(h,Q(x,2^i))\right)^q\right)^{\frac{1}{q}}\right\|_{L^p}.$$

This leads to the following minorant:

$$c_p^{-1}\left(\sum_{j\in J}\int_{\mathbb{R}^n}\left(\sum_{i:Q(x,2^i)\subset Q_j}\left(\frac{1}{|Q(x,2^i)|^{1+\frac{s}{n}}}E_k(f-g_{Q_j},Q(x,2^i))\right)^q\right)^{\frac{p}{q}}\right)^{\frac{1}{p}}$$
$$-\left(\sum_{j\in J}\int_{\mathbb{R}^n}\left(\sum_{i:Q(x,2^i)\subset Q_j}\left(\frac{1}{|Q(x,2^i)|^{1+\frac{s}{n}}}E_k(g-g_{Q_j},Q(x,2^i))\right)^q\right)^{\frac{p}{q}}\right)^{\frac{1}{p}}$$
$$\geq c_p^{-1}\left(t^p\sum_j|Q_j|^{\lambda p}\right)^{\frac{1}{p}}-\left(\varepsilon^p t^p\sum_j|Q_j|^{\lambda p}\right)^{\frac{1}{p}}=(c_p^{-1}-\varepsilon)t\left(\sum_j|Q_j|^{\lambda p}\right)^{\frac{1}{p}}.$$

Since $g\in B_{\dot{C}^{\lambda,k}(\dot{F}^{s,k}_{p,q})}(\varepsilon t)$ and the packing $\{Q_j\}$ are arbitrary, taking $\varepsilon>0$ sufficiently small (specifically, $\varepsilon\leq\frac{1}{2}c_p^{-1}$ suffices) we arrive at the inequality

$$\operatorname{dist}_{\dot{F}^{s,k}_{p,q}}(f,B_{\dot{C}^{\lambda,k}(\dot{F}^{s,k}_{p,q})}(\varepsilon t))\geq\varepsilon t\left(|\Omega_{f,t}|_{\lambda p}\right)^{\frac{1}{p}}.$$

The claim follows if we put $c=\frac{1}{\varepsilon}$. □

Consider an arbitrary WB-covering $\{K_i\}_{i\in I}$. Let $\{\psi_i\}_{i\in I}$ be a smooth partition of unity adjusted to this covering. Taking a function $h\in\dot{F}^{s,k}_{p,q}$, we decompose it as follows:

$$h=h_{\{K_i\}}+(h-h_{\{K_i\}}),$$

where

$$h_{\{K_i\}}=\sum_{i\in I}h_{K_i}\psi_i+h\chi_{\mathbb{R}^n/\cup K_i} \tag{A.39}$$

and the $h_{K_i}$ are polynomials of degree strictly smaller than $k$ that provide the best approximation to $h$ on the cubes $K_i$, $i\in I$, in the metric of $\dot{F}^{s,k}_{p,q}(K_i)$.

We prove the following analog of Lemma A.8.

**Lemma A.14.** *If $1\leq q$ and $s>0$, then*

$$\left\|\left(\sum_{i:Q(x,2^i)\ is\ big}\left(\frac{1}{|Q(x,2^i)|^{1+\frac{s}{n}}}\int_{Q(x,2^i)}|h-h_{\{K_i\}}|\right)^q\right)^{\frac{1}{q}}\right\|_{L^p} \tag{A.40}$$
$$\leq c\left(\sum_{i\in I}\left(\frac{1}{|K_i|^{1+\frac{s}{n}}}\int_{K_j}|h-h_{K_j}|\right)^p|K_i|\right)^{\frac{1}{p}}.$$

*Proof.* For arbitrary cube $Q$ we have

$$\int_Q|h-h_{\{K_i\}}|\leq\sum_{j:Q\cap K_j\neq\varnothing}\int_{K_j}|h-h_{K_j}|,$$

so that the left-hand side of (A.40) is dominated by the quantity

$$\left\| \left( \sum_{Q(x,2^i) \text{ is big}} \left( \frac{1}{|Q(x,2^i)|^{1+\frac{s}{n}}} \sum_{j:Q(x,2^i)\cap K_j \neq \varnothing} \int_{K_j} |h - h_{K_j}| \right)^q \right)^{\frac{1}{q}} \right\|_{L^p}.$$

We denote by $Q(x, 2^{i(j)})$ the smallest of the big cubes of the form $Q(x, 2^i)$ that intersect $K_j$. Since $q \geq 1$, we obtain

$$\begin{aligned}
&\left\| \left( \sum_{Q(x,2^i) \text{ is big}} \left( \frac{1}{|Q(x,2^i)|^{1+\frac{s}{n}}} \sum_{j:Q(x,2^i)\cap K_j \neq \varnothing} \int_{K_j} |h - h_{K_j}| \right)^q \right)^{\frac{1}{q}} \right\|_{L^p} \\
&\leq \left\| \sum_{Q(x,2^i) \text{ is big}} \sum_{j:Q(x,2^i)\cap K_j \neq \varnothing} \frac{1}{|Q(x,2^i)|^{1+\frac{s}{n}}} \int_{K_j} |h - h_{K_j}| \right\|_{L^p} \\
&\leq c \left\| \sum_j \frac{1}{|Q(x,2^{i(j)})|^{1+\frac{s}{n}}} \int_{K_j} |h - h_{K_j}| \right\|_{L^p}.
\end{aligned}$$

Since the cube $Q(x, 2^{i(j)})$ is big, Proposition A.7 (see (A.17)) shows that

$$K_j \subset \gamma Q(x, 2^{i(j)})$$

for some $\gamma$ depending only on the dimension $n$. Consequently (this is an important point of the proof), we have the following estimate in terms of the Hardy–Littlewood maximal function:

$$\frac{|K_j|}{|Q(x,2^{i(j)})|} = \frac{\gamma^n}{|\gamma Q(x,2^{i(j)})|} \int_{\gamma Q(x,2^{i(j)})} \chi_{K_j} dx \leq c M(\chi_{K_j})(x).$$

Therefore, the expression to be estimated can be rewritten in a form convenient for application of the Fefferman–Stein maximal theorem (Theorem 0.23):

$$\begin{aligned}
&\left\| \sum_j \frac{1}{|Q(x,2^{i(j)})|^{1+\frac{s}{n}}} \int_{K_j} |h - h_{K_j}| \right\|_{L^p} \qquad \text{(A.41)} \\
&= \left\| \sum_j \frac{|K_j|^{1+\frac{s}{n}}}{|Q(x,2^{i(j)})|^{1+\frac{s}{n}}} \cdot \frac{1}{|K_j|^{1+\frac{s}{n}}} \int_{K_j} |h - h_{K_j}| \right\|_{L^p} \\
&\leq c \left\| \sum_j (M\chi_{K_j})^{1+\frac{s}{n}} \frac{1}{|K_j|^{1+\frac{s}{n}}} \int_{K_j} |h - h_{K_j}| \right\|_{L^p}
\end{aligned}$$

$$= c\left(\left\|\left(\sum_j \left(M\left[\frac{1}{|K_j|}\left(\int_{K_j}|h-h_{K_j}|\right)^{\frac{1}{1+\frac{s}{n}}}\chi_{K_j}\right]\right)^{1+\frac{s}{n}}\right)^{\frac{1}{1+\frac{s}{n}}}\right\|_{L^{p(1+\frac{s}{n})}}\right)^{1+\frac{s}{n}}.$$

By (A.32), we have $p(1+\frac{s}{n}) > 1$. Therefore, the Fefferman–Stein maximal theorem (Theorem 0.23) is applicable, yielding

$$\left\|\sum_j \frac{1}{|Q(x,2^{i(j)})|^{1+\frac{s}{n}}}\int_{K_j}|h-h_{K_j}|\right\|_{L^p}$$
$$\le c\left(\left\|\left(\sum_j \left(\frac{1}{|K_j|}\left(\int_{K_j}|h-h_{K_j}|\right)^{\frac{1}{1+\frac{s}{n}}}\chi_{K_j}\right)^{1+\frac{s}{n}}\right)^{\frac{1}{1+\frac{s}{n}}}\right\|_{L^{p(1+\frac{s}{n})}}\right)^{1+\frac{s}{n}}.$$

The family $\{K_i\}$ being of finite multiplicity, we see that

$$\left\|\sum_j \frac{1}{|Q(x,2^{i(j)})|^{1+\frac{s}{n}}}\int_{K_j}|h-h_{K_j}|\right\|_{L^p}$$
$$\le c\left\|\sum_j \frac{1}{|K_j|^{1+\frac{s}{n}}}\left(\int_{K_j}|h-h_{K_j}|\right)\chi_{K_j}\right\|_{L^p}.$$

Again by finite multiplicity, we estimate the right-hand side by the quantity

$$c\left(\sum_{i\in I}\left(\frac{1}{|K_i|^{1+\frac{s}{n}}}\int_{K_j}|h-h_{K_j}|\right)^p |K_i|\right)^{\frac{1}{p}}. \qquad \square$$

The same argument yields the following statement, to be used below.

**Proposition A.15.** *If a cube $Q$ is big, then*

$$\left\|\left(\sum_{i:Q(x,2^i)\ is\ big,\ Q(x,2^i)\subset Q}\left(\frac{1}{|Q(x,2^i)|^{1+\frac{s}{n}}}\int_{Q(x,2^i)}|h-h_{\{K_i\}}|\right)^q\right)^{\frac{1}{q}}\right\|_{L^p}$$
$$\le c\left(\sum_{i:K_i\cap Q\neq\varnothing}\left(\frac{1}{|K_i|^{1+\frac{s}{n}}}\int_{K_i}|h-h_{K_i}|\right)^p |K_i|\right)^{\frac{1}{p}}.$$

We have also an analog of Lemma A.6.

**Lemma A.16.** *The following inequality is true:*

$$\left\| \left( \sum_{i:Q(x,2^i) \text{ is small}} \left( \frac{1}{|Q(x,2^i)|^{1+\frac{s}{n}}} \int_{Q(x,2^i)} \left| h_{\{K_i\}} - \left(h_{\{K_i\}}\right)_{Q(x,2^i)} \right| \right)^q \right)^{\frac{1}{q}} \right\|_{L^p}$$
$$\leq c \left( \sum_{i \in I} \left( \frac{1}{|K_i|^{1+\frac{s}{n}}} \int_{K_i} |h - h_{K_i}| \right)^p |K_i| \right)^{\frac{1}{p}}.$$

*Proof.* As in the proof of Lemma A.6, for every small cube $Q(x,2^i)$ we can write

$$\frac{1}{|Q(x,2^i)|^{1+\frac{s}{n}}} \int_{Q(x,2^i)} \left| h_{\{K_i\}} - (h_{\{K_i\}})_{Q(x,2^i)} \right|$$
$$\leq c \left| Q(x,2^i) \right|^{\frac{k-s}{n}} \max_{|m|=k} \left\| D^m (h_{\{K_i\}}) \right\|_{L^\infty(Q(x,2^i))}.$$

We denote by $i_x$ the largest $i$ such that the cube $Q(x,2^i)$ is small, and by $K_{j_x}$ a cube in the family $\{K_i\}$ with $Q(x,2^{i_x}) \subset K_{j_x}$.

Since $k - s > 0$ (see the definition (A.31) of the Triebel–Lizorkin spaces), we obtain

$$\left( \sum_{i \leq i_x} \left( \frac{1}{|Q(x,2^i)|^{1+\frac{s}{n}}} \int_{Q(x,2^i)} \left| h_{\{K_i\}} - (h_{\{K_i\}})_{Q(x,2^i)} \right| \right)^q \right)^{\frac{1}{q}}$$
$$\leq c \left( \sum_{i \leq i_x} \left( \left| Q(x,2^i) \right|^{\frac{k-s}{n}} \max_{|m|=k} \left\| D^m (h_{\{K_i\}}) \right\|_{L^\infty(Q(x,2^i))} \right)^q \right)^{\frac{1}{q}}$$
$$\leq c \left| K_{j_x} \right|^{\frac{k-s}{n}} \max_{|m|=k} \left\| D^m (h_{\{K_i\}}) \right\|_{L^\infty(K_{j_x})}.$$

As above (see (A.16)), we establish the inequality

$$\left\| D^m (h_{\{K_i\}}) \right\|_{L^\infty(K_{j_x})} \leq c \sum_{i: K_i \cap K_{j_x} \neq \varnothing} \frac{1}{|K_i|^{1+\frac{k}{n}}} \int_{K_i} |h - h_{K_i}|.$$

Therefore, the strong engagement property of the family $\{K_i\}$ implies

$$\left( \sum_{i \leq i_x} \left( \frac{1}{|Q(x,2^i)|^{1+\frac{s}{n}}} \int_{Q(x,2^i)} \left| h_{\{K_i\}} - (h_{\{K_i\}})_{Q(x,2^i)} \right| \right)^q \right)^{\frac{1}{q}}$$
$$\leq c \sum_{i: K_i \cap K_{j_x} \neq \varnothing} \frac{1}{|K_i|^{1+\frac{s}{n}}} \int_{K_i} \left| h - h_{K_j} \right|.$$

Invoking the finite multiplicity property of the family $\{K_i\}$ once again, we arrive at the required estimate

$$\left\| \left( \frac{1}{|K_i|^{1+\frac{s}{n}}} \sum_{K_i \cap K_{j_x} \neq \varnothing} \int_{K_i} \left| h - h_{K_j} \right| \right) \chi_{K_{j_x}} \right\|_{L^p} \\ \leq c \left( \sum_{i \in I} \left( \frac{1}{|K_i|^{1+\frac{s}{n}}} \int_{K_j} \left| h - h_{K_j} \right| \right)^p |K_i| \right)^{\frac{1}{p}}. \qquad \square$$

**Proposition A.17.** *Under the assumptions of Theorem* A.12, *we have*

$$\max(\|f_{\{K_i\}}\|_{\dot{F}^{s,k}_{p,q}}, \|f - f_{\{K_i\}}\|_{\dot{F}^{s,k}_{p,q}}) \leq c\|f\|_{\dot{F}^{s,k}_{p,q}},$$

*where $c > 0$ depends only on the dimension and the parameters $p, q, s, k$.*

*Proof.* It suffices to show that

$$\|f_{\{K_i\}}\|_{\dot{F}^{s,k}_{p,q}} \leq c\|f\|_{\dot{F}^{s,k}_{p,q}}.$$

As in the proof of Proposition A.9, the matter reduces to the estimate

$$\left( \sum_{i \in I} \left( \frac{1}{|K_i|^{1+\frac{s}{n}}} E_k(f, K_i) \right)^p |K_i| \right)^{\frac{1}{p}} \leq c\|f\|_{\dot{F}^{s,k}_{p,q}}. \tag{A.42}$$

This inequality is proved as follows. Since the family $\{K_i\}$ is of finite multiplicity, we may assume that it consists of mutually disjoint cubes. Now, we observe that for $x \in K_i$ we have

$$\frac{1}{|K_i|^{1+\frac{s}{n}}} E_k(f, K_i) \leq c \left( \sum_{i \in \mathbb{Z}} \left( \frac{1}{|Q(x, 2^i)|^{1+\frac{s}{n}}} E_k(f, Q(x, 2^i)) \right)^q \right)^{\frac{1}{q}}. \tag{A.43}$$

Indeed, take the smallest cube $Q(x, 2^j)$ with $K_i \subset Q(x, 2^j)$, then even

$$\frac{1}{|K_i|^{1+\frac{s}{n}}} E_k(f, K_i)\chi_{K_i} \leq c \frac{1}{|Q(x, 2^j)|^{1+\frac{s}{n}}} E_k(f, Q(x, 2^j))\chi_{K_i}.$$

Now, (A.43) can be rewritten as follows:

$$\sum_{i \in I} \frac{1}{|K_i|^{1+\frac{s}{n}}} E_k(f, K_i)\chi_{K_i}(x) \leq c \left( \sum_{i \in \mathbb{Z}} \left( \frac{1}{|Q(x, 2^i)|^{1+\frac{s}{n}}} E_k(f, Q(x, 2^i)) \right)^q \right)^{\frac{1}{q}}.$$

Taking the $L^p$-norm of both sides, we arrive at (A.42). $\square$

### A.3.2 Proof of the main result (Theorem A.12)

Largely, the arguments repeat those in the proof of Theorem A.3. For the reader's convenience, we present the details.

*Proof.* First, we show that

$$\left|\{Q_x\}_{x\in\Omega}\right|_{\lambda p} \le c\left|\Omega_{f,t}\right|_{\lambda p}. \tag{A.44}$$

Indeed, consider a packing $\pi = \left\{Q_{x_j}\right\}$ composed of cubes of the family $\{Q_x\}_{x\in\Omega}$. By (A.34), for every $Q_{x_j}$ there exists $\lambda_j \in \left[\frac{1}{2}, 1\right]$ with $\lambda_j Q_{x_j} \in \Omega_{f,t}$. As in the proof of Theorem A.3, this implies (A.44).

Now, (A.44) and Lemma A.13 show that the theorem will be proved if we establish the inequalities

$$\|f_t\|_{\dot{C}^{\lambda,k}(\dot{F}^{s,k}_{p,q})} \le ct \text{ and } \|f - f_t\|_{\dot{F}^{s,k}_{p,q}} \le ct\left(\left|\{Q_x\}_{x\in\Omega}\right|_{\lambda p}\right)^{\frac{1}{p}}. \tag{A.45}$$

We start with the second. By Lemma A.14 with $h = f$, we have

$$\left\|\left(\sum_{i:Q(x,2^i)\text{ is big}}\left(\frac{1}{|Q(x,2^i)|^{1+\frac{s}{n}}}\int_{Q(x,2^i)}|f-f_t|\right)^q\right)^{\frac{1}{q}}\right\|_{L^p}$$

$$\le c\left(\sum_{i\in I}\left(\frac{1}{|K_i|^{1+\frac{s}{n}}}\int_{K_j}|f - f_{K_j}|\right)^p|K_i|\right)^{\frac{1}{p}}. \tag{A.46}$$

Then the invocation of a controlling family $\pi = \left\{Q_{x_j}\right\}$ of cubes leads to the following (in the same way as in Theorem A.3):

$$\left(\sum_{K_i\subset\gamma Q_{x_j}}\left(\frac{1}{|K_i|^{1+\frac{s}{n}}}\int_{K_i}|f - f_{K_i}|\right)^p|K_i|\right)^{\frac{1}{p}} \le ct\left|Q_{x_j}\right|^{\lambda}.$$

Therefore,

$$\left(\sum_{i\in I}\left(\frac{1}{|K_i|^{1+\frac{s}{n}}}\int_{K_i}|f - f_{K_i}|\right)^p|K_i|\right)^{\frac{1}{p}} \le ct\left(\sum_{j\in J}\left|Q_{x_j}\right|^{\lambda p}\right)^{\frac{1}{p}}$$
$$\le ct\left(\left|\{Q_x\}_{x\in\Omega}\right|_{\lambda p}\right)^{\frac{1}{p}}.$$

The corresponding estimate over small cubes involves the inequality

$$\left\|\left(\sum_{i:Q(x,2^i)\text{ is small}}\left(\frac{1}{|Q(x,2^i)|^{1+\frac{s}{n}}}E_k(f - f_t, Q(x,2^i))\right)^q\right)^{\frac{1}{q}}\right\|_{L^p}$$

$$\leq c \left\| \left( \sum_{i:Q(x,2^i) \text{ is small}} \left( \frac{1}{|Q(x,2^i)|^{1+\frac{s}{n}}} E_k(f, Q(x,2^i)) \right)^q \right)^{\frac{1}{q}} \right\|_{L^p}$$

$$+ c \left\| \left( \sum_{i:Q(x,2^i) \text{ is small}} \left( \frac{1}{|Q(x,2^i)|^{1+\frac{s}{n}}} E_k(f_t, Q(x,2^i)) \right)^q \right)^{\frac{1}{q}} \right\|_{L^p}$$

$$\leq c \left\| \left( \sum_{i:Q(x,2^i) \text{ is small}} \left( \frac{1}{|Q(x,2^i)|^{1+\frac{s}{n}}} \int_{Q(x,2^i)} |f - f_{Q(x,2^i)}| \right)^q \right)^{\frac{1}{q}} \right\|_{L^p}$$

$$+ c \left\| \left( \sum_{i:Q(x,2^i) \text{ is small}} \left( \frac{1}{|Q(x,2^i)|^{1+\frac{s}{n}}} \int_{Q(x,2^i)} \left| f_t - (f_t)_{Q(x,2^i)} \right| \right)^q \right)^{\frac{1}{q}} \right\|_{L^p}.$$

With the help of Lemma A.16 with $h = f$, the second term in the last expression is dominated by the right-hand side of (A.46), which has already been estimated. So, it remains to control the quantity

$$\left\| \left( \sum_{i:Q(x,2^i) \text{ is small}} \left( \frac{1}{|Q(x,2^i)|^{1+\frac{s}{n}}} \int_{Q(x,2^i)} |f - f_{Q(x,2^i)}| \right)^q \right)^{\frac{1}{q}} \right\|_{L^p}.$$

Consider a family $\{Q_{x_j}\}_{j \in J}$ that controls the family $\{K_i\}_{i \in I}$. In relation to the former family, the cubes $Q(x, 2^i)$ will be split as follows. For each $x \in \bigcup K_i$, find $i = i(x)$ with the property that $Q(x, 2^i)$ is still small, but $Q(x, 2^{i+1})$ is big. Then $Q(x, 2^i)$ is included in $\gamma Q_{x_j}$, where $Q_{x_j}$ is a cube of the controlling family. Then

$$\left\| \left( \sum_{i:Q(x,2^i) \subset \gamma Q_{x_j}} \left( \frac{1}{|Q(x,2^i)|^{1+\frac{s}{n}}} \int_{Q(x,2^i)} |f - f_{Q(x,2^i)}| \right)^q \right)^{\frac{1}{q}} \right\|_{L^p}^p$$
$$\leq \left\| f - f_{\gamma Q_{x_j}} \right\|_{\dot{F}^{s,k}_{p,q}(\gamma Q_{x_j})}^p \leq ct^p \left| Q_{x_j} \right|^{\lambda p}$$

and therefore

$$\left\| \left( \sum_{i:Q(x,2^i) \text{ is small}} \left( \frac{1}{|Q(x,2^i)|^{1+\frac{s}{n}}} \int_{Q(x,2^i)} |f - f_{Q(x,2^i)}| \right)^q \right)^{\frac{1}{q}} \right\|_{L^p}^p$$
$$\leq ct^p \sum_j \left| Q_{x_j} \right|^{\lambda p} \leq ct^p \left| \{Q_x\}_{x \in \Omega} \right|_{\lambda p}.$$

This proves the inequality

$$\|f - f_t\|_{\dot{F}^{s,k}_{p,q}} \le ct \left( \left| \{Q_x\}_{x\in\Omega} \right|_{\lambda p} \right)^{\frac{1}{p}}.$$

Now, we verify the inequality

$$\|f_t\|_{\dot{C}^{\lambda,k}(\dot{F}^{s,k}_{p,q})} \le ct,$$

which means that

$$\frac{\|f_t - (f_t)_Q\|_{\dot{F}^{s,k}_{p,q}(Q)}}{|Q|^{\lambda}} \le ct.$$

for every cube $Q$.

Two cases are possible: either $Q$ is small or it is big. In the former case, $Q$ is a subset of some $K_{i_0}$.

By assumption, $\frac{1}{p} + \frac{k-s}{n} \ge \lambda$. Since

$$\max_{|m|=k} \|D^m(f_t)\|_{L^\infty(K_{i_0})} \le c \sum_{K_i \cap K_{i_0} \ne \varnothing} \frac{1}{|K_i|^{1+\frac{k}{n}}} \int_{K_i} |f - f_{K_i}|$$

and

$$\frac{1}{|K_i|^{1+\frac{s}{n}-\frac{1}{p}}} \int_{K_i} |f - f_{K_i}| \le |K_i|^{\lambda}\, t,$$

we obtain

$$\|f_t - (f_t)_Q\|_{\dot{F}^{s,k}_{p,q}(Q)} = \left\| \left( \sum_{Q(x,2^i)\subset Q} \left( \frac{1}{|Q(x,2^i)|^{1+\frac{s}{n}}} E_k(f_t, Q(x,2^i)) \right)^q \right)^{\frac{1}{q}} \right\|_{L^p}$$

$$\le c \left\| \left( \sum_{Q(x,2^i)\subset Q} \left( \frac{1}{|Q(x,2^i)|^{1+\frac{s}{n}}} |Q(x,2^i)|^{1+\frac{k}{n}} \max_{|m|=k} \|D^m(f_t)\|_{L^\infty(K_{i_0})} \right)^q \right)^{\frac{1}{q}} \right\|_{L^p}$$

$$\le c\,|Q|^{\frac{1}{p}+\frac{k-s}{n}} \max_{|m|=k} \|D^m(f_t)\|_{L^\infty(K_{i_0})}$$

$$\le c\,|Q|^{\frac{1}{p}+\frac{k-s}{n}} \sum_{K_i\cap K_{i_0}\ne\varnothing} \frac{1}{|K_i|^{1+\frac{k}{n}}} |K_i|^{1+\frac{s}{n}-\frac{1}{p}} |K_i|^{\lambda}\, t$$

$$\le c \frac{|Q|^{\frac{1}{p}+\frac{k-s}{n}-\lambda}}{|K_{i_0}|^{\frac{1}{p}+\frac{k-s}{n}-\lambda}} |Q|^{\lambda}\, t \le c\,|Q|^{\lambda}\, t.$$

Now, let $Q$ be big. Then

$$\|f_t - (f_t)_Q\|_{\dot{F}^{s,k}_{p,q}(Q)} = \left\| \left( \sum_{Q(x,2^i)\subset Q} \left( \frac{1}{|Q(x,2^i)|^{1+\frac{s}{n}}} E_k(f_t, Q(x,2^i)) \right)^q \right)^{\frac{1}{q}} \right\|_{L^p}$$

$$\leq c \left\| \left( \sum_{Q(x,2^i) \subset Q, Q(x,2^i) \text{ is small}} \left( \frac{1}{|Q(x,2^i)|^{1+\frac{s}{n}}} E_k(f_t, Q(x,2^i)) \right)^q \right)^{\frac{1}{q}} \right\|_{L^p}$$
$$+ c \left\| \left( \sum_{Q(x,2^i) \subset Q, Q(x,2^i) \text{ is big}} \left( \frac{1}{|Q(x,2^i)|^{1+\frac{s}{n}}} E_k(f_t, Q(x,2^i)) \right)^q \right)^{\frac{1}{q}} \right\|_{L_p}.$$

With the help of Lemma A.16, the first term on the right is dominated by the right-hand side of (A.46), which has already been estimated. The second term is treated as follows:

$$\left\| \left( \sum_{Q(x,2^i) \subset Q, Q(x,2^i) \text{ is big}} \left( \frac{1}{|Q(x,2^i)|^{1+\frac{s}{n}}} E_k(f_t, Q(x,2^i)) \right)^q \right)^{\frac{1}{q}} \right\|_{L^p} \tag{A.47}$$
$$\leq c \left\| \left( \sum_{Q(x,2^i) \subset Q, Q(x,2^i) \text{ is big}} \left( \frac{1}{|Q(x,2^i)|^{1+\frac{s}{n}}} E_k(f_t - f, Q(x,2^i)) \right)^q \right)^{\frac{1}{q}} \right\|_{L^p}$$
$$+ c \left\| \left( \sum_{Q(x,2^i) \subset Q, Q(x,2^i) \text{ is big}} \left( \frac{1}{|Q(x,2^i)|^{1+\frac{s}{n}}} E_k(f, Q(x,2^i)) \right)^q \right)^{\frac{1}{q}} \right\|_{L^p}.$$

The first norm on the right is estimated with the help of Propositions A.15 and A.7, and the observation that the cube $\gamma Q$ is big for $\gamma \geq 1$:

$$\left\| \left( \sum_{Q(x,2^i) \subset Q, Q(x,2^i) \text{ is big}} \left( \frac{1}{|Q(x,2^i)|^{1+\frac{s}{n}}} E_k(f_t - f, Q(x,2^i)) \right)^q \right)^{\frac{1}{q}} \right\|_{L^p}$$
$$\leq \left\| \left( \sum_{Q(x,2^i) \subset Q, Q(x,2^i) \text{ is big}} \left( \frac{1}{|Q(x,2^i)|^{1+\frac{s}{n}}} \int_{Q(x,2^i)} |f - f_t| \right)^q \right)^{\frac{1}{q}} \right\|_{L^p}$$
$$\leq c \left( \sum_{i: K_i \cap Q \neq \varnothing} \left( \frac{1}{|K_i|^{1+\frac{s}{n}}} \int_{K_i} |f - f_{K_i}| \right)^p |K_i| \right)^{\frac{1}{p}}$$
$$\leq c \left( \sum_{i: K_i \subset \gamma Q} \left( \frac{1}{|K_i|^{1+\frac{s}{n}}} \int_{K_i} |f - f_{K_i}| \right)^p |K_i| \right)^{\frac{1}{p}} \leq ct\,|Q|^{\lambda}.$$

The second term on the right in (A.47) is controlled directly by $t\,|Q|^{\lambda}$ because

$Q$ is big:

$$\left\| \left( \sum_{Q(x,2^i)\subset Q, Q(x,2^i) \text{ is big}} \left( \frac{1}{|Q(x,2^i)|^{1+\frac{s}{n}}} E_k(f, Q(x,2^i)) \right)^q \right)^{\frac{1}{q}} \right\|_{L^p}$$

$$\leq \left\| \left( \sum_{Q(x,2^i)\subset Q} \left( \frac{1}{|Q(x,2^i)|^{1+\frac{s}{n}}} \int_{Q(x,2^i)} \left| f - f_{Q(x,2^i)} \right| \right)^q \right)^{\frac{1}{q}} \right\|_{L^p} \leq ct\,|Q|^{\lambda}. \qquad \square$$

# Notes and remarks

This Appendix is a revision of an unpublished paper by Kruglyak and Kuznetsov. That work was highly influenced by discussions between Kruglyak and Lars-Inge Hedberg in the Fall of 2005 in Oberwolfach, Germany.

Generalized variation spaces were introduced by Brudnyi in [Br0]; see also the survey [Br1]. The seminorm on the Triebel–Lizorkin space employed here is not the usual seminorm (see [T]) but is equivalent to it (see §2.3.3 and especially Theorem (ii) in [T]); a similar seminorm was invoked in [HN].

# Bibliography


[A] D. R. Adams, *Lectures on $L_p$-potential theory*, Umea University, Math. Dept. Report, 2, 1981.

[AK] I. Asekritova and N. Kruglyak, *The Besicovitch covering theorem and near-minimizers for the couple* $(L^2, BV)$, Proc. Estonian Academy of Sciences, to appear.

[BDKPW] P. Bechler, R. DeVore, A. Kamot, G. Petrova, and P. Wojtaszczyk, *Greedy wavelet projections are bounded on* $BV$, Trans. AMS, 359 (2007), 619-635.

[BSh] C. Bennet and R. Sharpley, *Weak type inequalities on* $H^p$ *and* BMO, Proc. Symp. Pure Math., 35 (1979), 201-229.

[BSh1] C. Bennett and R. Sharpley, *Interpolation of operators*, Pure and Applied Mathematics, vol. 129, Academic Press, Orlando, 1988.

[BL] J. Bergh and J. Löfström, *Interpolation spaces. An introduction*, Springer-Verlag, Berlin-New York, 1976, 227pp.

[BP] J. Bergh and J. Peetre, *On the spaces* $V_p$ $(0 < p < \infty)$, Bolletino U. M. I. (4), 10 (1974), 632-648.

[B] A. S. Besicovitch, *A general form of the covering principle and relative differentiation of additive functions II*, Proc. Cambridge Philos. Soc., 42 (1946), 1-10.

[BS] M. Sh. Birman and M. Z. Solomyak, *Piecewise-polynomial approximations of functions of the class* $W_p^\alpha$, Mat. Sb., 73 (1967), no. 3, 331-355.

[Bo] J. Bourgain, *Some consequences of Pisier's approach to interpolation*, Isr. Math. J., 77 (1992), 165-185.

[BB] A. Brudnyi and Yu. Brudnyi, *Methods of geometric analysis in extension and trace problems*, Monographs in Mathematics, v. 102, 103, Springer Verlag, 2011.

[Br0] Yu. A. Brudnyi, *Spaces defined in terms of variation of local approximations*, Dokl. Akad. Nauk SSSR, 190 (1970), no. 4, 753-755 (Russian).

[Br1] Yu. A. Brudnyi, *Multivariate functions of bounded $(k,p)$-variation*, in: Banach spaces and their applications in analysis, B. Randrianantonina and N. Randrianantonina, eds., de Gruyter, 2007.

[Br] Yu. A. Brudnyi, *Spaces determined by local approximations*, Trudy Mosk. Matem. Ob-va, 24 (1971), 69-132.

[BrK] Yu. A. Brudnyi and B. Kotlyar, *A certain problem of combinatorial geometry*, Sibirsk. Mat. Zh. 11 (1970), no. 5, 1171-1173; English transl., Siberian Math. J., 11 (1970), 870-871.

[BK] Yu. A. Brudnyi and N. Krugljak, *Interpolation functors and interpolation spaces* 1, North Holland, Amsterdam, 1991, 722 pp.

[Ca] A. P. Calderon, *Estimates of singular integral operatros in terms of maximal functions*, Studia Math., 44 (1972), 167-186.

[CZ] A. P. Calderón and A. Zygmund, *On the existence of certain singular integrals*, Acta Math., 88 (1952), 85-139.

[CM] C. P. Calderón and M. Milman, *Interpolation of Sobolev spaces. The real method*, Indiana Math. J., 32 (1983), 801-808.

[C] S. Campanato, *Proprieta di holderianita di alcune classi di funzioni*, Ann. Scuola Norm. Sup. Pisa, 17 (1963), 175-188.

[CS] T. Chan and J. Shen, *Image processing and analysis: variational, PDE, wavelet, and stochastic methods*, SIAM, Philadelphia, 2005., 400 pp.

[CFM] F. Cobos, P. Fernándes-Martinez, and A. Martinez, *Interpolation of the measure of noncompactness by the real method*, Studia Math., 135 (1999), no. 1, 25-38.

[CK] F. Cobos and N. Kruglyak, *Exact minimizer for the couple $(L^\infty, BV)$ and one-dimensional analog of Rudin-Osher-Fatemi model*, Journal of Approximation Theory 163, 4 (2011), 481-490.

[CDDD] A. Cohen, W. Dahmen, I. Daubechies, and R. DeVore, *Harmonic analysis of the space $BV$*, Rev. Mat. Iberoamericana, 19 (2003), 235-263.

[CDPH] A. Cohen, R. DeVore, P. Petrushev, and H. Xu, *Nonlinear approximation and the space $BV(R^2)$*, Amer. J. Math., 121, no. 3 (1999), 587-628.

[CRW] R. Coifman, R. Rochberg, and G. Weiss, *Factorization theorems for Hardy spaces in several variables*, Ann. of Math., 103 (1976), 611-635.

[CNS] M. Cwikel, P. G. Nilsson, and G. Schechtman, *Interpolation of weighted Banach lattices. A characterization of relatively decomposable Banach lattices*, Mem. Amer. Math. Soc., 165, no. 787 (2003).

[Cw] M. Cwikel, *Real and complex interpolation and extrapolation of compact operators*, Duke Math. J., 65 (1992), no. 2, 333-343.

[CMW] M. Cwikel, J. E. McCarthy, and T. Wolff, *Interpolation between weighted Hardy spaces*, Proc. Amer. Math. Soc., 116 (1992), 381-388.

[D] I. Daubechies, *Ten lectures on wavelets*, CBMS-NSR Series in appl. math., SIAM, 1992.

[DeVSh] R. A. DeVore and R. C. Sharpley, *Maximal functions measuring smoothness*, Mem. Amer. Math. Soc., 47, no. 293 (1984).

[DeVSch] R. DeVore and K. Scherer, *Interpolation of linear operators on Sobolev spaces*, Ann. Math., 109 (1979), 583-599.

[Dou] J. Douandikoetxea, *Fourier analysis*, Graduate Studies in Math., 29, AMS, Providence, 2001, 222 pp.

[DS] N. Dunford and J. Schwartz, *Linear operators, vol. 2. Spectral theory*, Interscience, 1958.

[EE] D. E. Edmunds and W. D. Evans, *Spectral theory and differential operators*, The Clarendon Press, Oxford University Press, New York, 1987. xviii+574 pp.

[FS2] C. Fefferman, N. Rivière, and Y. Sagher, *Interpolation between $H^p$ spaces, the real method*, Trans. Amer. Math. Soc., 191 (1974), 75-81.

[FS] C. Fefferman and E. M. Stein, *Some maximal inequalities*, Amer. J. Math., 93 (1971), 107-115.

[FS1] C. Fefferman and E. M. Stein, *$H^p$ spaces of several variables*, Acta Math., 129 (1972), 137-193.

[FJ] M. Frazier and B. Jawerth, *A discrete transform and decompositions of distribution spaces*, J. Funct. Anal., 93 (1990), 34-170.

[GR] J. Garcia-Cuerva and J.L. Rubio de Francia, *Weighted norm inequalities and related topics*, North-Holland, Amsterdam, 1985, 604 pp.

[G] M. de Guzman, *Differentiation of integrals in $R^n$*, Lecture Notes in Mathematics, 481, Springer-Verlag, New York, 1975, 226 pp.

[HN] L. Hedberg and N. Netrusov, *Axiomatic approach to function spaces, spectral synthesis, and Luzin approximation*, Mem. Amer. Math. Soc., 188 (no. 882), 2007.

[H0] L. Hörmander, *Estimates for translation invariant operators in $L^p$ spaces*, Acta Math., 104 (1960), 93-140.

[H] L. Hörmander, *The analysis of linear partial differential operators* I, Springer-Verlag, Berlin etc., 1983.

[JRW] B. Jawerth, R. Rochberg, and G. Weiss, *Commutator and other second order estimates in real interpolation theory*, Ark. Mat., 24 (1986), 191-219.

[JN] F. John and L. Nirenberg, *On functions of bounded mean oscillation*, Comm. Pure. Appl. Math., 14 (1961), 415-426.

[Ki] A. Kirsch, *An introduction to the mathematical theory of inverse problems*, Springer, 1996, 282 pp.

[Kis0] S. V. Kisliakov, *A remark on the space of functions of bounded p-variation*, Math. Nachr., 119 (1984), 137-140.

[Kis] S. V. Kislyakov, *Interpolation of $H^p$-spaces: some recent developments*, Israel Math. Conf. Proceedings, 13 (1999), 102-140.

[Kis1] S. V. Kislyakov, *On* BMO*-regular lattices of measurable functions*, St. Petersburg Math. J., 14 (2003), no. 2, 273–286.

[Kis2] S. V. Kislyakov, *On* BMO*-regular couples of lattices of measurable functions*, Studia Math., 159 (2003), no. 2, 277–290.

[KK0] S. V. Kislyakov and N. Kruglyak, *Stability of approximation under singular integrals, and Calderón–Zygmund type decompositions*, PDMI Preprint No. 7, 2005, www.pdmi.ras.ru/preprint/2005/index.html.

[KK] S. V. Kislyakov and N. Kruglyak, *Stability of Approximation under the Action of Singular Integral Operators*, Funkts. Anal. Prilozh., 40, no. 4 (2006), 49-64.

[KX] S. V. Kislyakov and Q. Xu, *Real interpolation and singular integrals*, St. Petersburg Math. J., 8 (1997), no. 4, 593-615.

[KX1] S. V. Kisliakov and Q. Xu, *Partial retractions for weighted Hardy spaces*, Studia Math., 138 (2000), no. 3, 251–264.

[KX2] S. V. Kisliakov and Q. Xu, *Interpolation of weighted and vector-valued Hardy spaces*, Trans. Amer. Math. Soc., 343 (1994), no. 1, 1-34.

[Koo] P. Koosis, *Introduction to $H^p$-spaces. With an appendix on Wolff's proof of the corona theorem*, London Math. Soc. Lecture Notes Series, vol. 40, Cambridge University Press, Cambridge–New York, 1980.

[KPS] S. G. Krein, Ju. I. Petunin, and E. M. Semenov, *Interpolation of linear operators*, Translations of Mathematical Monographs, vol. 54, AMS, Providence, 1982.

[K] N. Kruglyak, *Investigations in the theory of real interpolation*, Steklov Mathematical Institute, St. Petersburg, Doctor of Science Thesis, 1996, 267 pp.

[K0] N. Kruglyak, *Quantitative theorems on Whitney-type coverings*, Zap. Nauchn. Sem. St.-Petersburg Otdel. Math. Inst. im. Steklova (POMI), 247 (1997), Issled. Po Linein. Oper. I Teor. Funkts. 25, 96-113, English translation in J. Math. Sci. (New York), 101 (2000), no. 3, 3104-3114.

[K1] N. Kruglyak, *Smooth analogs of Calderón-Zygmund decompositions, quantitative covering theorems and the K-functional for the couple* $(L^p, \dot{W}_q^k)$ (Russian), Algebra i Analiz, v. 8 (1996), no. 4, 110-160, English translation in St.-Petersburg Math. Journal, v. 8 (1997), no. 4, 617-649.

[K2] N. Kruglyak, *The K-functional and Calderón-Zygmund type decompositions*, Contemporary Mathematics, 446 (2007), 183-194.

[KKu] N. Kruglyak, E. Kuznetsov, *Smooth and nonsmooth Calderón-Zygmund decompositions for Morrey spaces*, Journal of Fourier Analysis and Applications, 11 (2005), no. 6, 697-714.

[KKu1] N. Kruglyak, E. Kuznetsov, *Sharp integral estimates for the fractional maximal function and interpolation*, Arkiv för Matematik, 44 (2006), no. 2, 309-326.

[KN] N. Kruglyak and D. Nevskii, *Almost optimality of algorithms of dyadic approximation and real interpolation* (Russian), Algebra i Analiz, 14 (2002), no. 4, 63-90, English translation in St. Petersburg Math. J., 14 (2003), no. 4.

[Ku] E. Kuznetsov, *Maximal theorems and Calderón-Zygmund decompositions for the fractional maximal function*, PhD Thesis, Lulea University of Technology, 2005.

[Maz] V. G. Maz'ya, *Sobolev spaces*, Leningrad State University Publishers, Leningrad, 1985 (Russian).

[Me] G. N. Meyers, *Mean oscillation over cubes and Hölder continuity*, Proc. Amer. Math. Soc., 15 (1964), 717-721.

[M1] Y. Meyer, *Wavelets and operators*, Cambridge Univ. Press, Cambridge, 1992.

[M] Y. Meyer, *Oscillating pattern in image processing and nonlinear evolution equations*, University Lecture Series, Vol. 22, AMS Providence, 2002, 122 pp.

[Mi] S. G. Mikhlin, *On multipliers of Fourier integrals*, Dokl. Akad. Nauk SSSR, 109 (1956), 701-703 (Russian).

[Mo] C. B. Morrey, *On the solutions of quasi-linear elliptic partial differential equations*, Trans. Amer. Math. Soc., 43 (1938), 126-166.

[NPS] I. Ya. Novikov, B. Yu. Protasov, and M. A. Skopina, *Theory of wavelets*, Fizmatlit, Moscow, 2005 (in Russian).

[Pe0] J. Peetre, *Espaces d'interpolation et théorème de Sobolev*, Ann. Inst. Fourier, 1966, 16, no.1, 279-317.

[Pe] J. Peetre, *A new approach in interpolation spaces*, Studia Math., 34 (1970), 23-42.

[Pe1] J. Peetre, *$H_p$-spaces*, Lund, 1974.

[P] G. Pisier, *Interpolation between $H^p$-spaces and noncomutative generalisation* 1, Pacific J. Math., 155 (1992), 341-368.

[R] F. Riesz, *Sur un théorème de maximum de MM. Hardy et Littlewood*, J. London Math. Soc., 7 (1932), 10-13.

[Ru] W. Rudin, *Function theory in polydiscs*, Mathematics lecture note series, no. 41, W. A. Benjamin, 1969.

[Rut] D. V. Rutskii, BMO *regularity in lattices of measurable functions on spaces of homogeneous type.* (Russian) Algebra i Analiz 23 (2011), no. 2, 248–295. English translation in: St. Petersburg Math. J., 23 (2012), no. 2.

[SGGHL] O. Scherzer, M. Grasmair, H. Grossauer, M. Haltmeier, and F. Lenzen, *Variational methods in imaging.* Applied Mathematical Sciences, vol. 167. Springer, New York, 2009.

[Sh] Z. Shen, *Boundary value problems in Morrey spaces for elliptic systems on Lipschitz domains*, Amer. J. Math., 125 (2003), 1079-1115.

[S] E. M. Stein, *Singular integrals and differentiability properties of functions*, Princeton Univ. Press, Princeton, N.J., 1970.

[S1] E. M. Stein, *Intégrales singulières et fonctions différentiables de plusieurs variables*, Lecture notes by Bachvan and A. Somen of a course given at Orsay, academic year 1966-67.

[SW] E. M. Stein and G. Weiss, *Introduction to Fourier analysis on euclidean spaces*, Princeton University Press, Princeton, NJ, 1971.

[ST] J.-O. Strömberg and A. Torchinsky, *Weighted Hardy spaces*, Lecture Notes Math., 1381, Springer-Verlag, Berlin etc., 1989.

[Ta] M. E. Taylor, *Analysis on Morrey spaces and applications to Navier–Stokes and other evolution equations*, Comm. Partial Differrential Equations, 17 (1992), 1407-1356.

[T] H. Triebel, *Theory of function spaces II*, Monographs in Mathematics, vol. 84, 1992.

[Xu] Q. Xu, *Some properties of the quotient space* $L^1(\mathbb{T})/H^1(\mathbb{T})$, Illinois J. Math., 37 (1993), no. 3, 437-454.

[Vol] A. Volberg, *Calderón–Zygmund capacities on nonhomogeneous spaces*, Reg. Conf. Ser. in Math., no. 100, Amer. Math. Soc., 2003.

[Wh] H. Whitney, *Analytic extensions of differentiable functions defined in closed sets*, Trans. Amer. Math. Soc., 36 (1934), 63-89.

[W] N. Wiener, *The ergodic theorem*, Duke Math. J., 5 (1939), 1-18.

[Wo] P. Wojtaszczyk, *Wavelets as unconditional bases in* $L_p(R)$, J. Fourier Anal. Appl., 5 (1999), no. 1, 73-85.

[Z] A. Zygmund, *Trigonometric series*, Cambridge Univ. Press, London, 1968.

# Index

# Monografie Matematyczne

[1] S. Banach, *Théorie des opérations linéaires*, 1932
[2] S. Saks, *Théorie de l'integrale*, 1933
[3] C. Kuratowski, *Topologie* I, 1933
[4] W. Sierpiński, *Hypothèse de continu*, 1934
[5] A. Zygmund, *Trigonometrical Series*, 1935
[6] S. Kaczmarz, H. Steinhaus, *Theorie der Orthogonalreihen*, 1935
[7] S. Saks, *Theory of the integral*, 1937
[8] S. Banach, *Mechanika*, T. I, 1947
[9] S. Banach, *Mechanika*, T. II, 1947
[10] S. Saks, A. Zygmund, *Funkcje analityczne*, 1948
[11] W. Sierpiński, *Zasady algebry wyższej*, 1946
[12] K. Borsuk, *Geometria analityczna w n wymiarach*, 1950
[13] W. Sierpiński, *Działania nieskończone*, 1948
[14] W. Sierpiński, *Rachunek różniczkowy poprzedzony badaniem funkcji elementarnych*, 1947
[15] K. Kuratowski, *Wykłady rachunku różniczkowego i całkowego*, T. I, 1948
[16] E. Otto, *Geometria wykreślna*, 1950
[17] S. Banach, *Wstęp do teorii funkcji rzeczywistych*, 1951
[18] A. Mostowski, *Logika matematyczna*, 1948
[19] W. Sierpiński, *Teoria liczb*, 1950
[20] C. Kuratowski, *Topologie* I, 1948
[21] C. Kuratowski, *Topologie* II, 1950
[22] W. Rubinowicz, *Wektory i tensory*, 1950
[23] W. Sierpiński, *Algèbre des ensembles*, 1951
[24] S. Banach, *Mechanics*, 1951
[25] W. Nikliborc, *Równania różniczkowe*, Cz. I, 1951
[26] M. Stark, *Geometria analityczna*, 1951
[27] K. Kuratowski, A. Mostowski, *Teoria mnogości*, 1952
[28] S. Saks, A. Zygmund, *Analytic functions*, 1952
[29] F. Leja, *Funkcje analityczne i harmoniczne*, Cz. I, 1952
[30] J. Mikusiński, *Rachunek operatorów*, 1953
* [31] W. Ślebodziński, *Formes extérieures et leurs applications*, 1954
[32] S. Mazurkiewicz, *Podstawy rachunku prawdopodobieństwa*, 1956
[33] A. Walfisz, *Gitterpunkte in mehrdimensionalen Kugeln*, 1957
[34] W. Sierpiński, *Cardinal and ordinal numbers*, 1965
[35] R. Sikorski, *Funkcje rzeczywiste*, 1958
[36] K. Maurin, *Metody przestrzeni Hilberta*, 1959
[37] R. Sikorski, *Funkcje rzeczywiste*, T. II, 1959
[38] W. Sierpiński, *Teoria liczb* II, 1959

* [39] J. Aczél, S. Gołąb, *Funktionalgleichungen der Theorie der geometrischen Objekte*, 1960
[40] W. Ślebodziński, *Formes extérieures et leurs applications*, II, 1963
[41] H. Rasiowa, R. Sikorski, *The mathematics of metamathematics*, 1963
[42] W. Sierpiński, *Elementary theory of numbers*, 1964
* [43] J. Szarski, *Differential inequalities*, 1965
[44] K. Borsuk, *Theory of retracts*, 1967
[45] K. Maurin, *Methods of Hilbert spaces*, 1967
[46] M. Kuczma, *Functional equations in a single variable*, 1967
[47] D. Przeworska-Rolewicz, S. Rolewicz, *Equations in linear spaces*, 1968
[48] K. Maurin, *General eigenfunction expansions and unitary representations of topological groups*, 1968
[49] A. Alexiewicz, *Analiza funkcjonalna*, 1969
* [50] K. Borsuk, *Multidimensional analytic geometry*, 1969
* [51] R. Sikorski, *Advanced calculus. Functions of several variables*, 1969
[52] W. Ślebodziński, *Exterior forms and their applications*, 1971
[53] M. Krzyżański, *Partial differential equations of second order*, vol. I, 1971
[54] M. Krzyżański, *Partial differential equations of second order*, vol. II, 1971
[55] Z. Semadeni, *Banach spaces of continuous functions*, 1971
[56] S. Rolewicz, *Metric linear spaces*, 1972
[57] W. Narkiewicz, *Elementary and analytic theory of algebraic numbers*, 1974
[58] Cz. Bessaga, A. Pełczyński, *Selected topics in infinite dimensional topology*, 1975
* [59] K. Borsuk, *Theory of shape*, 1975
[60] R. Engelking, *General topology*, 1977
[61] J. Dugundji, A. Granas, *Fixed point theory*, 1982
* [62] W. Narkiewicz, *Classical problems in number theory*, 1986

The volumes marked with * are available at the exchange department of the library of the Institute of Mathematics, Polish Academy of Sciences.

MIX
Papier aus verantwortungsvollen Quellen
Paper from responsible sources
FSC® C105338

If you have any concerns about our products,
you can contact us on
**ProductSafety@springernature.com**

In case Publisher is established outside the EU,
the EU authorized representative is:
**Springer Nature Customer Service Center GmbH**
**Europaplatz 3, 69115 Heidelberg, Germany**

Printed by Libri Plureos GmbH
in Hamburg, Germany